Linear Algebra

An Interactive Approach

S. K. Jain
Ohio University

A. D. Gunawardena
Carnegie Mellon University

THOMSON

BROOKS/COLE

Australia • Canada • Mexico • Singapore • Spain
United Kingdom • United States

THOMSON
BROOKS/COLE

Mathematics Editor: John-Paul Ramin
Assistant Editor: Molly Nance
Editorial Assistant: Lisa Chow
Marketing Manager: Tom Ziolkowski
Marketing Assistant: Stephanie Taylor
Project Manager, Editorial Production: Janet Hill
Print/Media Buyer: Vena Dyer

Production Service: Hearthside Publishing Services
Text Designer: John Edeen
Cover Designer: Laurie Albrecht
Cover Image: Eyewire
Compositor: ATLIS, Inc.
Printer: Phoenix Color Corp

For more information about our products, contact us at:
Thomson Learning Academic Resource Center
1-800-423-0563

For permission to use this material, contact us by:
Phone: 1-800-730-2214
Fax: 1-800-730-2215
Web: http://www.thomsonrights.com

MATLAB and Simulink are registered trademarks of
The MathWorks, Inc. Further information about MATLAB
and related publications may be obtained from:
The MathWorks, Inc.
3 Apple Hill Drive
Natick, MA 01760
(508) 647-700; Fax: (508) 647-7001
e-mail: info@mathworks.com
www.mathworks.com

Library of Congress Control Number: 2002113976

Student Edition: ISBN 0-534-40915-6

Brooks/Cole-Thomson Learning
10 Davis Drive
Belmont, CA 94402
USA

Asia
Thomson Learning
5 Shentom Way #01-01
UIC Building
Singapore 068808

Australia/New Zealand
Thomson Learning
102 Dodds Street
Southbank, Victoria
Australia 3006

Canada
Nelson
1120 Birchmount Road
Toronto, Ontario M1K 5G4
Canada

Europe/Middle East/Africa
Thomson Learning
High Holborn House
50/51 Bedford Row
London WC1R 4LR
United Kingdom

Latin America
Thomson Learning
Seneca 53
Colonia Polanco
11560 Mexico D.F. Mexico

Spain/Portugal
Parainfo
Calle Magallanes, 25
28015 Madrid
Spain

Contents

10 Determinants (Revisited) 247

Preface

Introduction

The purpose of this book is to provide an introduction to linear algebra, a branch of mathematics dealing with matrices and vector spaces. Matrices have been introduced here as a handy tool for solving systems of linear equations. But their utility goes far beyond this initial application. There is hardly any area of modern mathematics in which matrices do not have some application. They have also many applications in other disciplines, such as statistics, economics, engineering, physics, chemistry, biology, geology, and business.

The present text in linear algebra is designed for a general audience of sophomore-level students majoring in any area of art, science, or engineering. The only prerequisites are two or three years of high school mathematics with some knowledge of calculus. A special feature of this book is that it can be used in a course taught in a traditional manner as well as in a course using technology. Those using technology may refer to complete solutions of selected exercises (marked as drills) using Matlab at the end of the book, while others using Maple or Mathematica may refer to corresponding solutions on the web page. The readers would find the examples and solutions to drills using technology quite helpful and illustrative to solve similar problems. Concepts and practical methods for solving problems are illustrated through plenty of examples. The theorems and facts underlying these methods are clearly stated as they arise, but their proofs are provided in a separate section called "Proofs of Facts" near the end of the chapter.

Text Organization

The subject matter is laid out in a leisurely manner with plenty of examples to illustrate concepts and applications. Most of the sections contain a fairly large number of exercises, some of which relate to real-life problems. Chapters 1 and 2 deal with linear systems, leading naturally to matrices and to algebraic operations on matrices. Chapter 3 introduces the vector space F^n of n-tuples, linear dependence and independence of vectors. The concept of rank is introduced in Chapter 4 and is followed by more applications of elementary row operations in Chapter 5. Specifically, we show in Chapter 5 how to find the inverse of an invertible matrix, the LU-decomposition and full rank factorization of a matrix. Chapter 6 provides a working knowledge of determinants, which are later considered in a rigorous fashion in Chapter 10. Eigenvalue problems and inner product spaces are given in Chapters 7 and 8, respectively. An interesting feature of Chapter 7 is a method for finding eigenvalues without using determinants. Two methods for finding the least-squares solution of an inconsistent linear system are given in Chapter 8. The first method is geometric and uses the notion of shortest distance; the second is algebraic and uses the concept of generalized inverses. Vector spaces are revisited in Chapter 9, where a formal definition of a vector space is given and important examples of vector spaces of functions are considered.

In addition to the answers to all exercises, we have provided hints or solutions to selected ones. Complete Matlab solutions have been provided for the exercises that are marked as drills. The student's solution manual that contains solutions to the odd-numbered exercises is available separately as is the complete solution manual for the instructor.

Electronic Text on a CD

This text–CD package includes a CD that contains the entire contents of the book neatly organized into an electronic text. In addition, the CD also contains concept demonstrations, Matlab drills, solutions, projects, and chapter review questions. The electronic material is supported by a well-designed graphical user interface that allows the user to navigate to any part of the text by clicking the mouse. The ability to read the text electronically, find any topic you need, do an on-line test, or perform a drill activity using Matlab makes this electronic text a useful instrument for learning linear algebra. We believe that the use of technology in mathematics enriches the learning experience and encourages exploration of computationally hard problems that might not be easy to solve by hand. The experience of using technology to solve mathematical problems is an essential skill for today's graduates entering a high-tech dominated world. Although we have chosen Matlab as our technology tool, the printed text can also be used in a traditional setting. For the instructor who is interested in the use of technology in teaching, this CD contains a wealth of material for teaching linear algebra in a computer lab setting. It provides an interactive environment that encourages a hands-on approach.

Suggestions for Implementation

The book is suitable for a one-semester course in linear algebra. It may also be used for a one-quarter course by skipping certain sections in Chapters 5–10. We suggest the following guidelines for teaching the course in a computer lab setting. First, begin with demos in the electronic textbook to illustrate a new concept. Then advise the students to read related material in the text to reinforce the concept. Next, explain the Matlab operations needed in a given chapter. The list of basic Matlab operations is provided not only on the CD but also in the printed text. Pick a drill in a chapter and work it out fully in the class, and assign Matlab exercises to work in the lab or at home. Exercises marked as drills with a picture of a CD have Matlab solutions in the CD as well as at the end of the printed text. Solutions of drills using Mathematica/Maple are given on the Web page. Finally, encourage students to do the projects using Matlab.

Acknowledgments

We would like to thank, among others, Jonathan Golan, Dinh V. Huynh, Pramod Kanwar, Sergio R. Lopez-Permouth, S. R. Nagpaul, M. S. K. Sastry, and Charles Withee for using the manuscript of this book in their traditional or on-line classes and for their helpful comments. We would like to express special thanks to Pramod Kanwar for his assistance various ways, including the preparation of the CD.

A number of people have played an important role in the production of the book. Among them are John Tynan of Marietta College, Ohio; Yong UK Cho of Silla University (Korea); graduate students Adel Al-Ahmadi and Hussain Al-Hazmi who have been of great assistance in the tedious task of proofreading; and Larry E. Snyder and Roland Swardson for their help in editing some portions of the manuscript. We also mention with great pleasure the cooperation and assistance offered by individuals at Brooks/Cole, especially our publisher Bob Pirtle, editor John-Paul Ramin, and assistant editor Molly Nance.

S. K. Jain would like to thank his wife and his family; A. D. Gunawardena would like to thank his family, Sriya, Neomi, and Naoka, for their understanding and patience.

We wish to acknowledge the inspiration given by the late Professor P. B. Bhattacharya in initiating this project. We dedicate this book to his memory.

Finally, we are very grateful to the following who have reviewed this text and offered valuable comments: Roman Hilcher, Michigan State University; Michael Tsatsomeros, Washington State University; Jane Day, San Jose State Unversity; David Miller, William Paterson University; and Alan Schuchat, Wellesley College.

S. K. Jain
A. D. Gunawardena

Notes to Students

Chapter 1

An important activity that occurs throughout the book is reducing a system of linear equations to a system that is readily solvable. This is equivalent to reducing a certain matrix into a specific form, known as row echelon form, discussed in this chapter. Sections 1.1 and 1.2 guide the reader to the central theme of the book: reducing a matrix into row echelon form. Most instructors would come to Section 1.3 quickly after just giving a glimpse of the material discussed earlier. This is fine because it allows time to cover more topics. You will be learning how to solve a linear system of equations by Gauss elimination method. We suggest that you familiarize yourself with (1) the three elementary row operations, (2) the steps for reducing a matrix to row echelon form, and (3) the method for solving a linear system by using backward substitution, as discussed in Section 1.4. Reducing the augmented matrix into reduced echelon form is not necessary. However, if you do so, you will get the solution(s) without using the backward substitution method. In practice, doing backward substitution amounts to less work than reducing a row echelon matrix further to reduced row echelon form.

Chapter 2

Matrix addition, matrix subtraction, and scalar multiplication are defined naturally, and these operations are easy to apply. But matrix multiplication, although motivated by linear systems of equations for finding their solutions, does not appear natural, and you might find it difficult to appreciate the formula for multiplication of matrices. We suggest that you practice matrix multiplication problems and work on the many interesting applications to real-life problems given in worked-out examples and in exercises.

Chapter 3

You will find this chapter challenging. Some of you might not like the abstract definition of subspace and proving or disproving whether a given subset is a subspace or not. If you get an intuitive picture of a subspace, you may proceed and come back as needed. Linear independence and linear dependence are central concepts that are discussed in this chapter and will appear throughout the book. The definitions of these concepts are easy, but still you may have difficulty in reproducing the definitions correctly and applying them to problems. There is no magic prescription! You simply go through many worked-out examples in the book and solve exercises. Other concepts discussed in this chapter are basis and dimension. Finding a basis of a subspace may require long computations. We suggest that you work on the examples in the book and practice some problems. A much shorter method of finding a basis of a subspace is given in Chapter 4 after more machinery has been developed. The proofs of facts contained in Section 3.4 are nontrivial.

Chapter 4

Computing the rank of a matrix is easy once you are given its definition and the procedure to compute it. Computing the nullity of a matrix is straightforward but involves somewhat long computations. Nullity of a matrix turns out to be the number of arbitrary unknowns in the solution set of a homogeneous linear system whose coefficient matrix is the given matrix. The rank-nullity theorem connects the rank of a matrix with its nullity by stating "rank + nullity = number of columns." This fact has many applications in linear algebra. It can be used, for example, to compute the nullity once you know its rank unless you are required to compute it by using definition only. You will need to know the effect of performing elementary row or column operations on a matrix as given in the last section. This will be needed to understand the theoretical reason behind computing the inverse of a matrix and other factorizations discussed in Chapter 5. If you skip this section or forget the results contained therein, you might still be able to compute the inverse and perform factorizations correctly by following the steps given in the text but you wouldn't know the reasons behind your work.

Chapter 5

This chapter is straightforward. Read the steps to compute the inverse and to obtain various factorizations and practice some problems. Read the properties of the inverse, especially about the inverse of the product of invertible matrices.

Chapter 6

You will find this chapter as easy as Chapter 5. You will often be using the theorem on expanding the determinant with respect to any row or column (Theorem 6.3.3) and properties of determinants. To find the determinant of numerical matrices, you will mostly be using the method of reducing the matrix into row echelon form. You are urged to read and remember the properties of determinants, as they are heavily used in expanding determinants to reduce the amount of calculations.

Chapter 7

Computations of eigenvalues and eigenvectors are straightforward once you know their definitions. Two methods are presented in the book. One is traditional as given in Section 7.2; the other method given in Section 7.3 is not commonly given in texts, but we recommend that you read both methods and then choose which one you prefer. Indeed, you might find it helpful to know both techniques, because one technique may be superior to the other for a given problem. Read and remember the properties of eigenvalues and eigenvectors, in particular that eigenvectors corresponding to distinct eigenvalues are linearly independent. As a consequence, a matrix with distinct eigenvalues always is diagonalizable. This is a highly important fact that is used often. That every matrix is triangularizable is also proved and is illustrated with examples. This fact has many applications in matrix theory. But this topic might not be taught because of time. The Cayley-Hamilton Theorem also has many applications, but it might be skipped if time is short. We give a very short proof that we believe is new in the sense that it is not in the literature. It is extremely short and easy to follow. The traditional proof using adjoint of matrices is also given. If this theorem is taught and its applications are shown, you will find them interesting and easy to understand.

Chapter 8

This chapter has the closest connection with geometry. Memorize the Gram-Schmidt method of finding an orthonormal set. The important topics are diagonalization of symmetric matrias, singular value decomposition, applications of the principal-axis theorem, and best approximate solutions of inconsistent linear systems. Not all topics from this chapter are covered in a beginning-level course in linear algebra. However, these are very interesting topics and have applications to real-life problems. Most of these problems involve long computations, and it is helpful to use technology. You will find systemic steps given in the chapter for tackling a given problem on diagonalizing a symmetric matrix and finding a best approximate solution of an inconsistent linear system.

Chapter 9

This chapter is an abstract version of the concepts discussed in Chapter 3. It deals with abstract vector spaces and functions between them. You will appreciate it if you have a mathematical bent of mind for viewing concepts in a general setting. Not all schools might find this appropriate at your level or might not have time to teach abstract vector spaces. However, functions between concrete vector spaces known as linear mappings (also called linear transformations) are likely to be introduced at some stage. You should read the sections on linear mappings without being deterred by the abstract definition of vector space. An important fact in the theory of linear mappings is that there is one-to-one correspondence between linear mappings and matrices of suitable sizes (depending on the dimensions of vector spaces between which the linear mapping is defined). Follow the steps to work on the problems to compute a matrix of a linear mapping.

Chapter 10

This contains the abstract definition of the determinant of a square matrix introduced in Chapter 6. The proofs of various properties are given in this chapter. Not all schools might teach determinants in such a rigorous manner. Some instructors may combine Chapters 6 and 10. Read as advised. Those who are simply interested in applications of determinants can skip this chapter.

Suggested Syllabus

One-Quarter First Course in Linear Algebra

Chapter 1, Chapter 2, Chapter 3, Chapter 4, Chapter 5 (omit Sections 5.3–5.4), Chapter 6 (omit Section 6.4), Chapter 7 (Section 7.4 optional), Chapter 8 (omit Sections 8.4–8.6). Presenting all proofs may require omission of some material toward the end.

One-Semester First Course in Linear Algebra

One should be able to cover most of the material, possibly omitting some section or proofs to find more time for later chapters, and for the projects.

1 Linear Systems and Matrices

Introduction

Mathematical modeling of real-world problems often involves a large number of variables and a large number of constraints, giving rise to linear systems of equations. Solving these linear systems requires restructuring of equations and reduction into a form that is easily solvable. In this chapter, we discuss a very useful and practical method called Gauss elimination for solving such systems. The method involves reducing a linear system or, equivalently, the *augmented matrix* of a system into a form known as *row echelon form* that readily yields a solution or solutions if such solutions exist.

1.1 Linear Systems of Equations

In plane geometry, an equation of the form

$$ax + by = c,$$

where a, b are not both 0, denotes a line, and so it is called a linear equation in variables x and y. A linear equation in three variables corresponds to a plane in three-dimensional space. When we have to write a linear equation in a large number, say n, of variables, it is convenient to list them by subscripts $1, 2, \ldots, n$ attached to the letter x. Then a linear equation in $x_1, x_2, \ldots, x_n$, is of the form

$$a_1 x_1 + a_2 x_2 + \cdots + a_n x_n = b.$$

When we have to write more than one linear equation in $x_1, x_2, \ldots, x_n$, it is convenient to use double subscripts for the coefficients:

$$a_{11}x_1 + a_{12}x_2 + \cdots + a_{1n}x_n = b_1,$$
$$\vdots$$
$$a_{i1}x_1 + a_{i2}x_2 + \cdots + a_{ij}x_j + \cdots + a_{in}x_n = b_i, \qquad (1)$$
$$\vdots$$
$$a_{m1}x_1 + a_{m2}x_2 + \cdots + a_{mn}x_n = b_m.$$

1

In the above set of equations, a_{ij} is the coefficient in the ith equation of the jth variable and b_i is the constant term in that equation.

We do not read $a_{11}, a_{12}, \ldots$ as "a eleven," "a twelve," and so on; we read them as "a one-one," "a one-two," and so on, to mean that a_{11} is the coefficient in the first equation of the first variable x_1, a_{12} is the coefficient in the first equation of the second variable x_2, and so on. This notation a_{ij} is fundamental in linear algebra, and you are advised to become familiar with it.

The above set of m equations in n variables (also called unknowns), when considered as a single piece or entity, will be called an $m \times n$ linear system, or simply an $m \times n$ LS.

Solving an $m \times n$ linear system such as the one above amounts to finding a set of numbers $\alpha_1, \alpha_2, \ldots, \alpha_n$ such that when $\alpha_1, \alpha_2, \ldots, \alpha_n$ are substituted for $x_1, \ldots, x_n$, respectively, in Equations (1), we get the m equalities

$$a_{11}\alpha_1 + \cdots + a_{1n}\alpha_n = b_1,$$
$$\vdots$$
$$a_{m1}\alpha_1 + \cdots + a_{mn}\alpha_n = b_m.$$

Such a list $\alpha_1, \alpha_2, \ldots, \alpha_n$ of numbers is called a solution of the $m \times n$ LS (1). The answers to questions such as whether an LS has a solution and how to solve it in the best possible way form the foundation of the subject called linear algebra.

1.1.1
DEFINITION

Consistent and Inconsistent LS

If a linear system has a solution, it is called *consistent*; if it has no solution, it is called *inconsistent*.

For example, consider the following linear systems and their graphs.

1. $x_1 + x_2 = 1,$

$x_1 - x_2 = 0;$

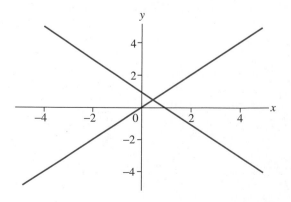

2. $x_1 + x_2 + x_3 = 0$;

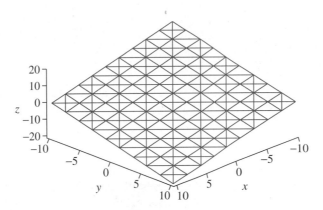

3. $x_1 - 2x_2 + 3x_3 = 1$,
$$2x_1 + x_3 = 0;$$

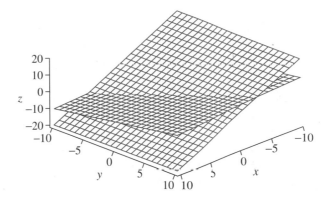

All three linear systems given above are consistent. Geometrically, the first LS represents two lines. The solution of this system is the point of intersection of the two lines in a plane. The second LS represents a plane, and every solution of the system is a point on the plane $x_1 + x_2 + x_3 = 0$. The solution of the third LS represents points on the line of intersection of the two planes. Both the second and third linear systems have infinitely many solutions.

Consider next the linear system

$$x_1 + x_2 = 1,$$
$$4x_2 = 1,$$
$$x_1 + 4x_2 = 1,$$

and its graph

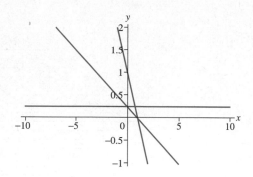

This linear system represents three lines, and there is no point common to all of them. Thus this LS is inconsistent.

Furthermore, since any two parallel lines or planes do not intersect, the linear systems of the types

$$x_1 + x_2 = 1, \qquad 2x_1 + 2x_2 = 1;$$

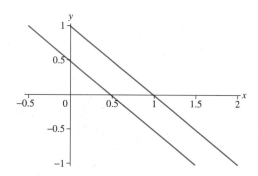

and

$$x_1 + x_2 + x_3 = 1, \qquad 2x_1 + 2x_2 + 2x_3 = 35;$$

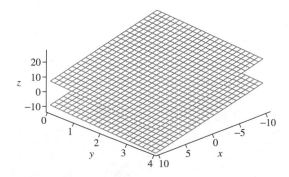

are inconsistent.

<table>
<tr><td>

1.1.2
DEFINITION

</td><td>

Equivalent Linear System
A consistent linear system S is said to be equivalent to a linear system S' if both linear systems have the same solutions.

</td></tr>
</table>

1.2   Elementary Operations and Gauss Elimination Method

The method described below to solve an LS is essentially the method of successive elimination of variables that you may have learned in high school. It consists in transforming a given $m \times n$ LS into an equivalent LS in which the first nonzero coefficient in each equation is to the right of the first nonzero coefficient in the preceding equation. Although it is not essential, it is helpful to make the first nonzero coefficient in each equation equal to 1. Such a system is said to be in *echelon form*.

Following are some examples of a linear system in echelon form:

a. $x_1 - x_2 + 3x_3 = 1,$
$$3x_2 + 2x_3 = -1,$$
$$4x_3 = 3;$$

b. $x_1 + 4x_2 - 3x_3 + 5x_4 = 7,$
$$2x_3 - x_4 = 3,$$
$$3x_4 = 1;$$

c. $3x_1 + 7x_2 - 5x_3 + 0x_4 - 2x_5 = -1,$
$$4x_2 - x_3 + 3x_4 + x_5 = 3,$$
$$x_4 + 2x_5 = 2,$$
$$3x_5 = -2.$$

The following are examples of linear systems that are not in echelon form:

d. $-x_2 + 3x_3 = 1,$
$$2x_1 + 3x_2 + 2x_3 = -1,$$
$$5x_2 + 4x_3 = 3;$$

e. $3x_1 + 7x_2 - 5x_3 + 0x_4 - 2x_5 = -1,$
$$4x_2 - x_3 + 3x_4 + x_5 = 3,$$
$$x_4 + 2x_5 = 2,$$
$$x_3 + x_4 - 3x_5 = 0.$$

The operations performed on a linear system to reduce it into echelon form are of the following three types.

> **1.2.1**
> **DEFINITION**
>
> **Elementary Operations on a Linear System**
>
> **I.** Interchange of two equations, say the ith and jth. This is denoted by $R_i \longleftrightarrow R_j$.
>
> **II.** Multiplying any equation, say the ith, by a nonzero number α. This is denoted by αR_i.
>
> **III.** Adding to the jth equation α times the ith equation. This is denoted by $R_j + \alpha R_i$. (This keeps the ith equation unchanged.)

These operations are called *elementary operations*. When performed successively on any linear system in any order, these operations transform it into a linear system that has the same solution.

We next describe the procedure to transform a given linear system into echelon form.

Suppose $x_1, \ldots, x_n$ are the variables in an $m \times n$ LS.

Step 1   Select the equation, say the ith, in which the coefficient of x_1 is not zero. If this is the first equation, then nothing needs to be done. Otherwise, interchange the first equation and the ith equation. This amounts to the elementary operation $R_1 \longleftrightarrow R_i$.

Step 2   Next perform the operation αR_1 to make the coefficient of x_1 equal to 1.

Step 3   Perform operations of the type $R_i + \alpha R_1$, $i > 1$, to eliminate x_1 from all equations other than the top equation.

Step 4   Ignoring temporarily the top equation, repeat Steps 1, 2, and 3 on the system of equations from which x_1 has been eliminated. Continue the process of elimination of variables until you come to a stage with only one equation. The process of elimination of variables must stop here.

Step 5   Solve the system by *backward substitution*. This means solving the system beginning with the solution of the last equation and then moving upward one by one, successively, to solve each equation.

> **1.2.2**
> **EXAMPLES**

a. Solve the LS

$$3x + y + z = 4, \tag{1}$$
$$-x + y - 2z = -15, \tag{2}$$
$$-2x + 2y + z = -5. \tag{3}$$

We would like to have on the top an equation whose coefficient of the first variable is 1. This can be achieved in several ways. We may either divide the first equation by 3 or add the third equation to the first equation. To avoid fractions, we may perform the latter operation to obtain a new linear system (LS)$'$:

$$x + 3y + 2z = -1, \tag{1$'$}$$
$$-x + y - 2z = -15 \tag{2$'$}$$
$$-2x + 2y + z = -5 \tag{3$'$}$$

Add Equation $(1')$ to Equation $(2')$. Also add 2 times Equation $(1')$ to Equation $(3')$. The new system $(LS)''$ is

$$x + 3y + 2z = -1, \qquad\qquad (1'')$$
$$4y = -16, \qquad\qquad (2'')$$
$$8y + 5z = -7. \qquad\qquad (3'')$$

Finally, add -2 times Equation $(2'')$ to Equation $(3'')$. This will transform the linear system $(LS)''$ to the system $(LS)'''$:

$$x + 3y + 2z = -1, \qquad\qquad (1''')$$
$$4y = -16, \qquad\qquad (2''')$$
$$5z = 25. \qquad\qquad (3''')$$

Solving backward, we get

$$z = 5,$$
$$y = -4,$$
$$x = -1 - 3(-4) - 2(5) = 1.$$

b. Solve the LS

$$2x_1 + 4x_2 - 6x_3 + x_4 = 2, \qquad\qquad (1)$$
$$x_1 - x_2 + 4x_3 + x_4 = 1, \qquad\qquad (2)$$
$$-x_1 + x_2 - x_3 + x_4 = 0. \qquad\qquad (3)$$

Interchange Equations (1) and (2) to obtain a new $(LS)'$. Note that this will make the coefficient of x_1 equal to 1 in the top equation of the new LS:

$$x_1 - x_2 + 4x_3 + x_4 = 1, \qquad\qquad (1')$$
$$2x_1 + 4x_2 - 6x_3 + x_4 = 2, \qquad\qquad (2')$$
$$-x_1 + x_2 - x_3 + x_4 = 0. \qquad\qquad (3')$$

Multiply Equation $(1')$ by -2 and add to Equation $(2')$. Also, add Equation $(1')$ to Equation $(3')$. This will eliminate x_2 in the new LS. The transformed $(LS)''$ is

$$x_1 - x_2 + 4x_3 + x_4 = 1, \qquad\qquad (1'')$$
$$6x_2 - 14x_3 - x_4 = 0, \qquad\qquad (2'')$$
$$3x_3 + 2x_4 = 1. \qquad\qquad (3'')$$

We are now ready to solve $(LS)''$ beginning from the last equation, $3x_3 + 2x_4 = 1$. Since this equation has two unknowns, we can choose one variable, say x_4, to be arbitrary. Putting $x_4 = t$, we get from the last equation $3x_3 + 2t = 1$. So by solving backward, we have

$$x_3 = \frac{1}{3}(1 - 2t).$$

Then moving up to the next equation,

$$6x_2 - 14x_3 - x_4 = 0,$$

we obtain

$$6x_2 - (14)\frac{1}{3}(1 - 2t) - t = 0,$$

which yields

$$x_2 = \frac{1}{18}(14 - 25t).$$

Finally, from the equation above the previous one, namely, $x_1 - x_2 + 4x_3 + x_4 = 1$, we obtain, by substituting the values of x_2, x_3, x_4,

$$x_1 = \frac{4}{9} - \frac{5}{18}t.$$

This shows that the given linear system has infinitely many solutions because t can take arbitrary values.

c. Solve the LS

$$x_1 - 2x_2 + 3x_3 = 4, \tag{1}$$
$$5x_1 - 4x_2 - x_3 = 5, \tag{2}$$
$$4x_1 - 2x_2 - 4x_3 = 9. \tag{3}$$

Add -5 times Equation (1) to Equation (2), and add -4 times Equation (1) to Equation (3). We get the following linear system (LS)′:

$$x_1 - 2x_2 + 3x_3 = 4, \tag{1′}$$
$$6x_2 - 16x_3 = -15, \tag{2′}$$
$$6x_2 - 16x_3 = -7. \tag{3′}$$

It immediately follows from Equations (2)′ and (3)′ that the linear system (LS)′, and hence (LS), is inconsistent. If we had proceeded further to make the coefficient of x_2 equal to 1 in (2)′ and had used (2)′ to eliminate x_2 in (3′), we would have ended with an absurdity, namely, the number 0 equal to some nonzero number.

Suppose you are interested in finding a solution of a linear system in which one of the coefficients, say a, is undetermined. One may ask the question "what are the values or the range of values of a that will make the system consistent?" As an example, suppose an engineer wants to design a stable bridge but one of the items may have a range of values depending on some physical constraints. The problem we would like to solve is to find all possible values of the item that would enable the engineer to design a stable bridge.

d. For what values of a is the following LS consistent?

$$3x_1 + 2x_2 + x_3 = 2,$$
$$x_1 - x_2 - 2x_3 = -3,$$
$$ax_1 - 2x_2 + 2x_3 = 6.$$

We perform elementary operations to reduce the given linear system to echelon form. Given below are equivalent linear systems obtained at each stage. The reader is advised to describe the type of elementary operation performed at every step. The given LS is equivalent to

$$x_1 - x_2 - 2x_3 = -3,$$
$$3x_1 + 2x_2 + x_3 = 2,$$
$$ax_1 - 2x_2 + 2x_3 = 6,$$

which is again equivalent to

$$x_1 - x_2 - 2x_3 = -3,$$
$$5x_2 + 7x_3 = 11,$$
$$(a-2)x_2 + (2a+2)x_3 = 3a + 6,$$

which is equivalent to

$$x_1 - x_2 - 2x_3 = -3,$$
$$x_2 + \frac{7}{5}x_3 = \frac{11}{5},$$
$$(a-2)x_2 + (2a+2)x_3 = 3a + 6,$$

which is equivalent to

$$x_1 - x_2 - 2x_3 = -3,$$
$$x_2 + \frac{7}{5}x_3 = \frac{11}{5},$$
$$\left[-\frac{7}{5}(a-2) + (2a+2) \right] x_3 = -\frac{11}{5}(a-2) + 3a + 6.$$

The given LS is inconsistent if

$$-\frac{7}{5}(a-2) + (2a+2) = 0$$

and

$$-\frac{11}{5}(a-2) + 3a + 6 \neq 0.$$

Now,

$$-\frac{7}{5}(a-2) + (2a+2) = 0$$

gives

$$-7(a-2) + 5(2a+2) = 0,$$

or

$$3a = -24, \quad \text{and} \quad \text{so } a = -8.$$

For the value $a = -8$,

$$-\frac{11}{5}(a-2) + 3a + 6 = -\frac{11}{5} \cdot 10 - 24 + 6 = 22 - 24 + 6 = 4,$$

which is not zero. Hence the given LS is consistent for all values of a except $a = -8$.   ■

1.2 Exercises

Solve the following linear systems, if consistent, by first reducing into echelon form.

1. $x_1 - 3x_2 + x_3 = -1,$
$3x_1 - 8x_2 + 2x_3 = 0,$
$x_1 - x_2 - 2x_3 = -4.$

2. $x_1 - 2x_2 + x_3 = 7,$
$2x_1 - 5x_2 + 2x_3 = 6,$
$3x_1 + 2x_2 - x_3 = 1.$

3. $x_1 + 2x_2 - 3x_3 + 4x_4 = 2,$
$2x_1 + 5x_2 - 2x_3 + x_4 = 1,$
$5x_1 + 12x_2 - 7x_3 + 6x_4 = -7.$

4. (Drill 1.1) $x + 2y - 2z = 1,$
$3x - y + 2z = 7,$
$2x - 3y - 4z = 5.$

5. (Drill 1.2) $2x - 5y + 3z - 4u + 2v = 4,$
$3x - 7y + 2z - 5u + 4v = 9,$
$5x - 10y - 5z - 4u + 7v = 22.$

6. For each of the following LS, find all values of a that make it consistent:

a. $x_1 + ax_2 = 5,$
$2x_1 + 5x_2 = 1.$

b. $x_1 + 3x_2 = 5,$
$2x_1 + 6x_2 = a.$

c. $x_1 + 2x_2 = -3,$
$ax_1 - 2x_2 = 5.$

In Exercises 7–10, recall that the linear system of equations is given by

$$a_{11}x_1 + a_{12}x_2 + \cdots + a_{1n}x_n = b_1,$$

$$\vdots$$

$$a_{i1}x_1 + a_{i2}x_2 + \cdots + a_{in}x_n = b_i,$$

$$\vdots$$

$$a_{m1}x_1 + a_{m2}x_2 + \cdots + a_{mn}x_n = b_m.$$

7. Solve the 3×3 LS in which

$$
\begin{array}{llll}
a_{11} = 1, & a_{12} = -1, & a_{13} = 1, & b_1 = 1, \\
a_{21} = 1, & a_{22} = 1, & a_{23} = -1, & b_2 = 1, \\
a_{31} = 2, & a_{32} = 0, & a_{33} = 3, & b_3 = 8.
\end{array}
$$

8. Is the following 2×3 LS consistent?

$$a_{11} = 1, \quad a_{12} = 1, \quad a_{13} = -1, \quad b_1 = 2,$$
$$a_{21} = -1, \quad a_{22} = -1, \quad a_{23} = 1, \quad b_2 = -2.$$

9. Is the following 3×3 LS consistent?

$$a_{11} = 1, \quad a_{12} = 2, \quad a_{13} = 4, \quad b_1 = 1,$$
$$a_{21} = 1, \quad a_{22} = 0, \quad a_{23} = 2, \quad b_2 = 0,$$
$$a_{31} = 2, \quad a_{32} = 3, \quad a_{33} = 7, \quad b_3 = 0.$$

10. Find all values of b_1 for which the following LS is consistent:

$$a_{11} = 1, \quad a_{12} = 4, \quad a_{13} = -1,$$
$$a_{21} = 1, \quad a_{22} = -2, \quad a_{23} = 1, \quad b_2 = 3,$$
$$a_{31} = 2, \quad a_{32} = -1, \quad a_{33} = 1, \quad b_3 = 0.$$

1.3 Homogeneous Linear Systems

1.3.1
DEFINITION

Homogeneous LS
If the right-hand side members $b_1, b_2, \ldots, b_m$ of the LS (1) in Section 1.1 are all zero, then it is called a *homogeneous* LS.

Thus a homogeneous LS looks like this:

$$a_{11}x_1 + \cdots + a_{1n}x_n = 0,$$
$$\vdots$$
$$a_{m1}x_1 + \cdots + a_{mn}x_n = 0.$$

It is obvious that $(0, 0, \ldots, 0)$ is a solution of any homogeneous LS. The solution $(0, 0, \ldots, 0)$ is called the *trivial solution*. Any other solution of a homogeneous LS is called *nontrivial*. In plane and solid geometry in particular, this means that a homogeneous linear system of equations represent lines or planes, all passing through the origin.

1.3.2
EXAMPLES

a. Solve the LS

$$x_1 + x_2 = 0, \tag{1}$$
$$x_1 - x_2 = 0. \tag{2}$$

whose graph is given by

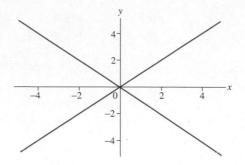

To eliminate x_1 from Equation (2), we add -1 times Equation (1) to Equation (2), yielding

$$x_1 + x_2 = 0, \tag{1'}$$
$$2x_2 = 0. \tag{2'}$$

Solving backward, we get $x_2 = 0$ and $x_1 = 0$. Thus the given LS has no nontrivial solution, as is also clear from the graph.

b. Solve the LS

$$x_1 + x_2 - x_3 = 0, \tag{1}$$
$$x_1 - x_2 - x_3 = 0. \tag{2}$$

whose graph is given by

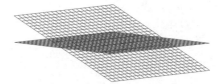

Add -1 times Equation (2) to Equation (1) to obtain

$$x_1 - x_2 - x_3 = 0, \tag{1'}$$
$$2x_2 = 0. \tag{2'}$$

Solving backward, we get $x_2 = 0$, and so (1') yields $x_1 - x_3 = 0$. Since there are two unknowns and one equation, we set one of the variables, say x_3, equal to t, where t is arbitrary. Then $x_1 = t$. Thus the given LS has infinitely many solutions consisting of the points on the line of intersection of the two planes. These are given by $x_1 = t$, $x_2 = 0$, $x_3 = t$, where t is any number.

c. Solve the LS

$$x_2 + 2x_3 = 0, \tag{1}$$
$$x_1 - x_2 + 3x_3 = 0, \tag{2}$$
$$2x_1 - x_2 + 8x_3 = 0. \tag{3}$$

We interchange Equations (1) and (2) to create a new LS in which the top equation is such that the coefficient of x_1 is 1:

$$x_1 - x_2 + 3x_3 = 0, \tag{1'}$$
$$x_2 + 2x_3 = 0, \tag{2'}$$
$$2x_1 - x_2 + 8x_3 = 0. \tag{3'}$$

To eliminate x_1 from (3'), we add -2 times (1') to (3'). This yields

$$x_1 - x_2 + 3x_3 = 0, \tag{1''}$$
$$x_2 + 2x_3 = 0, \tag{2''}$$
$$x_2 + 2x_3 = 0. \tag{3''}$$

Since (2'') and (3'') are the same, our new LS is simply

$$x_1 - x_2 + 3x_3 = 0, \tag{1''}$$
$$x_2 + 2x_3 = 0. \tag{2''}$$

If we had performed the usual operation to eliminate x_2 from (3'), we would have ended at

$$x_1 - x_2 + 3x_3 = 0, \tag{1'''}$$
$$x_2 + 2x_3 = 0, \tag{2'''}$$
$$0x_3 = 0. \tag{3'''}$$

Equation (3''') gives $x_3 = t$, where t is arbitrary. If we had ignored (3'''), then we would begin from (2'') and solve for x_2 and x_3. Because there is one Equation (2'') with two unknowns, we will set one of the variables, say x_3, as arbitrary and solve for the other variable x_2 in terms of the arbitrary value of x_3. Then $x_2 = -2t$. Moving upward to the next Equation (1''), namely, $x_1 - x_2 + 3x_3 = 0$, we obtain $x_1 = x_2 - 3x_3 = -2t - 3t = -5t$. Therefore the given LS has infinitely many solutions given by $x_1 = -5t$, $x_2 = -2t$, and $x_3 = t$, where t is an arbitrary number. In this example the solution consists of all points common to the three planes represented by given equations as shown in the following graph:

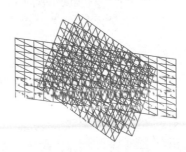

d. Given the 3×2 homogeneous LS

$$a_{11}x_1 + a_{12}x_2 = 0,$$
$$a_{21}x_1 + a_{22}x_2 = 0,$$
$$a_{31}x_1 + a_{32}x_2 = 0,$$

suppose $x_1 = c_1$, $x_2 = c_2$ and $x_1 = d_1$, $x_2 = d_2$ are two solutions. Then:

i. the sum of the two solutions, i.e., $x_1 = c_1 + d_1$, $x_2 = c_2 + d_2$, is also a solution, and

ii. any scalar multiple of a solution is also a solution, that is,

$$x_1 = kc_1, \ x_2 = kc_2 \text{ is also a solution,}$$

where k is an arbitrary number.

We have

$$a_{11}(c_1 + d_1) + a_{12}(c_2 + d_2) = (a_{11}c_1 + a_{12}c_2) + (a_{11}d_1 + a_{12}d_2)$$
$$= 0 + 0 = 0.$$

Similarly, $a_{21}(c_1 + d_1) + a_{22}(c_2 + d_2) = 0$, and so on. Thus $x_1 = c_1 + d_1$, $x_2 = c_2 + d_2$ is also a solution of the LS. Also,

$$a_{11}(kc_1) + a_{12}(kc_2) = k(a_{11}c_1 + a_{12}c_2) = k \cdot 0 = 0,$$

and so on. Thus $x_1 = kc_1$, $x_2 = kc_2$ is also a solution.

e. For what values of c does the following homogeneous linear system of equations have a nontrivial solution:

$$2x_1 + 3x_2 = 0,$$
$$cx_1 + x_2 = 0.$$

Geometrically, this means that besides the origin, there must be another point of intersection of the two lines represented by the given equations. But if two lines have two points in common, they must be same lines. Now the first equation can be rewritten as $x_2 = -\frac{2}{3}x_1$ and the second equation can be rewritten as $x_2 = -cx_1$. Thus $-\frac{2}{3}x_1 = -cx_1$, and so $x_1(c - \frac{2}{3}) = 0$. This implies that either $x_1 = 0$ or $c = \frac{2}{3}$. In case $c \neq \frac{2}{3}$, $x_1 = 0$. Then we get from any one of the given equations that $x_2 = 0$. But because we are looking for a condition that guarantees the existence of a solution other than the trivial solution, $x_1 \neq 0$, and so $c = \frac{2}{3}$ is the desired answer.

In this case the two equations are $2x_1 + 3x_2 = 0$ and $\frac{2}{3}x_1 + x_2 = 0$, and thus they reduce to one single equation $2x_1 + 3x_2 = 0$.

We close this section by giving a condition that guarantees the existence of a nontrivial solution of a homogenous linear system. ∎

1.3.3 THEOREM

(Theorem 1, Section 1.6) Every homogenous linear system with m equations in n unknowns has a nontrivial solution if $m < n$.

We may point out that the above statement does not give any condition that may hold for a linear system of homogeneous equations in which the number of equations is equal to or greater than the number of variables. Such a system may or may not have a nontrivial solution (see *Example 1.3.2(c)*).

The interested reader may look up the proof of the above fact in Section 1.6.

1.3 Exercises

1. Which of the following are homogeneous systems? Which homogeneous LS has a nontrivial solution?

a. $x_1 + x_2 = 0,$
$\quad 2x_1 - x_2 = 0.$

b. $x_1 + x_2 - x_3 = 0,$
$\quad x_1 + 2x_2 + x_3 = 4.$

c. $x_1 + x_2 = 0,$
$\quad\quad x_2 = 1.$

d. $x_1 - x_2 + 2x_3 = 0,$
$\quad x_1 + x_2 - 3x_3 = 0.$

2. Find nontrivial solutions for each of the following homogeneous LS:

a. $x_1 + x_2 - x_3 = 0,$
$\quad x_1 - x_2 + x_3 = 0.$

b. $x_1 + x_2 - x_3 = 0,$
$\quad x_1 + x_2 - x_3 = 0.$

c. $\quad x_1 + x_2 + x_3 = 0,$
$\quad 3x_1 + 3x_2 + 3x_3 = 0.$

d. $\quad x_1 + 2x_2 + x_3 + 3x_4 = 0,$
$\quad 2x_1 + 5x_2 + 2x_3 + x_4 = 0,$
$\quad x_1 + 3x_2 + x_3 - x_4 = 0.$

Also write down a solution, if possible, by choosing $x_3 = 1$.

3. Find a nontrivial solution (if it exists) for each of the following homogeneous LS:

a. $\quad\quad x_2 + 2x_3 = 0,$
$\quad x_1 - x_2 + 3x_3 = 0,$
$\quad 2x_1 - x_2 + 8x_3 = 0.$

b. $\quad x_1 + 2x_2 + x_3 = 0,$
$\quad -x_1 + 3x_2 + 4x_3 = 0,$
$\quad -x_1 + 8x_2 + x_3 = 0.$

c. $3x_1 + 2x_2 + 16x_3 + 5x_4 = 0,$
$\quad 2x_2 + 10x_3 + 8x_4 = 0,$
$\quad x_1 + x_2 + 7x_3 + 3x_4 = 0.$

d. $\quad\quad 2x_1 + 4x_3 = 0,$
$\quad\quad 2x_2 + 4x_3 = 0,$
$\quad -4x_1 + 4x_2 + 3x_3 = 0.$

e. $\quad x_1 + x_2 + 2x_3 = x_3,$
$\quad 5x_1 + 3x_2 + 3x_3 = 2x_1.$

4. Without computing, state which of the following LS has nontrivial solutions. Give the reason.

a. $x_1 + 2x_2 + x_3 + 3x_4 = 0.$

b. $\quad x_1 + 2x_2 + x_3 + 3x_4 = 0,$
$\quad 2x_1 + 5x_2 + 2x_3 + x_4 = 0.$

c. $\quad x_1 + 2x_2 + x_3 + 3x_4 = 0,$
$\quad 2x_1 + 5x_2 + 2x_3 + x_4 = 0,$
$\quad x_1 + 3x_2 + x_3 + x_4 = 0.$

d. $x_1 + 2x_2 + x_3 + 3x_4 + 4x_5 = 0,$
$\quad 2x_1 + 5x_2 + 2x_3 + x_4 = 0,$
$\quad x_1 + 3x_2 + x_3 - x_4 + 4x_5 = 0,$
$\quad 3x_1 + 6x_2 + 3x_3 + 9x_4 = 0$

5. (Drill 1.3) Find a nontrivial solution, if it exists, for the following homogeneous LS by first reducing it to row echelon form:

$$x_1 + 2x_2 + x_3 + 2x_4 = 0,$$
$$-x_1 + 3x_2 + 4x_3 - x_4 = 0,$$
$$4x_1 + 3x_2 + 4x_3 + 3x_4 = 0,$$
$$3x_1 + 3x_2 + 2x_3 + x_4 = 0.$$

6. (Drill 1.4) Find the general solution for the following homogeneous LS by first reducing it to row echelon form:

$$x_1 + 2x_2 + x_3 + 3x_4 + 2x_5 = 0,$$
$$2x_1 + 5x_2 + 2x_3 + x_4 + 2x_5 = 0,$$
$$x_1 + 3x_2 + x_3 - 2x_4 + x_5 = 0,$$
$$3x_1 + 2x_2 + 3x_3 - x_4 + x_5 = 0.$$

7. For each of the following LS, find a value of the constant c such that the LS has nontrivial solutions:

a. $2x_1 + 5x_2 = 0,$
$\quad 3x_1 - cx_2 = 0.$

b. $cx_1 + x_2 = 0,$
$\quad 2x_1 + 3x_2 = 0.$

c. $\quad x_1 + cx_2 = 0,$
$\quad -5x_1 - x_2 = 0.$

Solve each of the following LS by reducing it to echelon form. Also verify that the sum of any two solutions is a solution and any scalar multiple of a solution is also a solution (see Example 1.3.2(d)).

8. $x_1 + 2x_2 + x_3 + 3x_4 = 0,$
$\quad 2x_1 + 5x_2 + 2x_3 + x_4 = 0,$
$\quad x_1 + 3x_2 + x_3 - x_4 = 0.$

9. $x_1 - x_2 + 2x_3 = 0,$
$\quad x_1 + x_2 - 3x_3 = 0.$

10. $x_1 + x_2 + x_3 + x_4 = 0,$
$\quad x_1 - x_2 + x_3 - x_4 = 0.$

11. $\quad x_1 - 2x_2 + x_3 = 0,$
$\quad\quad x_1 + 3x_3 = 0,$
$\quad x_1 + 4x_2 + 7x_3 = 0.$

12. Show that a linear system, if consistent, has either a unique solution or infinitely many solutions. In other words, no consistent linear systen can have $k \geqslant 2$ solutions, where k is a positive integer.

1.4 Introduction to Matrices and the Matrix of a Linear System

Let

$$a_{11}x_1 + \cdots + a_{1n}x_n = b_1,$$
$$\vdots$$
$$a_{m1}x_1 + \cdots + a_{mn}x_n = b_m,$$

be an $m \times n$ LS in which a_{ij} is the coefficient of the jth variable in the ith equation (it may possibly be 0). If we suppress the variables $x_1, \ldots, x_n$ as well as the symbols $+$ and $=$ in the above LS, we are left with a rectangular array of numbers:

$$\begin{bmatrix} a_{11} & \cdots & a_{1n} & b_1 \\ a_{21} & \cdots & a_{2n} & b_2 \\ \vdots & & \vdots & \vdots \\ a_{m1} & \cdots & a_{mn} & b_m \end{bmatrix},$$

which we usually enclose within brackets or parentheses. This array is called the *augmented matrix* of the LS. It consists of m rows and $n + 1$ columns. If the last column is deleted, we get the $m \times n$ rectangular array

$$\begin{bmatrix} a_{11} & \cdots & a_{1n} \\ \vdots & & \vdots \\ a_{m1} & \cdots & a_{mn} \end{bmatrix},$$

which is called the *coefficient matrix* of the LS.

1.4.1 EXAMPLE The augmented matrix for the 4×3 LS

$$7x_1 - 2x_2 + 3x_3 = 2, \tag{1}$$

$$x_1 + \frac{3}{2}x_2 = -1, \tag{2}$$

$$-x_1 + 3x_2 - \frac{4}{3}x_3 = 0, \tag{3}$$

$$-3x_2 + 4x_3 = -2, \tag{4}$$

is

$$\begin{bmatrix} 7 & -2 & 3 & 2 \\ 1 & \frac{3}{2} & 0 & -1 \\ -1 & 3 & -\frac{4}{3} & 0 \\ 0 & -3 & 4 & -2 \end{bmatrix},$$

in which we have inserted a zero in place of any missing unknown. This matrix has four rows and four columns. If we delete the last column, we get the coefficient matrix

$$\begin{bmatrix} 7 & -2 & 3 \\ 1 & \frac{3}{2} & 0 \\ -1 & 3 & -\frac{4}{3} \\ 0 & -3 & 4 \end{bmatrix}.$$

∎

1.4.2 DEFINITION

Matrix
An array of numbers (or symbols) in m rows and n columns is called an $m \times n$ *matrix*.

The notation

$$A = \begin{pmatrix} a_{11} & a_{12} & \cdots & a_{1n} \\ a_{21} & a_{22} & \cdots & a_{2n} \\ \vdots & \vdots & & \vdots \\ a_{m1} & a_{m2} & \cdots & a_{mn} \end{pmatrix} \quad \text{or} \quad A = \begin{bmatrix} a_{11} & a_{12} & \cdots & a_{1n} \\ a_{21} & a_{22} & \cdots & a_{2n} \\ \vdots & \vdots & & \vdots \\ a_{m1} & a_{m2} & \cdots & a_{mn} \end{bmatrix}$$

represents an $m \times n$ matrix A.

We shall write an $m \times n$ matrix A as $(a_{ij})_{m \times n}$, or simply (a_{ij}). The entry that appears at the intersection of the ith row and jth column is called the (i, j) entry. The symbol a_{ij} denotes the (i, j) entry of the matrix $A = (a_{ij})$, $i = 1, \ldots, m$ and $j = 1, \ldots, n$. Sometimes it is convenient to write the (i, j) entry of A as $(A)_{ij}$.

An $n \times n$ matrix is called a *square* matrix of order n. In other words, the number of rows in a square matrix is equal to the number of columns. If the number of rows is not equal to the number of columns, then the matrix is called *rectangular*.

A matrix obtained from a given matrix A by deleting some (but of course not all) of its rows and columns is called the submatrix of A.

The term *matrix* was first introduced by the British mathematician James Joseph Sylvester in 1890. The word "matrix" is derived from the Indo-European root *mater*, meaning "mother." Matrices are indeed the core of linear algebra.

A linear system is determined by its augmented matrix. If there are two linear systems, they will be equivalent if their augmented matrices are identical, which are defined to be equal in the following sense.

1.4.3 Remark Since any number a may be looked upon as the 1×1 matrix $[a]$, it is customary to identify $[a]$ with a.

1.4.4 DEFINITION

Equality of Matrices

Let $A = (a_{ij})_{m \times n}$, $B = (b_{ij})_{p \times q}$. Then A is said to be *equal* to B if

i. A and B have the same number of rows and the same number of columns, that is, $m = p, n = q$, and

ii. the corresponding entries of A and B are equal, that is, $a_{ij} = b_{ij}$ for each pair of subscripts (i, j).

Many real-life problems can be conveniently described by means of a diagram consisting of a set of points together with lines joining certain pairs of these points. For example, the points could be cities, and the lines joining the points could be airline routes. In such diagrams, what is important is whether or not two points are joined. The manner in which they are joined is immaterial.

1.4.5 DEFINITION

Graphs, Vertices, and Edges

By a *graph*, we mean a pair (V, E), where V is a finite set and E is a list (possibly empty) of 1×2 matrices whose entries belong to V.

The elements of V are called *vertices*, and those of E are called *edges* or *paths*. If $e = [a \ \ b]$ is in E, then e is called an edge (or a path of length 1) from a to b. If e appears in the list E twice, we say that there are two edges (or paths of lengths 1) from a to b.

Graphs that have the additional property that if there are k edges from a to b, then there are also k edges from b to a, are called *undirected graphs*. In undirected graphs, the direction of the line segments between the vertices is immaterial, and therefore it is not indicated.

Graphs, in general, defined in *Definition 1.4.5* are known as *directed graphs*.

To represent a graph pictorially, we indicate each vertex by a point and each edge by a line, not necessarily a straight line, joining two vertices. There is no unique way of representing a graph or assigning positions of vertices. A graph simply shows some relation that holds between its vertices and edges.

1.4.6 EXAMPLE Let $V = \{a, b, c, d\}$ be the set of vertices, and let $E = [a \ b], [a \ c], [b \ d], [a \ a]$. Pictorially, the directed and undirected graphs corresponding to the given data are respectively

$$\begin{array}{ccc} @\!\!\longrightarrow & b & \\ & \searrow & \downarrow, \\ & c & d \end{array} \quad \text{and} \quad \begin{array}{ccc} @\!\!\longrightarrow & b \\ & \backslash & | \\ & c & d \end{array}.$$

To any graph G with n vertices, there corresponds an $n \times n$ matrix called the vertex or adjacency matrix of G. ∎

1.4.7 DEFINITION

> **Vertex or Adjacency Matrix**
> Let G be a graph with n vertices $P_1, \ldots, P_n$. The *vertex* (or *adjacency*) matrix A of the graph G is the $n \times n$ matrix whose (i, j) entry is the number of edges (paths) from P_i to P_j.

1.4.8 EXAMPLE The vertex (or adjacency) matrices of the graphs in the above example are, respectively,

$$\begin{bmatrix} 1 & 1 & 1 & 0 \\ 0 & 0 & 0 & 1 \\ 0 & 0 & 0 & 0 \\ 0 & 0 & 0 & 0 \\ 0 & 0 & 0 & 0 \end{bmatrix} \quad \text{and} \quad \begin{bmatrix} 1 & 1 & 1 & 0 \\ 1 & 0 & 0 & 1 \\ 1 & 0 & 0 & 0 \\ 0 & 1 & 0 & 0 \end{bmatrix}.$$

∎

Note that the $(1, 1)$, $(1, 2)$, $(1, 3)$, and $(1, 4)$ entries of the vertex matrices are 1, 1, 1, and 0, respectively, because there is one path from vertex 1 to vertices 1, 2, 3; no path from vertex 1 to vertex 4; and so on. In the second matrix that corresponds to the undirected graph, if the (i, j) entry is nonzero, then the (j, i) entry is also nonzero because if there is a path from i to j, then it is implied that there is also a path from j to i.

1.4.9 DEFINITION

> **Path of Length k**
> We say that there is a path of length $k \geqslant 1$ from a vertex a to a vertex b if there exist $k - 1$ vertices $v_1, \ldots, v_{k-1}$, possibly repeating, such that there is an edge from a to v_1, v_1 to v_2, v_2 to v_3, $\ldots$, v_{k-2} to v_{k-1}, and v_{k-1} to b.

Speaking informally, we say that there is a path of length k from a to b if you pass through exactly $k - 1$ vertices (possibly repeating) to come from a to b. Path of length 1 means there is an edge from a to b.

Thus the entries of the vertex (or adjacency) matrix count the number of paths of length 1 between vertices of the graph. If there is exactly one vertex j of the graph between initial vertex i and vertex k, we say that there is a path of length 2 from i to k. In general, if there are exactly $k - 1$ vertices of the graph between vertex i and vertex j, we say that there is a path of length k from i to j. (Repeated vertices are counted as many times as their multiplicity.)

1.4 Exercises

1. Find the coefficient matrix and the augmented matrix for each of the following LS:

a. $2x_1 - 3x_2 + x_3 = 0,$
$5x_1 + 6x_3 = 1,$
$x_1 - x_2 + x_3 = 5.$

b. $5x_1 + 6x_2 + 7x_3 + 8x_4 = 9.$

d. $2x_1 + 3x_2 + 4x_3 + 7x_4 = 1,$
$5x_1 + 6x_2 + 7x_3 + 8x_4 = 2,$
$x_2 - 6x_4 = 0.$

c. $-x_1 + 2x_2 + 6x_3 = 0,$
$x_1 - x_2 + 7x_3 = 0.$

2. Find the LS whose augmented matrix is

a. $\begin{bmatrix} 1 & 2 & 3 & 4 \\ 5 & 6 & 7 & 8 \end{bmatrix}.$

b. $\begin{bmatrix} 0 & 1 & 2 \\ 1 & 0 & 3 \\ 2 & 1 & 8 \end{bmatrix}.$

c. $\begin{bmatrix} -1 & -2 & 3 & 4 \\ 0 & 5 & 1 & 1 \\ 7 & 6 & 5 & 3 \end{bmatrix}.$

d. $\begin{bmatrix} 2 & 4 & 6 & 8 & 10 \\ 1 & 3 & 5 & 7 & 9 \\ -1 & -2 & 0 & 0 & 1 \\ 1 & 0 & 0 & 6 & 5 \end{bmatrix}.$

e. What is the coefficient matrix for each of the linear systems obtained in (a)–(d)?

3. Find the values of x, y, and z in the following equations:

a. $\begin{bmatrix} x^2 & 1 \\ 2 & 3 \end{bmatrix} = \begin{bmatrix} 5x - 6 & 1 \\ 2 & 3 \end{bmatrix}.$

b. $\begin{bmatrix} x & 2 \\ y & z \end{bmatrix} = \begin{bmatrix} y+1 & y \\ 2 & z \end{bmatrix}.$

c. $\begin{bmatrix} y & x & -x \\ 0 & 1 & x \\ 2 & 0 & 0 \end{bmatrix} = \begin{bmatrix} 4-x & x^2 & -1 \\ 0 & x & x \\ 2 & 0 & 0 \end{bmatrix}.$

d. $\begin{bmatrix} x & y \\ z & 1 \end{bmatrix} = \begin{bmatrix} y & z \\ 1 & 1 \end{bmatrix}.$

e. $\begin{bmatrix} x & y \\ y & 0 \end{bmatrix} = \begin{bmatrix} 1+y & 1+x \\ y & 0 \end{bmatrix}.$

4. Get a flight route map of your favorite airline. Write down the vertex (or adjacency) matrix A of the flights between 10 of the cities to which the airline flies.

5. Let $G = (V, E)$ be a directed graph, where $V = \{1, 2, 3, 4\}$ and $E = [1\ 2], [2\ 3], [3\ 3], [3\ 4], [2\ 4],$ and $[3\ 4].$

a. Give a pictorial representation of the graph.

b. Write down the adjacency matrix of the graph.

6. Let $G = (V, E)$ be a directed graph, where $V = \{1, 2, 3, 4, 5\}$ and $E = [1\ 1], [1\ 2],$ $[3\ 4], [1\ 4], [4\ 5], [4\ 5],$ and $[5\ 2]$.

    a. Give a pictorial representation of the graph.
    b. Write down the adjacency matrix of the graph.

7. Give a pictorial representation of the graphs whose adjacency matrices are given below. Find whether the graph is directed or undirected.

$$\textbf{a. } A = \begin{bmatrix} 0 & 2 & 1 & 1 \\ 2 & 0 & 1 & 0 \\ 1 & 1 & 0 & 1 \\ 1 & 0 & 1 & 1 \end{bmatrix}, \qquad \textbf{b. } B = \begin{bmatrix} 0 & 0 & 1 & 1 & 0 \\ 0 & 0 & 1 & 1 & 1 \\ 1 & 1 & 0 & 0 & 1 \\ 1 & 1 & 0 & 0 & 2 \\ 0 & 1 & 1 & 2 & 0 \end{bmatrix}.$$

1.5 Elementary Row Operations on a Matrix

We showed earlier that the Gauss elimination method for solving a linear system consists of performing suitable elementary operations $R_i \longleftrightarrow R_j$, αR_i, and $R_j + \alpha R_i$. Now, for every elementary operation on an LS we can define, in a natural way, the corresponding operation on the augmented matrix of the LS, to be denoted by the same symbol. We illustrate this by the following example, in which the linear system S is written on the left and the augmented matrix M is written on the right.

<table>
<tr><th>Linear System S</th><th>Augmented Matrix M</th></tr>
</table>

$$\begin{array}{l} 3x_1 + 4x_2 - 6x_3 = 4, \\ x_1 - x_2 + 4x_3 = 1, \\ -x_1 + 2x_2 - 7x_3 = 0; \end{array} \qquad \begin{bmatrix} 3 & 4 & -6 & 4 \\ 1 & -1 & 4 & 1 \\ -1 & 2 & -7 & 0 \end{bmatrix};$$

$$\downarrow R_1 \longleftrightarrow R_2; \qquad\qquad\qquad \downarrow R_1 \longleftrightarrow R_2;$$

$$\begin{array}{l} x_1 - x_2 + 4x_3 = 1, \\ 3x_1 + 4x_2 - 6x_3 = 4, \\ -x_1 + 2x_2 - 7x_3 = 0; \end{array} \qquad \begin{bmatrix} 1 & -1 & 4 & 1 \\ 3 & 4 & -6 & 4 \\ -1 & 2 & -7 & 0 \end{bmatrix};$$

$$\left\downarrow \begin{array}{l} R_2 + (-3)R_1 \text{ and} \\ R_3 + 1R_1; \end{array} \right. \qquad \left\downarrow \begin{array}{l} R_2 + (-3)R_1 \text{ and} \\ R_3 + 1R_1; \end{array} \right.$$

$$\begin{array}{l} x_1 - x_2 + 4x_3 = 1, \\ 7x_2 - 18x_3 = 1, \\ x_2 - 3x_3 = 1; \end{array} \qquad \begin{bmatrix} 1 & 1 & 1 & 1 \\ 0 & 7 & -18 & 1 \\ 0 & 1 & -3 & 1 \end{bmatrix};$$

$$\downarrow \tfrac{1}{7}R_2; \qquad\qquad\qquad \downarrow \tfrac{1}{7}R_2;$$

$$\begin{array}{l} x_1 - x_2 + 4x_3 = 1, \\ x_2 - \tfrac{18}{7}x_3 = \tfrac{1}{7}, \\ x_2 - 3x_3 = 1; \end{array} \qquad \begin{bmatrix} 1 & -1 & 4 & 1 \\ 0 & 1 & -\tfrac{18}{7} & \tfrac{1}{7} \\ 0 & 1 & -3 & 1 \end{bmatrix};$$

Linear System S Augmented Matrix M

$$\downarrow R_3 + (-1)R_2;$$

$$\downarrow R_3 + (-1)R_2;$$

$$\begin{aligned} x_1 - x_2 + 4x_3 &= 1, \\ x_2 - \tfrac{18}{7}x_3 &= \tfrac{1}{7}, \\ -\tfrac{3}{7}x_3 &= \tfrac{6}{7}; \end{aligned}$$

$$\begin{bmatrix} 1 & -1 & 4 & 1 \\ 0 & 1 & -\tfrac{18}{7} & \tfrac{1}{7} \\ 0 & 0 & -\tfrac{3}{7} & \tfrac{6}{7} \end{bmatrix}.$$

Solving backward, we obtain

$$x_3 = -2,$$
$$x_2 = \frac{18}{7}(-2) + \frac{1}{7} = -\frac{35}{7} = -5,$$
$$x_1 = -5 - 4(-2) + 1 = 4.$$

We note the important fact that the augmented matrix of the reduced linear system is the matrix to its right. Such a matrix is called an echelon matrix—or rather a row echelon matrix.

1.5.1
DEFINITION

Row Echelon Form

A matrix is said to be in *row echelon form*, or simply an *echelon matrix*, if

  i. the zero rows, if any, are below all nonzero rows and

  ii. the first nonzero entry in any row is to the right of the first nonzero entry of the previous row. (This statement puts no restriction on the first row.)

1.5.2
Remark

Some authors require that the first entry in any row of a matrix in row echelon form be 1. This is not necessary, but sometimes it is helpful to have the first nonzero entry of each row equal to 1.

1.5.3
DEFINITION

Reduced Row Echelon Form

A matrix is said to be in *reduced row echelon form* if

  i. the matrix is in row echelon form with 1 as the first nonzero entry in each nonzero row and

  ii. the first nonzero entry in each row is the only nonzero entry of its column.

For example, the matrices

$$\begin{bmatrix} 1 & 2 \\ 0 & 3 \end{bmatrix}, \quad \begin{bmatrix} 1 & 2 \\ 0 & 0 \end{bmatrix}, \quad \begin{bmatrix} 2 & 3 & 7 \\ 0 & 0 & 1 \\ 0 & 0 & 0 \end{bmatrix}, \quad \begin{bmatrix} -1 & 0 & 0 \\ 0 & 1 & 0 \\ 0 & 0 & 5 \end{bmatrix}, \quad \begin{bmatrix} 1 & 2 & 3 & 4 \\ 0 & 2 & 5 & 6 \\ 0 & 0 & 0 & 1 \end{bmatrix}$$

are in row echelon form, but none except the second matrix is in reduced row echelon form.

The operations on a matrix described in the example at the beginning of this section are called elementary row operations.

1.5.4 **DEFINITION**	**Elementary Row Operations on a matrix** The elementary row operations on a matrix are of three types. **I.** $R_i \longleftrightarrow R_j$: Interchange row i and row j. **II.** αR_i: Multiply the ith row by a nonzero scalar α. **III.** $R_j + \alpha R_i$: Add α times the ith row to the jth row $(i \neq j)$.

We now describe the procedure to transform a given matrix into row echelon form or reduced row echelon form by performing suitable elementary operations.

- Select one of the rows that has its first entry nonzero. If this happens to be the first row, then nothing needs to be done. Otherwise, interchange the selected row, say the ith, and the first row. This amounts to the elementary operation $R_1 \longleftrightarrow R_i$.
- Next, perform the operation αR_1 on the transformed matrix to make the first entry of the first row equal to 1.
- Perform operations $R_i + \alpha R_1$, $i > 1$, to make all entries that are below the first entry of the first row equal to 0.
- Ignoring temporarily the first row and first column, repeat the above process on the matrix obtained by disregarding the first row and first column. This means that we move to the right and pick the first nonzero column. Suppose this column has a nonzero entry in the jth row (this row need not be the first row containing the nonzero entry of that column). Bring the jth row to the top by interchange of rows (type I operation), and make the nonzero entry 1 by performing elementary row operation αR_1 of type II. As before, with the help of this entry 1, make all entries of the column below it 0. We then ignore the first two rows and the two columns we had worked with to create ones and zeros below 1.
- Continue the process until we arrive at row echelon form.

To transform the matrix in row echelon form to reduced row echelon form, we proceed as follows:

- Begin from the first nonzero row from the bottom. Get hold of the first nonzero entry of that row. Perform elementary row operations to make all entries in the column to which that entry belongs as zero. Then move upward to the next row and repeat the process, continuing to the top.

Application of the Gauss elimination method amounts to the following steps:

Step 1 Write the augmented matrix of the LS.

Step 2 By elementary row operations, reduce the augmented matrix to an echelon matrix.

Step 3 Write the linear system that corresponds to the echelon matrix obtained in Step 2. This linear system has the same solution as the original linear system.

Step 4 Solve the new system by backward substitution. This means that we solve the system beginning with the solution of the last equation and then move upward one by one, successively, to solve each equation.

1.5.5
EXAMPLES

a. Solve

$$x_1 + x_2 + 2x_3 = 6,$$
$$3x_1 + 4x_2 - x_3 = 5,$$
$$5x_1 + 6x_2 + 3x_3 = 17.$$

The augmented matrix is

$$\begin{bmatrix} 1 & 1 & 2 & 6 \\ 3 & 4 & -1 & 5 \\ 5 & 6 & 3 & 17 \end{bmatrix} \quad \overset{R_2 + (-3)R_1,\ R_3 + (-5)R_1}{\longrightarrow} \quad \begin{bmatrix} 1 & 1 & 2 & 6 \\ 0 & 1 & -7 & -13 \\ 0 & 1 & -7 & -13 \end{bmatrix}$$

$$\overset{R_3 + (-1)R_2}{\longrightarrow} \quad \begin{bmatrix} 1 & 1 & 2 & 6 \\ 0 & 1 & -7 & -13 \\ 0 & 0 & 0 & 0 \end{bmatrix}.$$

The new LS is

$$x_1 + x_2 + 2x_3 = 6,$$
$$x_2 - 7x_3 = -13.$$

Since the last equation has two unknowns, we set one of them, say x_3, equal to t. Then $x_2 = 7t - 13$, and so $x_1 = 6 - x_2 - 2x_3 = 6 - 7t + 13 - 2t = 19 - 9t$.

We may, if we wish, reduce the augmented matrix further to reduced row echelon form. We note first that the first nonzero entry in each of the nonzero rows of row echelon form is 1. We perform $R_1 - R_2$ to obtain

$$\begin{bmatrix} 1 & 0 & 9 & 19 \\ 0 & 1 & -7 & -13 \end{bmatrix}.$$

This yields

$$x_2 - 7x_3 = -13,$$
$$x_1 + 9x_3 = 10,$$

which gives the same answer as above.

b. Solve

$$2x_1 - 3x_2 + 7x_3 + 5x_4 = 7,$$
$$x_1 - x_2 + x_3 - x_4 = 1,$$
$$x_1 + x_2 - 8x_3 + x_4 = 0,$$
$$4x_1 - 3x_2 + 5x_4 = 0.$$

The augmented matrix is

$$
\begin{bmatrix}
2 & -3 & 7 & 5 & 7 \\
1 & -1 & 1 & -1 & 1 \\
1 & 1 & -8 & 1 & 0 \\
4 & -3 & 0 & 5 & 0
\end{bmatrix}
\begin{array}{l} R_1 \longleftrightarrow R_2 \\ \longrightarrow \end{array}
\begin{bmatrix}
1 & -1 & 1 & -1 & 1 \\
2 & -3 & 7 & 5 & 7 \\
1 & 1 & -8 & 1 & 0 \\
4 & -3 & 0 & 5 & 0
\end{bmatrix}
\begin{array}{l} R_2 + (-2)R_1 \\ \longrightarrow \\ R_3 + (-1)R_1 \\ R_4 + (-4)R_1 \end{array}
$$

$$
\begin{bmatrix}
1 & -1 & 1 & -1 & 1 \\
0 & -1 & 5 & 7 & 5 \\
0 & 2 & -9 & 2 & -1 \\
0 & 1 & -4 & 9 & -4
\end{bmatrix}
\begin{array}{l} R_3 + 2R_2 \\ \longrightarrow \\ R_4 + 1R_2 \end{array}
\begin{bmatrix}
1 & -1 & 1 & -1 & 1 \\
0 & -1 & 5 & 7 & 5 \\
0 & 0 & 1 & 16 & 9 \\
0 & 0 & 1 & 16 & 1
\end{bmatrix}
\begin{array}{l} R_4 + (-1)R_3 \\ \longrightarrow \end{array}
$$

$$
\begin{bmatrix}
1 & -1 & 1 & -1 & 1 \\
0 & -1 & 5 & 7 & 5 \\
0 & 0 & 1 & 16 & 9 \\
0 & 0 & 0 & 0 & -8
\end{bmatrix}.
$$

The corresponding LS is

$$
\begin{aligned}
x_1 - x_2 + x_3 - x_4 &= 1, \\
-x_2 + 5x_3 + 7x_4 &= 5, \\
x_3 + 16x_4 &= 9, \\
0 &= -8.
\end{aligned}
$$

Since the last equation is absurd, the given LS is inconsistent.

c. A veterinarian recommends that a certain pet's diet should contain 100 units of protein, 200 units of carbohydrates, and 50 units of fat daily. A store's pet food department contains four varieties of foods with the following composition of protein, carbohydrate, and fat (in units) per ounce in foods A, B, C, and D:

Food	Protein	Carbohydrates	Fat
A	5	20	2
B	4	25	2
C	7	10	10
D	10	5	6

Find, if possible, the amounts of foods A, B, C, and D that can be included in the pet's diet to conform to the veterinarian's recommendation.

Let x_1, x_2, x_3, and x_4 be amounts in ounces of foods A, B, C, and D, respectively, to come up with the diet recommended. Then we have the system of equations

$$
\begin{aligned}
5x_1 + 4x_2 + 7x_3 + 10x_4 &= 100, \\
20x_1 + 25x_2 + 10x_3 + 5x_4 &= 200, \\
2x_1 + 2x_2 + 10x_3 + 6x_4 &= 50.
\end{aligned}
$$

The augmented matrix of the linear system is

$$\left[\begin{array}{ccccc|c} 5 & 4 & 7 & 10 & 100 \\ 20 & 25 & 10 & 5 & 200 \\ 2 & 2 & 10 & 6 & 50 \end{array}\right] \begin{array}{l} \frac{1}{2}R_3, \frac{1}{5}R_2 \\ \longrightarrow \\ R_1 \longleftrightarrow R_3 \end{array}$$

$$\left[\begin{array}{ccccc} 1 & 1 & 5 & 3 & 25 \\ 4 & 5 & 2 & 1 & 40 \\ 5 & 4 & 7 & 10 & 100 \end{array}\right] \begin{array}{l} R_2 - 4R_1 \\ \longrightarrow \\ R_3 - 5R_1 \end{array}$$

$$\left[\begin{array}{ccccc} 1 & 1 & 5 & 3 & 25 \\ 0 & 1 & -18 & -11 & -60 \\ 0 & -1 & -18 & -5 & -25 \end{array}\right] \begin{array}{l} R_2 + R_3 \\ \longrightarrow \\ (-1)R_3 \end{array}$$

$$\left[\begin{array}{ccccc} 1 & 1 & 5 & 3 & 25 \\ 0 & 1 & -18 & -11 & -60 \\ 0 & 0 & 36 & 16 & 85 \end{array}\right]$$

The above matrix is in row echelon form, and the corresponding linear system of equations is

$$x_1 + x_2 + 5x_3 + 3x_4 = 25,$$
$$x_2 - 18x_3 - 11x_4 = -60,$$
$$36x_3 + 16x_4 = 85.$$

We now solve backward with the understanding that only nonnegative values of the variables are acceptable. Since the last equation has two unknowns, we as usual set one of the variables, say x_4, to have arbitrary value t. Then from the last equation, $36x_3 + 16x_4 = 85$, by plugging in $x_4 = t$, we get $x_3 = \frac{1}{36}(85 - 16t)$. Moving up to the next equation, $x_2 - 18x_3 - 11x_4 = -60$, by plugging in the values of x_3 and x_4, we obtain $x_2 - 18\left[\frac{1}{36}(85 - 16t)\right] - 11t = -60$. On simplifying, we get $x_2 - \frac{85}{2} + 8t - 11t = -60$, whence $x_2 = 3t + \frac{85}{2} - 60$, and so $x_2 = 3t - \frac{35}{2}$. Finally, we can find the value of x_1 by moving up to the next equation, $x_1 + x_2 + 5x_3 + 3x_4 = 25$, and plugging in the values of x_2, x_3, and x_4. But just pause and think whether one can choose nonnegative solutions of the variables whose values have been determined thus far.

First, t must be nonnegative because $x_4 = t$. Next, because $x_3 = 85 - 16t$ is nonnegative, t must be less than or equal to $\frac{85}{16}$. In order for $x_2 = 3t - \frac{35}{2}$ to be nonnegative, t must be greater than or equal to $\frac{35}{6}$. This yields $\frac{35}{6} \leqslant t \leqslant \frac{85}{16}$, that is, $\frac{280}{48} \leqslant t \leqslant \frac{255}{48}$, which is absurd. Thus it is not possible to find a solution for which all the variables have nonnegative values. Equivalently speaking, we cannot form a diet out of the four foods available in the store. However, we must emphasize that the system of equations is consistent and has infinitely many solutions. ∎

1.5 Exercises

1. Solve each of the following linear systems, if consistent, and give the general solution if it has more than one solution.

a. $x_1 + 2x_2 - 3x_3 = 1,$
$\quad x_2 - 2x_3 = 2,$
$\quad 2x_2 - 4x_3 = 4.$

b. $\quad 2x + y + w = 2,$
$\quad 3x + 3y + 3z + 5w = 4,$
$\quad 3x - 3y - 2w = 3.$

 2. (Drill 1.5) Construct the augmented matrix of the following linear system, and reduce the augmented matrix to reduced row echelon form:

$$x_1 + 2x_2 + x_3 + 3x_4 = 0,$$
$$2x_1 + 4x_2 + 3x_3 + x_4 = 0,$$
$$3x_1 + 6x_2 + 6x_3 + 2x_4 = 0.$$

3. Solve each of the following linear systems, if consistent:

a. $\quad x + 2y - z = 4,$
$\quad 2x + 4y + 3z = 5,$
$\quad x + 2y - 6z = 7.$

 b. (Drill 1.6) $\quad 2x_1 + 4x_2 + 3x_3 + 2x_4 = 2,$
$\quad 3x_1 + 6x_2 + 5x_3 + 2x_4 = 2,$
$\quad 2x_1 + 5x_2 + 2x_3 - 3x_4 = 3,$
$\quad 4x_1 + 5x_2 + 14x_3 + 14x_4 = 11.$

c. $\quad 3x_1 - x_2 + 2x_3 = -4,$
$\quad 2x_1 + x_2 + x_3 = -1,$
$\quad x_1 + 3x_2 = 2.$

d. $\quad x_1 - x_2 - x_3 - x_4 = 5,$
$\quad x_1 + 2x_2 + 3x_3 + x_4 = -2,$
$\quad 3x_1 + x_2 + 2x_4 = 1,$
$\quad 2x_1 + 2x_3 + 3x_4 = 3.$

 4. (Drill 1.7) Consider the following linear system:

$$4x - y - z = 1,$$
$$-x + 4y - z - m = 2,$$
$$-x - y + 4z - m - n = 3,$$
$$-y - z + 4m - n = 4,$$
$$-z - m + 4n = 5.$$

a. Construct the augmented matrix, and solve the linear system.
b. Change the constraint vector [1 2 3 4 5] to [1 2 3 4.1 5], and solve the system. For the LS $Ax = b$ the vector b is sometimes referred to as constraint vector or output.
c. Find the difference vector, that is, the solution in(a) minus the solution in (b).

5. (Drill 1.8) A dietitian is to arrange a special diet composed of four foods A B, C, and D. The diet is to include 70 units of calcium, 35 units of iron, 35 units of vitamin A, and 50 units of vitamin B. The following table shows the amount of calcium, iron, vitamin A, and vitamin B (in units) per ounce in foods A, B, C, and D:

Food	Calcium	Iron	Vitamin A	Vitamin B
A	20	5	5	8
B	10	5	15	10
C	10	10	5	10
D	15	15	10	20

a. Write down a system of linear equations that models the given dietary requirement.
b. Solve the system using Matlab.

6. Jan Ritter hires three types of laborers, I, II, and III, and pays them \$8, \$6, and \$5 per hour, respectively. If the total amount paid is \$2800 for a total of 500 hours of work, find the possible number of hours put in by the three categories of workers if the category III workers must put in the maximum amount of hours.

7. A manufacturing company produces three products, I, II, and III. It uses three machines, A, B, and C, for 350, 150, and 100 hours, respectively. Making one thousand items of type I requires 30, 10, and 5 hours of machines A, B, and C, respectively. Making one thousand items of type II requires 20, 10, and 10 hours on machines A, B, and C, respectively. Making one thousand items of type III requires 30, 30, and 5 hours on machines A, B, and C, respectively. Find the number of items of each type of product that can be produced if the machines are used at full capacity.

8. The traffic flow of vehicles at a given rush hour in a town at the intersections of 1st Avenue, 2nd Avenue, 5th Street, and 6th Street is given in the following diagram. Find all possible solutions, and discuss the following cases:

a. When 5th Street is closed between intersections A and D.
b. When 2nd Avenue is closed between intersections C and D.

1.6 Proofs of Facts

THEOREM 1

Every homogenous linear system with m equations in n unknowns has a nontrivial solution if $m < n$.

Proof Reduce the LS to echelon form:

$$c_{11}x_1 + c_{21}x_2 + \cdots + c_{1n}x_n = 0,$$
$$c_{22}x_2 + \cdots + c_{1n}x_n = 0,$$
$$\vdots$$
$$c_{rr}x_r + \cdots + c_{1n}x_n = 0.$$

(Some of $c_{22}, \ldots, c_{rr}$, where $1 \leqslant r \leqslant m < n$, may be 0.)

Case 1
If the bottom equation contains more than one variable with nonzero coefficients, then by assigning parameters (arbitrary numbers) to all but one of these variables and solving backward, we get a nontrivial solution.

Case 2
If the bottom equation has just one variable with nonzero coefficient, the value of the corresponding variable in any possible solution of the LS is 0. We then proceed to the next upper equation and eliminate the variable whose value is to be 0. If this also contains just one variable with a nonzero coefficient, the value of two variables in any possible solution of the LS is 0. We will proceed to the next and eliminate variables whose values are to be 0. There are r steps corresponding to the r equations. So in this way we can eliminate at most r variables whose values are to be 0 in any possible solution. Since $(n - r) > 1$, we are bound to come to an equation with at least two variables with nonzero coefficients whose values cannot be 0 in all possible solutions of the LS. This brings us to Case 1, and we get a nontrivial solution. ∎

1.7   Chapter Review Questions and Project

1. What is meant by consistent and inconsistent linear systems? Give examples. (Section 1.1)

2. Write down three types of elementary (row) operations on a linear system of equations (matrix). (Sections 1.2 and 1.5)

3. What are the coefficient matrix and augmented matrix of a linear system? Explain by means of an example. (Section 1.4)

4. **a.** Carefully define row echelon form and reduced row echelon form of a matrix. (Section 1.5)
   b. Write in your own words steps to reduce a matrix into (i) row echelon form and (ii) reduced row echelon form. (Section 1.5)

c. Reduce the following matrices into (i) row echelon form and (ii) reduced row echelon form:

i. $\begin{bmatrix} 2 & 3 & 4 & 5 \\ 1 & 0 & -1 & 0 \\ 3 & 1 & 0 & 0 \end{bmatrix}$,   **ii.** $\begin{bmatrix} 0 & 0 & 0 \\ 0 & 1 & 2 \\ 0 & 0 & 1 \end{bmatrix}$.

(Section 1.5)

5. a. What is meant by the statement that a homogeneous linear system has a nontrivial solution? (Section 1.3)

b. Can you give a sufficient condition for a homogeneous linear system of m equations in n unknowns to possess a nontrivial solution? (Section 1.3)

c. Find, by actually solving the system, whether the following homogeneous system of equations has a nontrivial solution:

$$x_1 + x_2 + x_3 = 0$$
$$2x_1 - x_2 + 3x_3 = 0$$
$$x_1 - 2x_2 + 2x_3 = 0$$

What conclusion can be drawn with respect to the condition in (b) and the above LS? (Section 1.3)

6. Solve the following linear systems of equations, if consistent:

a. $x_1 - x_2 - 4x_3 = -11$
$-3x_2 - 8x_3 = -23$
$x_1 - 6x_2 - 17x_3 = -51$
$x_2 + 2x_3 = 7$

b. $x_1 - x_2 + x_3 - 2x_4 = 6,$
$-x_1 + x_2 - x_3 + 3x_4 = 2,$
$x_2 - 6x_4 = 0.$

c. $2x_1 + 4x_2 - 6x_3 = 14,$
$x_1 + 2x_2 - 5x_3 + 8x_4 = 9,$
$-x_1 - 2x_2 + x_3 - 6x_4 = 0.$

d. $x_1 - x_2 + x_3 + x_4 = 7$
$4x_1 - 3x_2 + 5x_4 = 5$
$2x_1 - x_2 + x_3 - x_4 = 1$
$x_1 - x_2 + 2x_3 - 2x_4 = 1$

Chapter Project

Consider a traffic flow problem. The (i, j)th entry represents (in thousands) the number of vehicles flowing from junction i to junction j.

		Into Junction						
		1	2	3	4	5	6	Other
	1	0	x_1	0	x_2	0	0	10
	2	x_3	0	x_2	0	0	0	35
From	3	0	x_3	0	0	0	x_6	10
Junction	4	x_4	0	0	0	x_3	0	0
	5	0	x_5	0	x_2	0	x_3	0
	6	0	0	x_3	0	0	0	20
	Other	20	15	0	0	30	10	0

a. Write down the system of equations so that the inflow equals the outflow at any junction.

b. Solve the linear system obtained in part (a).

c. Explain why the flow from junction 4 to junction 1 cannot handle more than 40,000 cars.

d. If the traffic from junction 4 to junction 1 decreases to 5000 vehicles, we make changes so that some streets are one-way or two-way as shown below.

$$
\begin{array}{ccccc}
1 & \longrightarrow & 2 & \longrightarrow & 3 \\
\updownarrow & & \updownarrow & & \updownarrow \\
4 & \longrightarrow & 5 & \longrightarrow & 6
\end{array}
$$

Then the corresponding table of inflow-outflow traffic is given by the following matrix:

		Into Junction						
		1	2	3	4	5	6	Other
	1	0	x_1	0	x_2	0	0	1
	2	0	0	x_3	0	x_4	0	10
From	3	0	0	0	0	0	x_6	20
Junction	4	5	0	0	0	0	0	10
	5	0	x_5	0	0	0	x_8	20
	6	0	0	x_7	0	0	0	69
	Other	10	55	40	5	10	10	0

Write down the system of equations so that the inflow equals the outflow at each junction, and give all possible solutions.

Key Words

Linear Systems
Equivalent Linear Systems
Echelon Form of a Linear System
Gauss Elimination Method
Backward Substitution
Homogeneous Linear Systems
Nontrivial Solution of a Homogenous
   Linear System
Matrix

Equality of Matrices
Coefficient Matrix of a Linear System
Augmented Matrix of a Linear
   System
Elementary Row Operations
Row Echelon Matrix
Reduced Row Echelon Matrix
Graphs, Vertices, Edges
Vertex or Adjacency Matrix

Key Phrases

■ Every matrix can be reduced to row echelon form as well as to reduced row echelon form by performing appropriate elementary row operations.

■ A homogeneous linear system in which the number of unknowns is more than the number of equations always has a nontrivial solution.

2 Algebra of Matrices

Introduction

To process effectively the information embedded in matrices, we must define algebraic operations such as addition, subtraction, and multiplication of matrices. In this chapter, we discuss, among others, definitions of addition and multiplication of matrices, along with a number of related applications to real-life problems.

2.1 Scalar Multiplication and Addition of Matrices

For any matrix $A = (a_{ij})_{m \times n}$, we consider the ith row of A, for each i, as a $1 \times n$ matrix

$$[a_{i1} \ a_{i2} \ \cdots \ a_{in}],$$

and the jth column of A, for each j, as an $m \times 1$ matrix

$$\begin{bmatrix} a_{1j} \\ a_{2j} \\ \vdots \\ a_{mj} \end{bmatrix}.$$

In the process of reducing a matrix A to row echelon form we perform certain operations on the rows of A, that is, on $1 \times n$ matrices. For example,

i. we multiply the ith row by α as follows:

$$\alpha[a_{i1} \ a_{i2} \ \cdots \ a_{in}] = [\alpha a_{i1} \ \alpha a_{i2} \ \cdots \ \alpha a_{in}],$$

and

ii. we add the jth row to the kth row in the following way:

$$[a_{j1} \ a_{j2} \ \cdots \ a_{jn}] + [a_{k1} \ a_{k2} \ \cdots \ a_{kn}] = [a_{j1} + a_{k1} \ a_{j2} + a_{k2} \ \cdots \ a_{jn} + a_{kn}].$$

Such operations on $1 \times n$ matrices can be extended in a natural way to more general matrices.

2.1.1 DEFINITION

Scalar Multiplication

Let α be a number, and let $A = (a_{ij})_{m \times n}$ be a matrix. Then

$$\alpha A = A\alpha = (\alpha a_{ij})_{m \times n}.$$

This means that αA is the matrix obtained by multiplying each entry of A by the same number α.

2.1.2 EXAMPLE

$$3 \begin{bmatrix} 1 & 2 & 5 \\ 0 & 6 & 7 \end{bmatrix} = \begin{bmatrix} 3 & 6 & 15 \\ 0 & 18 & 21 \end{bmatrix}.$$

∎

2.1.3 DEFINITION

Addition of Matrices

Let $A = (a_{ij})_{m \times n}$ and $B = (b_{ij})_{m \times n}$ be $m \times n$ matrices. Then

$$A + B = (c_{ij})_{m \times n}, \quad \text{where } c_{ij} = a_{ij} + b_{ij} \text{ for each pair } (i, j).$$

We note that

i. If A is an $m \times n$ matrix and B is a $p \times q$ matrix, then $A + B$ is defined only if $m = p$ and $n = q$; and

ii. Addition of matrices is componentwise, that is, we add the (i, j) entries of A and B to get the (i, j) entry of $A + B$.

2.1.4 EXAMPLE

$$\begin{bmatrix} 1 & 2 & -1 \\ 3 & 4 & 6 \end{bmatrix} + \begin{bmatrix} 2 & 5 & 7 \\ 0 & 1 & 3 \end{bmatrix} = \begin{bmatrix} 3 & 5 & 6 \\ 3 & 5 & 9 \end{bmatrix}.$$

Matrix addition is both commutative and associative, that is, for $m \times n$ matrices A, B, C,

a. $A + B = B + A$ (commutative law of addition).

b. $(A + B) + C = A + (B + C)$ (associative law of addition).

The reader who is interested in their proofs may go to *Section 2.4 (Theorem 1)*.

∎

2.1.5 DEFINITION

Negative of a Matrix

Let $A = (a_{ij})$ be an $m \times n$ matrix. The *negative* of the matrix A is the $m \times n$ matrix $B = (b_{ij})$ such that $b_{ij} = -a_{ij}$ for all i, j.

The negative of A is written as $-A$.

2.1.6
EXAMPLE   If

$$A = \begin{bmatrix} 1 & 2 & 3 \\ 4 & 5 & 6 \end{bmatrix}, \qquad \text{then} \qquad -A = \begin{bmatrix} -1 & -2 & -3 \\ -4 & -5 & -6 \end{bmatrix}.$$

■

2.1.7
DEFINITION

Zero Matrix
The $m \times n$ matrix in which every entry is 0 is called the $m \times n$ *zero matrix* and is denoted by $0_{m \times n}$ or simply 0.

Note that

a. $A + (-A) = 0$,
b. $(-A) + A = 0$.

We usually write $A + (-B)$ as $A - B$.

2.1.8
EXAMPLE   Let $A = \begin{bmatrix} 1 & 2 \\ 3 & 4 \end{bmatrix}$ and $B = \begin{bmatrix} 5 & 6 \\ 7 & 8 \end{bmatrix}$. We find X such that $4A + 2X = B$.

By adding $-4A$ on both sides of the given matrix equation, we obtain $-4A + (4A + 2X) = -4A + B$. This yields (using associativity and the property of negative of a matrix)

$$2X = -4A + B = -4 \begin{bmatrix} 1 & 2 \\ 3 & 4 \end{bmatrix} + \begin{bmatrix} 5 & 6 \\ 7 & 8 \end{bmatrix}$$

$$= \begin{bmatrix} -4 & -8 \\ -12 & -16 \end{bmatrix} + \begin{bmatrix} 5 & 6 \\ 7 & 8 \end{bmatrix}$$

$$= \begin{bmatrix} -4+5 & -8+6 \\ -12+7 & -16+8 \end{bmatrix} = \begin{bmatrix} 1 & -2 \\ -5 & -8 \end{bmatrix}.$$

Thus

$$X = \frac{1}{2} \begin{bmatrix} 1 & -2 \\ -5 & -8 \end{bmatrix} = \begin{bmatrix} \frac{1}{2} & -1 \\ -\frac{5}{2} & -4 \end{bmatrix}.$$

■

2.1 Exercises

1. Let the matrices A, B, and I be given by

$$A = \begin{bmatrix} 2 & 1 & 2 \\ 3 & 1 & 3 \\ 1 & 2 & 1 \end{bmatrix}, \qquad B = \begin{bmatrix} 2 & 9 & 2 \\ -1 & -6 & 3 \\ 2 & 2 & 1 \end{bmatrix}, \qquad I = \begin{bmatrix} 1 & 0 & 0 \\ 0 & 1 & 0 \\ 0 & 0 & 1 \end{bmatrix}.$$

a. Find $A - 2B$ and $6A + 7B$.
b. Find a matrix X such that $A - B + 10I + X = 0$ (0 is the 3×3 zero matrix).

2. Find a matrix X such that

   a. $4A + 3X = B$,
   b. $B - 5X = A$,
   c. $2A - 3B = 7X$,

   where

$$A = \begin{bmatrix} 1 & 2 & 3 \\ 0 & 1 & 0 \\ -1 & 1 & 1 \end{bmatrix}, \quad \text{and} \quad B = \begin{bmatrix} 0 & 1 & 0 \\ 6 & 7 & -1 \\ -2 & 0 & -1 \end{bmatrix}.$$

3. Find (i) $A + 2B + 3C$, (ii) $5A + 3B - C$, and (iii) $B + C - A$, where

$$A = \begin{bmatrix} 2 & 1 & 0 \\ 1 & 0 & 0 \end{bmatrix}, \quad B = \begin{bmatrix} -1 & 6 & -2 \\ 0 & 1 & 1 \end{bmatrix}, \quad C = \begin{bmatrix} 1 & 2 & 3 \\ 4 & 5 & 6 \end{bmatrix}.$$

4. For each pair of matrices A, B given below, find a matrix X such that $A + X = B$.

   a. $A = \begin{bmatrix} 2 & 3 \\ -1 & 5 \end{bmatrix}, \quad B = \begin{bmatrix} -1 & 0 \\ 0 & 0 \end{bmatrix}.$

   b. $A = \begin{bmatrix} 1 & 2 & 3 \\ 2 & -1 & 5 \end{bmatrix}, \quad B = \begin{bmatrix} 0 & 0 & 0 \\ 0 & 0 & 0 \end{bmatrix}.$

   c. $A = \begin{bmatrix} 0 & 0 & 0 \\ 0 & 0 & 0 \\ 0 & 0 & 0 \end{bmatrix}, \quad B = \begin{bmatrix} -1 & -1 & -1 \\ 2 & 2 & 2 \\ -2 & -2 & -2 \end{bmatrix}.$

5. Find a number α such that $B = \alpha A$.

   a. $B = \begin{bmatrix} 1 & 1 & 1 \\ 0 & 1 & 2 \end{bmatrix}, \quad A = \begin{bmatrix} \frac{1}{3} & \frac{1}{3} & \frac{1}{3} \\ 0 & \frac{1}{3} & \frac{2}{3} \end{bmatrix}.$

   b. $B = \begin{bmatrix} 3 & 6 & -12 \\ 15 & 30 & -3 \\ 0 & 0 & 0 \end{bmatrix}, \quad A = \begin{bmatrix} 1 & 2 & -4 \\ 5 & 10 & -1 \\ 0 & 0 & 0 \end{bmatrix}.$

6. (Drill 2.1) The matrices A and B represent the number of freshman males and females from the United States and abroad during 1989 and 1990, respectively, enrolled in an Ohio college.

$$A = \begin{array}{c} \\ \text{male} \\ \text{female} \end{array} \begin{array}{cc} \text{Domestic} & \text{Foreign} \\ \begin{bmatrix} 2000 & 500 \\ 1300 & 100 \end{bmatrix}, \end{array}$$

$$B = \begin{array}{c} \\ \text{male} \\ \text{female} \end{array} \begin{array}{cc} \text{Domestic} & \text{Foreign} \\ \begin{bmatrix} 2500 & 300 \\ 1400 & 200 \end{bmatrix}. \end{array}$$

a. Find the matrix showing the increase or decrease of students in 1990 over 1989.

b. If from 1990 to 1991 the student population changed in the opposite direction by 75% of the population change from 1989 to 1990, find the population matrix for 1991.

7. (Drill 2.2) The production of goods of types I, II, and III in January and February is given (in thousands) by the matrix

$$
\begin{array}{c} \\ \text{Jan.} \\ \text{Feb.} \end{array}
\begin{array}{ccc} \text{I} & \text{II} & \text{III} \end{array} \\
\left[\begin{array}{ccc} 7 & 6 & 4 \\ 8 & 5 & 3 \end{array} \right].
$$

The quality control manager rejects a percentage of items as given in the following matrix:

$$
\begin{array}{c} \\ \text{Jan.} \\ \text{Feb.} \end{array}
\begin{array}{ccc} \text{I} & \text{II} & \text{III} \end{array} \\
\left[\begin{array}{ccc} 0.5 & 0.3 & 0.4 \\ 0.5 & 0.3 & 0.3 \end{array} \right].
$$

Write down the matrix of shippable items for sale.

8. (Drill 2.3) Consider the following two tables from Nasdaq market history. The matrices A and B show Nasdaq data for the last 10 years for August 1 and September 1, respectively.

$$
A =
\begin{array}{c}
\text{Percent} \\
\text{change} \quad \text{Change} \quad \text{Volatility} \\
(\%) \qquad\qquad\qquad (\%)
\end{array}
\left[\begin{array}{ccc}
0.89 & 2.7 & 0.95 \\
-0.19 & -0.7 & 0.57 \\
0.18 & 0.7 & 0.31 \\
-0.11 & -0.5 & 0.49 \\
-0.53 & -2.3 & 0.8 \\
0.42 & 2.11 & 0.51 \\
0.37 & 2.69 & 0.31 \\
-1.02 & -10.1 & 1.57 \\
1.66 & 18.26 & 1.68 \\
0.03 & 0.52 & 1.16
\end{array} \right],
\quad
B =
\begin{array}{c}
\text{Percent} \\
\text{change} \quad \text{Change} \quad \text{Volatility} \\
(\%) \qquad\qquad\qquad (\%)
\end{array}
\left[\begin{array}{ccc}
-0.55 & -2.5 & 0.84 \\
-0.97 & -3.6 & 1.18 \\
0.42 & 2.0 & 0.4 \\
0.44 & 2.49 & 0.58 \\
0.44 & 3.31 & 0.56 \\
-0.88 & -6.67 & 0.73 \\
0.06 & 0.64 & 0.36 \\
4.81 & 75.84 & 6.48 \\
0.42 & 11.45 & 1.07 \\
0.66 & 27.98 & 1.58
\end{array} \right].
$$

a. Find the average matrix $\frac{1}{2}(A + B)$, and explain the meaning of entries in this matrix.

b. Find the average volatility on September 1 and also on August 1 over the last 10 years.

9. Let $A = (a_{ij})_{m \times n}$ and $B = (b_{ij})_{n \times n}$ be two matrices. Define $A \leqslant B$ if $a_{ij} \leqslant b_{ij}$ for all i and j. Show the following:

a. For any matrix $C = (c_{ij})_{m \times n}$, if $A \leqslant B$, then $A + C \leqslant B + C$.

b. If the sum of entries of each of the rows of the matrices A and B is equal to 1, show that $A \leqslant B$ implies $A = B$.

2.2 Matrix Multiplication and Its Properties

The product of two matrices A and B (in that order) is defined if A is $m \times n$ and B is $n \times p$, that is, the number of columns in A is equal to the number of rows in B. We first give the rule for multiplication in the special case in which A is $1 \times n$ and B is $n \times 1$. (Recall that a 1×1 matrix $[a]$ is identified with the number a itself.)

Let

$$x = [x_1 \quad x_2 \cdots x_n] \qquad \text{and} \qquad y = \begin{bmatrix} y_1 \\ y_2 \\ \vdots \\ y_n \end{bmatrix}$$

be $1 \times n$ and $n \times 1$ matrices, respectively. Then their product xy is defined to be the 1×1 matrix

$$xy = x_1 y_1 + x_2 y_2 + \cdots + x_n y_n.$$

For example, let

$$x = [1 \quad 2 \quad 3] \qquad \text{and} \qquad y = \begin{bmatrix} 4 \\ 3 \\ 6 \end{bmatrix}.$$

Then $xy = (1)(4) + (2)(3) + (3)(6) = 4 + 6 + 18 = 28$.

Now we give the rule for multiplication of matrices in general.

2.2.1 DEFINITION

Multiplication of Matrices

Let $A = (a_{ij})_{m \times n}$ and $B = (b_{ij})_{n \times p}$ be two matrices. Then

$$AB = (c_{ij})_{m \times p},$$

where c_{ij} is the product of the ith row $[\, a_{i1} \quad a_{i2} \quad \cdots \quad a_{in} \,]$ of A with the jth

column $\begin{bmatrix} b_{1j} \\ b_{2j} \\ \vdots \\ b_{nj} \end{bmatrix}$ of B, that is, $c_{ij} = a_{i1}b_{1j} + a_{i2}b_{2j} + \cdots + a_{in}b_{nj}$.

Note that if A is an $m \times n$ matrix and B is an $n \times p$ matrix, then the product AB is an $m \times p$ matrix.

Note also that AB is defined if and only if the number of columns of A is equal to the number of rows of B.

2.2.2 EXAMPLES

a. $[1 \ 2 \ 3 \ 4] \begin{bmatrix} 5 \\ 6 \\ 7 \\ 8 \end{bmatrix} = [5 + 12 + 21 + 32] = [70]$.

b. $\begin{bmatrix} 1 & 2 \\ 3 & 4 \end{bmatrix} \begin{bmatrix} 5 \\ 6 \end{bmatrix} = \begin{bmatrix} 5 + 12 \\ 15 + 24 \end{bmatrix} = \begin{bmatrix} 17 \\ 39 \end{bmatrix}.$

c. $\begin{bmatrix} 1 & 0 \\ 0 & 1 \end{bmatrix} \begin{bmatrix} 6 & 7 \\ 8 & 9 \end{bmatrix} = \begin{bmatrix} 6 & 7 \\ 8 & 9 \end{bmatrix}.$

d. $\begin{bmatrix} a & b \\ c & d \end{bmatrix} \begin{bmatrix} x \\ y \end{bmatrix} = \begin{bmatrix} ax + by \\ cx + dy \end{bmatrix}.$

e. Let us compute AB, where

$$A = \begin{bmatrix} -2 & 1 & 2 & 0 \\ 4 & -1 & 0 & 1 \\ 3 & 1 & -2 & 1 \end{bmatrix} \quad \text{and} \quad B = \begin{bmatrix} 1 & -4 & 2 \\ 0 & -3 & 1 \\ -2 & -1 & 0 \\ 1 & 0 & 1 \end{bmatrix}.$$

Since A and B are 3×4 and 4×3 matrices, respectively, AB is a 3×3 matrix. Now,

$$(AB)_{11} = [-2 \quad 1 \quad 2 \quad 0] \begin{bmatrix} 1 \\ 0 \\ -2 \\ 1 \end{bmatrix}$$

$$= -2 + 0 - 4 + 0 = -6.$$

Similarly,

$$(AB)_{12} = [-2 \quad 1 \quad 2 \quad 0] \begin{bmatrix} -4 \\ -3 \\ -1 \\ 0 \end{bmatrix}$$

$$= 8 - 3 - 2 + 0 = 3,$$

$$(AB)_{13} = [-2 \quad 1 \quad 2 \quad 0] \begin{bmatrix} 2 \\ 1 \\ 0 \\ 1 \end{bmatrix}$$

$$= -4 + 1 + 0 + 0 = -3.$$

Carrying out the same procedure for the remaining entries in AB, we get

$$(AB)_{21} = 4 + 0 + 0 + 1 = 5,$$
$$(AB)_{22} = -16 + 3 + 0 + 0 = -13,$$
$$(AB)_{23} = 8 - 1 + 0 + 1 = 8,$$
$$(AB)_{31} = 3 + 0 + 4 + 1 = 8,$$
$$(AB)_{32} = -12 - 3 + 2 + 0 = -13,$$
$$(AB)_{33} = 6 + 1 + 0 + 1 = 8.$$

Hence

$$AB = \begin{bmatrix} -6 & 3 & -3 \\ 5 & -13 & 8 \\ 8 & -13 & 8 \end{bmatrix}.$$

Matrices are useful for storing information and using it to solve certain problems in a concise way. The following example illustrates this.

f. Suppose the rows of the matrix $\begin{bmatrix} 150 & 70 & 20 \\ 100 & 50 & 40 \\ 70 & 60 & 10 \end{bmatrix}$ represent the number of books

purchased by a bookstore in Humanities I (H-I), Humanities II (H-II), and General Science (GS), in the fall, winter, and spring quarters. Let $\begin{bmatrix} 40 \\ 50 \\ 60 \end{bmatrix}$ and $\begin{bmatrix} 50 \\ 65 \\ 80 \end{bmatrix}$

represent the cost price and sale price of the books for H-I, H-II, and GS. Find the revenue earned by the bookstore on the books sold, assuming that it sold only 70%, 90%, and 80% of the books for H-I, H-II, and GS, respectively.

The matrix representing the sale of books is

$$A = \begin{array}{c} \\ \\ \end{array} \begin{array}{ccc} \text{H-I} & \text{H-II} & \text{GS} \\ \begin{bmatrix} 105 & 63 & 16 \\ 70 & 45 & 32 \\ 49 & 54 & 8 \end{bmatrix} & & \begin{array}{c} \text{Fall} \\ \text{Winter} \\ \text{Spring} \end{array} \end{array}$$

The matrix representing the revenue earned on each type of book is

$$A \begin{bmatrix} 50 \\ 65 \\ 80 \end{bmatrix} - A \begin{bmatrix} 40 \\ 50 \\ 60 \end{bmatrix} = \begin{bmatrix} 10\,625 \\ 8985 \\ 6600 \end{bmatrix} - \begin{bmatrix} 8310 \\ 6970 \\ 5140 \end{bmatrix} = \begin{bmatrix} 2315 \\ 2015 \\ 1460 \end{bmatrix}.$$

Total revenue = $5, 790.

g. It is interesting to note that any LS, say

$$ax + by + cz = d,$$
$$lx + my + nz = p,$$
$$sx + ty + uz = v,$$

can be written as a single matrix equation:

$$\begin{bmatrix} ax + by + cz \\ lx + my + nz \\ sx + ty + uz \end{bmatrix} = \begin{bmatrix} d \\ p \\ v \end{bmatrix},$$

that is,

$$\begin{bmatrix} a & b & c \\ l & m & n \\ s & t & u \end{bmatrix} \begin{bmatrix} x \\ y \\ z \end{bmatrix} = \begin{bmatrix} d \\ p \\ v \end{bmatrix}.$$

In general, the $m \times n$ LS

$$a_{11}x_1 + \cdots + a_{1n}x_n = b_1,$$
$$\vdots$$
$$a_{m1}x_1 + \cdots + a_{mn}x_n = b_m,$$

can be written as

$$\begin{bmatrix} a_{11} & a_{12} & \cdots & a_{1n} \\ a_{21} & a_{22} & \cdots & a_{2n} \\ \vdots & \vdots & & \vdots \\ a_{m1} & a_{m2} & \cdots & a_{mn} \end{bmatrix} \begin{bmatrix} x_1 \\ x_2 \\ \vdots \\ x_n \end{bmatrix} = \begin{bmatrix} b_1 \\ b_2 \\ \vdots \\ b_m \end{bmatrix}.$$

■

**2.2.3
Remark** We note that matrix multiplication may seem to be artificially contrived. However, *Example 2.2.2(g)* illustrates an important motivation behind the definition of matrix multiplication.

Let us now consider whether matrix multiplication is commutative. Let A and B be matrices. The products AB and BA both exist if and only if for some m, n, A and B are $m \times n$ and $n \times m$ matrices, respectively. But then AB is an $m \times m$ matrix and BA is an $n \times n$ matrix, and therefore AB and BA have the same size if and only if $m = n$. But even when A and B are both $n \times n$ matrices, AB and BA are not necessarily equal. This is easily proved by the following example. Let

$$A = \begin{bmatrix} 1 & 1 & 1 \\ 0 & 0 & 0 \\ 0 & 0 & 0 \end{bmatrix} \quad \text{and} \quad B = \begin{bmatrix} 1 & 0 & 0 \\ 1 & 0 & 0 \\ 1 & 0 & 0 \end{bmatrix}$$

be 3×3 matrices. In A every entry in the first row is 1, and every other row is zero. In B every entry in the first column is 1, and every other column is zero. (A row (column) in which every entry is zero is called a zero row (column).) It is clear that

$$AB = \begin{bmatrix} 3 & 0 & 0 \\ 0 & 0 & 0 \\ 0 & 0 & 0 \end{bmatrix} \quad \text{and} \quad BA = \begin{bmatrix} 1 & 1 & 1 \\ 1 & 1 & 1 \\ 1 & 1 & 1 \end{bmatrix}.$$

Hence $AB \neq BA$. This proves that multiplication of matrices is not generally commutative.

The matrix multiplication is, however, associative and distributive over addition. To put it formally, this means that for matrices A, B, and C, we have the following properties:

a. If A, B can be multiplied and B, C can be multiplied, then

$$A(BC) = (AB)C \text{ (associative law of multiplication)}.$$

b. If A, B can be multiplied and B, C can be added, then

$$A(B + C) = AB + AC \text{ (left distributive law)}.$$

c. If A, B can be added and B, C can be multiplied, then

$$(A + B)C = AC + BC \text{ (right distributive law)}.$$

The reader who is interested in proofs of the above properties may go to *Section 2.4* (*Theorem 2*).

In view of the associative property $(AB)C = A(BC)$, we define

$$A^2 = AA, \qquad A^3 = AAA,$$

and so on for any nth power of A, where n is a positive integer.

2.2.4
DEFINITION

Diagonal of a Matrix

If A is an $n \times n$ matrix, then the line joining the $(1, 1)$ entry, $(2, 2)$ entry, ..., (n, n) entry is called the *diagonal* (or the *main diagonal*) of the matrix.

2.2.5
DEFINITION

Diagonal Matrix

An $n \times n$ matrix D is called a *diagonal matrix* if each entry that is not on the diagonal is 0.

Thus an $n \times n$ diagonal matrix looks like this:

$$D = \begin{bmatrix} d_1 & 0 & \ldots & 0 \\ 0 & d_2 & \ldots & 0 \\ \vdots & \vdots & \ldots & \vdots \\ 0 & 0 & \ldots & d_n \end{bmatrix}.$$

The diagonal matrices form an important class of matrices in matrix theory. These are easy to work with. For example, multiplying two diagonal matrices of the same size simply amounts to multiplying their corresponding diagonal entries. Matrices that can be transformed to diagonal matrices by premultiplying and postmultiplying with suitable matrices are of special importance, and these will be considered in *Chapters 7 and 8*.

2.2.6
DEFINITION

Scalar Matrix

An $n \times n$ matrix is called a *scalar matrix* if it is a diagonal matrix and all the entries on the diagonal are equal. For example,

$$\begin{bmatrix} c & 0 \\ 0 & c \end{bmatrix} \quad \text{and} \quad \begin{bmatrix} c & 0 & 0 \\ 0 & c & 0 \\ 0 & 0 & c \end{bmatrix}$$

are scalar matrices.

We note that multiplying a matrix A with a scalar matrix each of whose diagonal entry is c amounts to scalar multiplication of the number c with the matrix A.

2.2.7
DEFINITION

Identity Matrix

An $n \times n$ matrix is called an *identity matrix* if it is a diagonal matrix and all the entries on the diagonal are equal to 1. For example,

$$\begin{bmatrix} 1 & 0 & 0 \\ 0 & 1 & 0 \\ 0 & 0 & 1 \end{bmatrix} \quad \text{and} \quad \begin{bmatrix} 1 & 0 & 0 & 0 \\ 0 & 1 & 0 & 0 \\ 0 & 0 & 1 & 0 \\ 0 & 0 & 0 & 1 \end{bmatrix}$$

are identity matrices. An $n \times n$ identity matrix is denoted by I_n. But it is common to drop the subscript and denote it by I when the size of the matrix is clear from the context.

An important property of the $n \times n$ identity matrix is

$$A I_n = A = I_n A,$$

where A is any $n \times n$ matrix.

Furthermore, we remark that any scalar matrix is a scalar multiple of the identity matrix.

By convention, for any nonzero square matrix A, A^0 is defined to be the identity matrix. This is in conformity with the usual definition of $a^0 = 1$ in algebra, where a is any nonzero number. This, along with the definition of positive powers of A, defines A^n for all nonnegative integers n.

2.2.8
DEFINITION

Upper (Lower) Triangular Matrix

An $n \times n$ matrix A is called an *upper (lower) triangular matrix* if all the entries below (above) the diagonal are zero. For example,

$$\begin{bmatrix} a & b \\ 0 & c \end{bmatrix} \quad \text{and} \quad \begin{bmatrix} a & b & c \\ 0 & d & e \\ 0 & 0 & f \end{bmatrix}$$

are 2×2 and 3×3 upper trianglar matrices, respectively. Indeed, any square matrix in row echelon form is upper triangular.

An example of a lower triangular matrix is

$$\begin{bmatrix} a & 0 & 0 \\ b & c & 0 \\ d & e & f \end{bmatrix}.$$

It is common practice to describe upper and lower triangular matrices as

$$\begin{bmatrix} 0 & \diagdown & * \end{bmatrix} \quad \text{and} \quad \begin{bmatrix} & \diagdown & 0 \\ * & & \end{bmatrix},$$

respectively, where $*$ denotes arbitrary entries and 0 stand for block with all entries zero.

Like diagonal matrices, triangular matrices form an important class of matrices. Square matrices in row echelon form are upper triangular matrices, and these matrices, as we saw in *Chapter 1*, play a vital role in solving a system of linear equations. One can define triangular matrices that are not necessarily square that look like matrices in row echelon form. However, we do not intend to go into such detailed discussion of rectangular triangular matrices. Triangular matrices have other useful properties, which are discussed in *Chapters 6 and 7*.

We close this section with an example that points to an application of matrix theory to real-life problems.

2.2.9 EXAMPLE   Consider the three major long-distance telephone companies AT&T, MCI, and Sprint, which serve a given area with a total population of N (in millions). Through their constant advertising directed toward attracting customers, each of them has been gaining as well as losing customers. Assume that there is no increase or decrease in the total number of customers being served. Furthermore, it is observed that at the end of each month, the number of customers changing companies is as follows: 20% of AT&T customers leave AT&T, but 10% of MCI customers and 10% of Sprint customers join AT&T. MCI loses 30% of its customers and attracts 10% of AT&T customers and 20% of Sprint customers. Sprint loses 30% of its customers and gains 10% of AT&T customers plus 20% of MCI customers.

Suppose x_0, y_0, and z_0 are the present numbers of customers of AT&T, MCI, and Sprint, respectively. Find the number of customers of each company after one year. ∎

Solution   If x_1, y_1, and z_1 are the customers of AT&T, MCI, and Sprint, respectively, at the end of the first month, then

$$x_1 = 0.8\, x_0 + 0.1\, y_0 + 0.1\, z_0,$$
$$y_1 = 0.1\, x_0 + 0.7\, y_0 + 0.2\, z_0,$$
$$z_1 = 0.1\, x_0 + 0.2\, y_0 + 0.7\, z_0.$$

Thus the initial matrix of customers, namely, $\begin{bmatrix} x_0 \\ y_0 \\ z_0 \end{bmatrix}$, becomes $A \begin{bmatrix} x_0 \\ y_0 \\ z_0 \end{bmatrix}$, where

$$A = \begin{bmatrix} 0.8 & 0.1 & 0.1 \\ 0.1 & 0.7 & 0.2 \\ 0.1 & 0.2 & 0.7 \end{bmatrix}$$

at the end of the first month.

Therefore, if we know $\begin{bmatrix} x_0 \\ y_0 \\ z_0 \end{bmatrix}$, we can find $\begin{bmatrix} x_1 \\ y_1 \\ z_1 \end{bmatrix}$ by computing $A \begin{bmatrix} x_0 \\ y_0 \\ z_0 \end{bmatrix}$.

Similarly,

$$\begin{bmatrix} x_2 \\ y_2 \\ z_2 \end{bmatrix} = A \begin{bmatrix} x_1 \\ y_1 \\ z_1 \end{bmatrix} = A^2 \begin{bmatrix} x_0 \\ y_0 \\ z_0 \end{bmatrix},$$

and so on. Thus by computing powers of A, we can find the number of customers each company has at the end of each month if we know the present number of customers served by each company. Note that at the end of one year, if x, y, and z are the number of customers of AT&T, MCI, and Sprint, respectively, then

$$\begin{bmatrix} x \\ y \\ z \end{bmatrix} = A^{12} \begin{bmatrix} x_0 \\ y_0 \\ z_0 \end{bmatrix}.$$

In *Chapter 7*, we will give a simple method to compute powers of a matrix.

2.2 Exercises

1. Find the product of each of the following pairs of matrices:

a. $[-1 \quad 2], \begin{bmatrix} 0 \\ -1 \end{bmatrix}.$

b. $[-2 \quad 3 \quad 5], \begin{bmatrix} 1 \\ 5 \\ 2 \end{bmatrix}.$

c. $[0 \quad 2 \quad 1 \quad -2], \begin{bmatrix} 1 \\ -2 \\ 1 \\ 0 \end{bmatrix}.$

2. Evaluate

a. $[1 \quad -1 \quad 2] \begin{bmatrix} 0 \\ 1 \\ 2 \end{bmatrix} + [1 \quad -1 \quad 2] \begin{bmatrix} 1 \\ -1 \\ 0 \end{bmatrix} + [1 \quad -1 \quad 2] \begin{bmatrix} 5 \\ 0 \\ -1 \end{bmatrix}.$

b. $[1 \quad 0 \quad -1 \quad 1] \begin{bmatrix} 2 \\ 0 \\ 1 \\ 3 \end{bmatrix} + [1 \quad 0 \quad -1 \quad 1] \begin{bmatrix} -1 \\ 1 \\ 0 \\ 2 \end{bmatrix} + [1 \quad 0 \quad -1 \quad 1] \begin{bmatrix} 1 \\ 6 \\ 2 \\ 0 \end{bmatrix}.$

3. Solve the following systems of equations for x, y, z. Recall that $[a]$ is identified with a.

a. $[1 \quad -1] \begin{bmatrix} x \\ y \end{bmatrix} = 2,$ **b.** $[2 \quad 3] \begin{bmatrix} x \\ y \end{bmatrix} = 5,$

$[2 \quad -1] \begin{bmatrix} x \\ y \end{bmatrix} = 0.$ $[-1 \quad 2] \begin{bmatrix} x \\ y \end{bmatrix} = -1.$

c. $[1 \quad 3 \quad -1] \begin{bmatrix} x \\ y \\ z \end{bmatrix} = 1,$ **d.** $[1 \quad 1 \quad -3] \begin{bmatrix} x \\ y \\ z \end{bmatrix} = 2,$

$[2 \quad 5 \quad 1] \begin{bmatrix} x \\ y \\ z \end{bmatrix} = 5,$ $[2 \quad 0 \quad -1] \begin{bmatrix} x \\ y \\ z \end{bmatrix} = 1.$

$[1 \quad -1 \quad 3] \begin{bmatrix} x \\ y \\ z \end{bmatrix} = 2.$

4. (Drill 2.4) Find AB and BA if

$$A = \begin{bmatrix} 1 & 3 & 0 \\ 2 & 1 & 1 \\ -1 & -2 & 0 \end{bmatrix}, \qquad B = \begin{bmatrix} 3 & 5 & 7 \\ 9 & 11 & 1 \\ 0 & 0 & 1 \end{bmatrix}.$$

5. Find CD if

$$C = \begin{bmatrix} 1 & 2 \\ 3 & 4 \end{bmatrix}, \qquad D = \begin{bmatrix} 5 & 6 & 7 \\ 8 & 9 & 10 \end{bmatrix}.$$

Is DC defined?

6. Find 2×2 matrices A, B such that $AB = 0$ but $A \neq 0$ and $B \neq 0$.

7. (Drill 2.5) Using 3×3, 4×4, and 5×5 matrices with random integral entries, verify the associative property for multiplication of matrices. That is, for any given matrices A, B, and C, show that $(AB)C = A(BC)$.

8. (Drill 2.6) For the matrices

$$A = \begin{bmatrix} 2 & 6 \\ 3 & 9 \end{bmatrix}, \qquad B = \begin{bmatrix} 3 & -1 \\ -1 & 2 \end{bmatrix}, \qquad C = \begin{bmatrix} 3 & 2 \\ 1 & 0 \end{bmatrix},$$

verify that

$$A(B + C) = AB + AC$$

and

$$(B + C)A = BA + CA.$$

Do the same for three random matrices A, B, and C each of size 6×6.

9. Write the following linear systems as matrix equations:

a. $x + 2y + 3z = 1,$
$5x + y + 6z = 2.$

b. $\begin{aligned} x_1 - 2x_2 + 3x_3 + 4x_4 &= 1, \\ 2x_2 - x_3 + x_4 &= 2, \\ 5x_1 + 6x_2 + 7x_3 + 8x_4 &= 9. \end{aligned}$

c. $\begin{aligned} 2u + 3v - 5w &= 1, \\ u + v - 7w &= 6. \end{aligned}$

10. (Drill 2.7) If matrix $A = \begin{bmatrix} 1 & 1.5 & 0.75 \\ 0.75 & 1 & 1.50 \end{bmatrix}$ represents in thousands the numbers of Toyotas, Hondas, and Dodges sold in January and February by dealer I and matrix $B = \begin{bmatrix} 0.15 & 0.2 & 0.2 \\ 0.12 & 0.2 & 0.3 \end{bmatrix}$ represents in thousands the numbers of Lexuses, Lincolns, and Cadillacs sold by dealer II during January and February, write down the matrix representing the total number of cars sold in January and February. Also, if the sale prices of Toyotas, Hondas, Dodges, Cadillacs, Lexuses, and Lincolns are $15,000, $16,000, $15,000, $35,000, $45,000, and $35,000, respectively, find which dealer had the higher revenue.

11. Consider a sequence $a_0, a_1, a_2, \ldots$ of numbers such that any number a_k, for $k \geqslant 3$, is the sum of the previous three numbers, that is, $a_{k+3} = a_{k+2} + a_{k+1} + a_k$. Find the matrix A such that $\begin{bmatrix} a_{k+3} \\ a_{k+2} \\ a_{k+1} \end{bmatrix} = A \begin{bmatrix} a_{k+2} \\ a_{k+1} \\ a_k \end{bmatrix}$.

12. Suppose $a_0, a_1, a_2, \ldots$ is a sequence of numbers such that $a_{k+1} = a_k + 2a_{k-1}$ for all $k \geqslant 1$. Find a_4 by computing powers of a suitable matrix A connecting $\begin{bmatrix} a_{k+1} \\ a_k \end{bmatrix}$ and $\begin{bmatrix} a_k \\ a_{k-1} \end{bmatrix}$ if $a_0 = 0$ and $a_1 = 1$. (We will revisit this problem in *Chapter 7*, since further development of the subject will enable us to write down an explicit expression for a_k.)

13. The Fibonacci numbers $a_0, a_1, a_2, \ldots$ are defined by the formula $a_{k+1} = a_k + a_{k-1}$. Find a matrix A such that $\begin{bmatrix} a_{k+1} \\ a_k \end{bmatrix} = A \begin{bmatrix} a_k \\ a_{k-1} \end{bmatrix}$. (An explicit expression for a_k will be derived in *Chapter 7*.)

14. (Drill 2.8) Consider two neighboring cities A and B whose populations at the end of the year t are represented by A_t and B_t, respectively. If because of movements of the residents, births, and deaths, the population of the cities at the end of year $t + 1$ are $0.6A_t + 0.7B_t$ and $0.1A_t + 1.2 B_t$, respectively, write down the recursion formula (difference equation) connecting $\begin{bmatrix} A_{t+1} \\ B_{t+1} \end{bmatrix}$ and $\begin{bmatrix} A_t \\ B_t \end{bmatrix}$ using an appropriate matrix. Assuming that the populations in the cities at year 0 are 50,000 and 100,000, find the populations after 1, 2, 3, 4, and 5 years. Obtain a general formula for the year n populations.

15. Consider the Pacific Ocean near Vancouver, British Columbia, where the rates of growth of populations of types A and B of fish change each year. It is found that if the type A fish kill the type B fish at a certain rate, both the types will continue

to exist. Observations suggest that

$$S_{i+1} = 0.7\ S_i + 0.4\ F_i,$$
$$F_{i+1} = -k\ S_i + 1.2\ F_i,$$

where S_i and F_i represent the population of fish of types A and B respectively, at the end of i years and k is the kill rate of type B by type A. If $S_0 = 1000$ and $F_0 = 50{,}000$ are the present populations, find S_i, F_i for $i = 5, 6$ by choosing (i) $k = 0.02$ and (ii) $k = 0.2$.

16. Four work-study students, Tom, Dick, David, and Stephanie, put in certain hours of work on each of the five working days of a week. They are paid $5, $5, $6, and $6, respectively, per hour. If the (i, j) entry of the matrix

$$\begin{bmatrix} 2 & 3 & 2 & 1 \\ 3 & 2 & 2 & 2 \\ 2 & 2 & 1 & 2 \\ 1.5 & 1 & 2 & 2.5 \\ 1 & 2 & 2 & 2 \end{bmatrix}$$

gives on the ith day of the week the number of hours put in by the jth student, find the total amount to be budgeted for biweekly checks.

17. Show that the product of two diagonal matrices is a diagonal matrix.

18. Show that the product of two upper (lower) triangular matrices is an upper (lower) triangular matrix.

19. Let

$$A = \begin{bmatrix} \cos x & \sin x \\ -\sin x & \cos x \end{bmatrix} \quad \text{and} \quad B = \begin{bmatrix} \cos y & \sin y \\ -\sin y & \cos y \end{bmatrix}.$$

Show that

$$AB = \begin{bmatrix} \cos(x + y) & \sin(x + y) \\ -\sin(x + y) & \cos(x + y) \end{bmatrix} = BA.$$

20. The trace of an $n \times n$ matrix $A = (a_{ij})$ is defined to be

$$\mathrm{tr}(A) = \sum_{i=1}^{n} a_{ii}.$$

Show that

a. $\mathrm{tr}(A + B) = \mathrm{tr}(A) + \mathrm{tr}(B)$.
b. $\mathrm{tr}(AB) = \mathrm{tr}(BA)$.

21. (Drill 2.9) Let A be the vertex matrix of a graph. It is known that the (i, j) entry of A^n gives the number of paths of length n from vertex i to vertex j of the graph. Consider the route map of Vanguard Airlines given below.

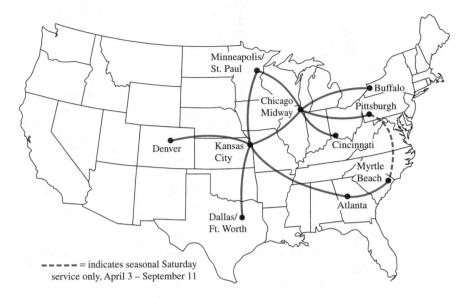

-----= indicates seasonal Saturday
service only, April 3 – September 11

Write down the vertex (or adjacency) matrix A of the flights between 10 cities.

a. Find the number of paths of lengths 1, 2, and 3 between Kansas City and Chicago by considering powers of A.

b. Find the lengths of the shortest paths between each pair of cities.

2.3 Transpose

2.3.1
DEFINITION

Transpose of a Matrix
Let $A = (a_{ij})$ be an $m \times n$ matrix. Then the *transpose* of A, written A^T, is the $n \times m$ matrix whose (i, j) entry is a_{ji} for all i, j.

In other words, the ith row of A^T is the ith column of A for all i.

2.3.2
EXAMPLES

a. For $A = [1 \quad 2 \quad 3]$, $A^T = \begin{bmatrix} 1 \\ 2 \\ 3 \end{bmatrix}$.

b. For $A = \begin{bmatrix} 5 & 6 \\ 7 & 8 \end{bmatrix}$, $A^T = \begin{bmatrix} 5 & 7 \\ 6 & 8 \end{bmatrix}$.

c. For $A = \begin{bmatrix} 1 & 7 \\ 9 & 0 \end{bmatrix}$, $B = \begin{bmatrix} 1 & -1 \\ 6 & 0 \end{bmatrix}$,

$$(A + B)^T = \begin{bmatrix} 2 & 6 \\ 15 & 0 \end{bmatrix}^T = \begin{bmatrix} 2 & 15 \\ 6 & 0 \end{bmatrix}$$

$$= \begin{bmatrix} 1 & 9 \\ 7 & 0 \end{bmatrix} + \begin{bmatrix} 1 & 6 \\ -1 & 0 \end{bmatrix} = A^T + B^T.$$

d. For $A = \begin{bmatrix} 2 & 0 \\ 1 & 5 \\ -1 & 1 \end{bmatrix}$, $B = \begin{bmatrix} 1 & 1 \\ 6 & 3 \end{bmatrix}$,

$$(AB)^T = \begin{bmatrix} 2 & 2 \\ 31 & 16 \\ 5 & 2 \end{bmatrix}^T = \begin{bmatrix} 2 & 31 & 5 \\ 2 & 16 & 2 \end{bmatrix},$$

$$B^T A^T = \begin{bmatrix} 1 & 6 \\ 1 & 3 \end{bmatrix} \begin{bmatrix} 2 & 1 & -1 \\ 0 & 5 & 1 \end{bmatrix} = \begin{bmatrix} 2 & 31 & 5 \\ 2 & 16 & 2 \end{bmatrix} = (AB)^T.$$

So $(AB)^T = B^T A^T$. (Note that $(AB)^T$ need not be equal to $A^T B^T$.)   ■

2.3.3
DEFINITION

Symmetric Matrix
An $n \times n$ matrix A is called *symmetric* if $A = A^T$. In other words, $A = (a_{ij})$ is symmetric if $a_{ij} = a_{ji}$ for all i, j.

For example, the matrices $\begin{bmatrix} 1 & 2 \\ 2 & 3 \end{bmatrix}$ and $\begin{bmatrix} 2 & 4 & 0 \\ 4 & 1 & 6 \\ 0 & 6 & 3 \end{bmatrix}$ are symmetric.

2.3.4
EXAMPLE

The quadratic function $ax_1^2 + 2hx_1x_2 + bx_2^2$ can be expressed as $X^T A X$, where $X = \begin{bmatrix} x_1 \\ x_2 \end{bmatrix}$ is a 2×1 matrix and A is a 2×2 symmetric matrix whose diagonal entries are the coefficients of x_1^2 and x_2^2 and the nondiagonal $(1, 2)$ entry as well as the $(2, 1)$ entry is half of the coefficient of the term x_1x_2.

The quadratic function $ax_1^2 + bx_2^2 + cx_3^2 + 2hx_1x_2 + 2gx_1x_3 + 2fx_2x_3$ can similarly be expressed as $X^T A X$, where $X = \begin{bmatrix} x_1 \\ x_2 \\ x_3 \end{bmatrix}$ is a 3×1 matrix and A is a 3×3 symmetric matrix whose diagonal entries are the coefficients of x_1^2, x_2^2, x_3^2 and the nondiagonal (i, j) entry is half of the coefficient of the term $x_i x_j \ (= x_j x_i)$.

In particular, consider an expression $5x_1^2 + 16x_1x_2 + 6x_2^2$. Choose

$$X = \begin{bmatrix} x_1 \\ x_2 \end{bmatrix} \qquad \text{and} \qquad A = \begin{bmatrix} 5 & 8 \\ 8 & 6 \end{bmatrix}.$$

Verify that $X^T A X = 5x_1^2 + 16x_1x_2 + 6x_2^2$.

Consider another expression $2x_1^2 - 6x_2^2 + 3x_3^2 + 2x_1x_2 + 20x_2x_3$. Here,

$$X = \begin{bmatrix} x_1 \\ x_2 \\ x_3 \end{bmatrix}, \qquad \text{and} \qquad A = \begin{bmatrix} 2 & 1 & 0 \\ 1 & -6 & 10 \\ 0 & 10 & 3 \end{bmatrix}.$$

One can verify that $X^T A X = 2x_1^2 - 6x_2^2 + 3x_3^2 + 2x_1x_2 + 20x_2x_3$.   ■

2.3.5 Properties of Transpose

1. If A, B are two matrices that can be added, then $(A + B)^T = A^T + B^T$.
2. For any matrix A and any number α, $(\alpha A)^T = \alpha A^T$.
3. If A and B are two matrices that can be multiplied in this order, then $(AB)^T = B^T A^T$.
4. For any matrix A, $(A^T)^T = A$.

The reader who is interested in the proof of these properties may refer to *Theorem 3* in *Section 2.4*.

2.3 | Exercises

1. (Drill 2.10) For the matrices

$$A = \begin{bmatrix} 1 & 2 & 3 \\ 0 & 1 & 0 \\ 1 & 0 & 1 \end{bmatrix}, \qquad B = \begin{bmatrix} -1 & 0 & 1 \\ 1 & 2 & 0 \\ 0 & 0 & 1 \end{bmatrix},$$

verify that $(A + B)^T = A^T + B^T$ and $(AB)^T = B^T A^T$. Also, verify the same properties by choosing two random matrices.

2. Prove that if A is a diagonal matrix, then $A = A^T$.
3. Write a nondiagonal matrix A such that $A = A^T$. Give the general form of a 3×3 matrix A such that $A = A^T$.
4. Find the symmetric matrix A such that

$$X^T A X = -x_1^2 + 2x_1 x_2 + x_2^2,$$

where $X = \begin{bmatrix} x_1 \\ x_2 \end{bmatrix}$.

5. Find the symmetric matrix A such that $X^T A X = 5x^2 + 6xy + 7y^2$, where $X = \begin{bmatrix} x \\ y \end{bmatrix}$.

6. Write $ax^2 + 2hxy + by^2 + 2gx + 2fy + c$ as $X^T A X$, where $X = \begin{bmatrix} x \\ y \\ 1 \end{bmatrix}$ and A is a symmetric matrix.

7. **a.** Prove that the sum of 2×2 symmetric matrices

$$\begin{bmatrix} a & b \\ b & a \end{bmatrix}, \quad \begin{bmatrix} x & y \\ y & x \end{bmatrix}$$

is a symmetric matrix.
   b. Generalize that the sum of two $n \times n$ symmetric matrices is a symmetric matrix.
   c. Give 2×2 symmetric matrices A and B such that AB is not symmetric.

8. Let A be a symmetric matrix. Show that
   a. $-A$ is a symmetric matrix;
   b. $(\alpha A)^T$ is a symmetric matrix, where α is a scalar.

9. **a.** Give 2×2 nondiagonal matrices A and B such that their product is symmetric.
   b. Show that for any square matrix A, $A^T + A$ and AA^T are symmetric.

10. Create a matrix $C = \begin{bmatrix} A & A \\ B & B \end{bmatrix}$, where A and B are 3×3 random matrices. When is C symmetric? Compute $C - C^T$.

11. Show that the ith column of the product AB of matrices A and B is the product of A with the ith column of B. State and prove the corresponding result for the ith row of the product AB of two matrices.

2.4 Proofs of Facts

THEOREM 1

Let A, B, C be $m \times n$ matrices. Then

a. $A + B = B + A$ (commutative law of addition).
b. $(A + B) + C = A + (B + C)$ (associative law of addition).

Proof For any matrix X, let $(X)_{ij}$ denote the (i, j) entry of X.

a. By definition of the sum of matrices, $A + B$ and $B + A$ are both $m \times n$ matrices and
$$(A + B)_{ij} = (A)_{ij} + (B)_{ij} = (B)_{ij} + (A)_{ij} = (B + A)_{ij}$$
for each (i, j). Hence $A + B = B + A$.

b. Again, $A + (B + C)$ and $(A + B) + C$ are both $m \times n$ matrices, and
$$\begin{aligned}
(A + (B + C))_{ij} &= A_{ij} + (B + C)_{ij} \\
&= A_{ij} + (B_{ij} + C_{ij}) \\
&= (A_{ij} + B_{ij}) + C_{ij} \\
&= (A + B)_{ij} + C_{ij} \\
&= ((A + B) + C)_{ij}.
\end{aligned}$$
∎

THEOREM 2

Let A, B, C be matrices. Then the following statements hold:

a. If A, B can be multiplied and B, C can be multiplied, then
$$A(BC) = (AB)C \text{ (associative law of multiplication).}$$

b. If A, B can be multiplied and B, C can be added, then
$$A(B + C) = AB + AC \text{ (left distributive law)}.$$

c. If A, B can be added, and B, C can be multiplied, then
$$(A + B)C = AC + BC \text{ (right distributive law)}.$$

Proof

a. The products $A(BC)$ and $(AB)C$ are possible if and only if A, B, and C are $m \times n$, $n \times p$ and $p \times q$ matrices, respectively. That being so, $A(BC)$ and $(AB)C$ are both $m \times q$ matrices. Recall that $(X)_{ij}$ denotes the (i, j) entry of matrix X. Then

$$(A(BC))_{ij} = \sum_{k=1}^{n} (A)_{ik}(BC)_{kj}$$

$$= \sum_{k=1}^{n} (A)_{ik} \left[\sum_{l=1}^{p} (B)_{kl}(C)_{lj} \right]$$

$$= \sum_{l=1}^{p} \sum_{k=1}^{n} ((A)_{ik}(B)_{kl})(C)_{lj}$$

$$= \sum_{l=1}^{p} (AB)_{il}(C)_{lj}$$

$$= ((AB)C)_{ij}.$$

This proves that $A(BC) = (AB)C$.

b. The condition holds if and only if A, B, and C are $m \times n$, $n \times p$, and $n \times p$ matrices, respectively. In that case, $A(B + C)$ and $AB + AC$ are both $m \times p$ matrices, and for every (i, j),

$$(A(B + C))_{ij} = \sum_{k=1}^{n} (A)_{ik}(B + C)_{kj}$$

$$= \sum_{k=1}^{n} (A)_{ik}((B)_{kj} + (C)_{kj})$$

$$= \sum_{k=1}^{n} (A)_{ik}(B)_{kj} + \sum_{k=1}^{n} (A)_{ik}(C)_{kj}$$

$$= (AB)_{ij} + (AC)_{ij}$$

$$= (AB + AC)_{ij}.$$

Hence

$$A(B + C) = AB + AC.$$

The proof for (c) is similar to the one for (b) and is left as an exercise for the reader.   ∎

THEOREM 3

a. If A, B are two matrices that can be added, then

$$(A + B)^T = A^T + B^T.$$

b. For any matrix A and any number α,

$$(\alpha A)^T = \alpha A^T.$$

c. If A and B are two matrices that can be multiplied in this order, then

$$(AB)^T = B^T A^T.$$

d. For any matrix A,
$$(A^T)^T = A.$$

Proof Throughout, let $A = (a_{ij})$ and $B = (b_{ij})$.

a. Let A and B be $m \times n$ matrices. By the definition of transpose, for any (i, j),

$$\begin{aligned}
(i, j) \text{ entry of } (A + B)^T &= (j, i) \text{ entry of } A + B \\
&= a_{ji} + b_{ji} \text{ (by definition of sum)} \\
&= (i, j) \text{ entry of } A^T + (i, j) \text{ entry of } B^T.
\end{aligned}$$

Hence

$$(A + B)^T = A^T + B^T.$$

b. Let A be an $m \times n$ matrix, and let $B = \alpha A$. Then $b_{ij} = \alpha a_{ij}$. By the definition of transpose, for any (i, j),

$$\begin{aligned}
(i, j) \text{ entry of } (\alpha A)^T &= (j, i) \text{ entry of } \alpha A \\
&= b_{ji} \\
&= \alpha a_{ji} \\
&= \alpha (i, j) \text{ entry of } A^T.
\end{aligned}$$

c. Let A and B be $m \times n$, $n \times p$ matrices, respectively. Then A^T, B^T are $n \times m$, $p \times n$ matrices, respectively. Therefore $B^T A^T$ is a $p \times m$ matrix.

Also, $(AB)^T$ is a $p \times m$ matrix. To prove $(AB)^T = B^T A^T$, we consider

$$((AB)^T)_{ij} = (AB)_{ji}$$

$$= \sum_{k=1}^{n} (A)_{jk}(B)_{ki}$$

$$= \sum_{k=1}^{n} (A^T)_{kj}(B^T)_{ik}$$

$$= \sum_{k=1}^{n} (B^T)_{ik}(A^T)_{kj}$$

$$= (B^T A^T)_{ij}.$$

Hence

$$(AB)^T = B^T A^T.$$

d. Let A be an $m \times n$ matrix. Set $B = A^T$. Further, let $C = B^T$, and so $C = (A^T)^T$. Suppose $C = (c_{ij})$. Then $c_{ij} = b_{ji}$, because $C = B^T$. Also, because $B = A^T$, $b_{ji} = a_{ij}$. Therefore $c_{ij} = a_{ij}$ for all i, j. This proves that $(A^T)^T = A$. ∎

2.5 Chapter Review Questions and Projects

1. Under what conditions do we define addition and subtraction of two matrices? Write down the rule for defining addition of matrices. (Section 2.1)

2. Is addition of matrices commutative and associative? Justify by outlining the reason. (Section 2.1)

3. Define scalar multiplication of a matrix. Explain by an example. (Section 2.1)

4. If A is an $m \times n$ matrix and B is a $k \times l$ matrix, write down the conditions under which (i) AB is defined, (ii) BA is defined, (iii) AB is defined but BA is not defined. (Section 2.2)

5. Is matrix multiplication commutative? Explain by an example. (Section 2.2)

6. Is matrix multiplication associative? (Section 2.2)

7. Is matrix multiplication distributive over addition? Explain by an example. (Section 2.2)

8. Define the transpose of a matrix. For two $n \times n$ matrices A and B, write down formulae for the transpose of $A + B$ and AB. (Section 2.3)

9. Give examples of symmetric matrices. For two $n \times n$ symmetric matrices A and B, find which of the matrices are symmetric: A^2, $A + B$, AB. (Section 2.3)

10. What is the vertex matrix of a graph? If A is the vertex matrix of the directed graph given below, find A^2 and A^3, and interpret the (2, 3) and (1, 3) entries. (Section 2.2)

$$
\begin{array}{ccc}
1 & \rightarrow & 2 \\
\downarrow & \swarrow & \uparrow \\
4 & \leftarrow & 3
\end{array}
$$

Chapter Projects

1. A group of five workers in a fertilizer company handles a hazardous chemical. If a worker gets contaminated owing to mishandling, the worker can spread the contamination through direct contact with his or her coworkers. Let $A = (a_{ij})$ be the vertex (adjacency) matrix of the directed graph of the contacts that workers have with one another, defined by the rule that

$$
a_{ij} = \begin{cases} 1, & \text{if } i \text{ can contaminate } j, \\ 0, & \text{if } i \text{ cannot contaminate } j. \end{cases}
$$

Assume that

$$
A = \begin{bmatrix}
0 & 1 & 0 & 1 & 1 \\
0 & 0 & 1 & 0 & 0 \\
1 & 0 & 0 & 0 & 1 \\
0 & 1 & 1 & 0 & 0 \\
0 & 1 & 0 & 1 & 0
\end{bmatrix}
$$

and recall that the (i, j) entry of A^n gives the number of paths of length n from vertex i to vertex j (see *Exercise 21, Section 2.2*).

a. What can you conclude if all the entries of some power n of the matrix A are positive (expressed symbolically as $A^n > 0$)?

b. Find the smallest positive integer n such that $A^n > 0$, and find which workers have the greatest potential for spreading contamination.

c. Which workers have the least potential for spreading contamination?

d. What interpretation can be made for the largest column sum of $A + A^2 + A^3$?

2. Additional Project (Markov Chain)

A Gambler's Ruin: It is still one week to payday, and Bill has only $5 left. He figures that he needs to have $10 for going to a movie tomorrow night with Cathy. So Bill convinces his roommate Eric to wager $1 on each toss of a coin. Bill plans to continue betting until he has accumulated enough money for the movie or until he goes broke, whichever comes first.

We are going to set up a mathematical model called a Markov chain to study Bill's chances for success or failure.

First, let us make some observations about how Bill's fortune may vary as the game progresses. After the first toss, it is equally likely that Bill will have $4 or $6. After two tosses, his fortune will consist of either $3, $5, or $7 with probabilities 1/4, 1/2, and 1/4, respectively. After three tosses, he can have $2, $4, $6, or $8 with probabilities

1/8, 3/8, 3/8, and 1/8, respectively. The following tree can help us to see how to arrive at these probabilities. (Each branch of the tree has probabilitiy 1/2.)

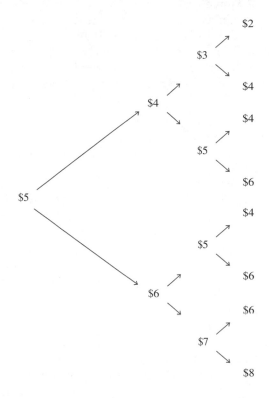

Throughout the game, Bill's fortune may assume any of the values $0, $1, $2, ..., $10. Let $p^{(0)}$ be the vector with 11 entries to denote the initial probabilities for each of these possible values:

$$p^{(0)} = [0\ 0\ 0\ 0\ 0\ 1\ 0\ 0\ 0\ 0\ 0].$$

Let $p^{(1)}$ be the vector with 11 entries to denote the probabilities for each of those possible fortunes after 1 step:

$$p^{(1)} = [0\ 0\ 0\ 0\ 0.5\ 0\ 0.5\ 0\ 0\ 0\ 0].$$

In general, if $p^{(k)}$ is the vector with 11 entries to denote the probabilities for each of these possible fortunes after k steps, then for $k = 2, 3, 4$, we have

$$p^{(2)} = [0\ 0\ 0\ 0.25\ 0\ 0.5\ 0\ 0.25\ 0\ 0\ 0],$$
$$p^{(3)} = [0\ 0\ 0.125\ 0\ 0.375\ 0\ 0.375\ 0\ 0.125\ 0\ 0],$$
$$p^{(4)} = [0\ 0.0625\ 0\ 0.25\ 0\ 0.375\ 0\ 0.25\ 0\ 0.0625\ 0].$$

Thus the probability that Bill has $9 after four steps of the game is the tenth entry, 0.0625, of the vector $p^{(4)}$.

The calculation of these probabilities might seem cumbersome. But we will consider an 11×11 matrix $T = (t_{ij})$, called the transition matrix, in which the (i, j) entry t_{ij} represents the probability of moving from the state i to the next state j, that is, the probability of accumulating the amount $\$(j - 1)$ in the next step from the present amount of $\$(i - 1)$. Since Bill will stop playing if he has no money, $t_{12} = 0, t_{13} = 0, \ldots,$ $t_{1,11} = 0$. However, the probability of changing from $\$0$ to $\$0$ is clearly 1. So $t_{11} = 1$.

Furthermore, $t_{21} = 0.5$, $t_{22} = 0$, $t_{23} = 0.5$, $t_{24} = 0$, and so on. Note that the change in the winnings at any step is always $\pm\$1$. Writing explicitly, we have

$$
T =
\begin{bmatrix}
1 & 0 & 0 & 0 & 0 & 0 & 0 & 0 & 0 & 0 & 0 \\
0.5 & 0 & 0.5 & 0 & 0 & 0 & 0 & 0 & 0 & 0 & 0 \\
0 & 0.5 & 0 & 0.5 & 0 & 0 & 0 & 0 & 0 & 0 & 0 \\
0 & 0 & 0.5 & 0 & 0.5 & 0 & 0 & 0 & 0 & 0 & 0 \\
0 & 0 & 0 & 0.5 & 0 & 0.5 & 0 & 0 & 0 & 0 & 0 \\
0 & 0 & 0 & 0 & 0.5 & 0 & 0.5 & 0 & 0 & 0 & 0 \\
0 & 0 & 0 & 0 & 0 & 0.5 & 0 & 0.5 & 0 & 0 & 0 \\
0 & 0 & 0 & 0 & 0 & 0 & 0.5 & 0 & 0.5 & 0 & 0 \\
0 & 0 & 0 & 0 & 0 & 0 & 0 & 0.5 & 0 & 0.5 & 0 \\
0 & 0 & 0 & 0 & 0 & 0 & 0 & 0 & 0.5 & 0 & 0.5 \\
0 & 0 & 0 & 0 & 0 & 0 & 0 & 0 & 0 & 0 & 1
\end{bmatrix}.
$$

a. Compute $p_1 = p_0 T$, $p_2 = p_1 T$, $p_3 = p_2 T, \ldots$ What is the relation between $p^{(i)}$ and p_i?

b. Show that $p_1 = p_0 T$, $p_2 = p_0 T^2$, $p_3 = p_0 T^3$.

c. The entry in the sixth row of T^k represents the probabilities of accumulating $\$0, \$1,$ $\$2, \ldots, \$10$ in k steps if Bill had $\$5$ to begin with. We can see how Bill's chances vary as the game progresses by looking at higher and higher powers of T. Thus we notice that Bill has a probability slightly better than 0.25 of having $\$10$ after 20 steps. But if he had started with $\$6$, then the probability of having $\$10$ after 20 steps is slightly better than 0.38. Why?

d. Compute T^{50}, T^{100}, and T^{200}. What does T^{200} tell us about Bill's chances of going to the movie?

e. Suppose Bill started with $\$6$. What is the probability then that he can go to a movie with Cathy?

Key Words

Row Matrix	Upper and Lower Triangular Matrices
Column Matrix	Multiplication of a Matrix
Square Matrix	by a Scalar
Diagonal of a Matrix	Addition of Matrices
Diagonal Matrix	Negative of a Matrix
Scalar Matrix	Multiplication of Matrices
Identity Matrix	Transpose of a Matrix
Zero Matrix	Symmetric Matrix

Key Phrases

- Matrix addition is commutative and associative.
- Matrix multiplication is associative but not commutative.
- The product of nonzero matrices can be zero.
- The transpose of the product of matrices is the product of the transposes of the matrices in reverse order.

3 Subspaces

Introduction

The concept of a vector in a plane or space, which originated in the study of mechanics, can be looked on as a 1×2 or 1×3 matrix representing the coordinates of the terminal point with respect to the initial point, the origin. It is also customary to represent vectors as 2×1 or 3×1 matrices. In this chapter, we introduce a vector as an $m \times 1$ column matrix or a $1 \times m$ row matrix. The vector algebra allows the formation of linear combinations of vectors. The idea of a *minimal generating set* of vectors in a subspace leads to the concepts of *basis* and *dimension* of a subspace. It can be seen that any vector in the xy plane $\mathbf{R}^2$ is a sum of scalar multiples of any two given perpendicular vectors. For example, $[a \quad b] = a[1 \quad 0] + b[0 \quad 1]$. We call the set of vectors $[1 \quad 0]$ and $[0 \quad 1]$ a basis of the subspace $\mathbf{R}^2$.

3.1 Linear Combination of Vectors

In geometry, we draw a line from the origin to a suitable point P in the plane to convey the idea of direction and magnitude of an object such as displacement or velocity:

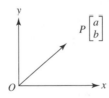

A line segment OP, joining O to P, is determined by the ordered pair of numbers (a, b), called the *coordinates* of P. The line segment OP is denoted by the column matrix $\begin{bmatrix} a \\ b \end{bmatrix}$ and is called a *vector*, since it conveys the idea of both direction and magnitude. The term "vector" is derived from the Latin word *vehere*, which means "to convey."

Let F denote the set of real numbers or the set of complex numbers. Elements of F are also called *scalars*.

**3.1.1
DEFINITION**

Column Vector and Row Vector

An $n \times 1$ matrix $\begin{bmatrix} a_1 \\ a_2 \\ \vdots \\ a_n \end{bmatrix}$ is called an *n-dimensional column vector* or simply a

vector.

A $1 \times n$ matrix $[a_1 \ a_2 \ \cdots \ a_n]$ is called an *n-dimensional row vector*.

A column vector or a row vector is known as an *n-tuple*.

F^n will denote the set of all $n \times 1$ or $1 \times n$ matrices. The context will generally make it clear in which sense F^n is used.

If $[\alpha_1 \ \alpha_2 \ \cdots \ \alpha_m]$ is a row vector, then its transpose $[\alpha_1 \ \alpha_2 \ \cdots \ \alpha_m]^T$, is a column vector. To save space, we will sometimes write a column vector as the transpose of a row vector.

Since an n-dimensional column vector is an $n \times 1$ matrix and an n-dimensional row vector is a $1 \times n$ matrix, we will frequently use the properties of addition, multiplication, and scalar multiplication of matrices given in *Chapter 2*. These are recorded below for convenience.

1. Addition of matrices is commutative and associative:

$$A + B = B + A, \qquad (A + B) + C = A + (B + C).$$

2. Zero matrix and negative of a matrix:

$$A + 0 = A = 0 + A, \qquad A + (-A) = 0 = (-A) + A.$$

3. Multiplication of matrices is associative and distributive over addition:

$$A(BC) = (AB)C, \qquad A(B + C) = AB + AC, \qquad (A + B)C = AC + BC.$$

4. Identity matrix I:

$$AI = A = IA.$$

5. In any product of scalars and matrices, a scalar can be shifted right or left:

$$(A\alpha)B = \alpha(AB) = (AB)\alpha.$$

6. Multiplication of matrices with scalars is distributive over addition:

$$(a + b)A = aA + bA = Aa + Ab.$$

**3.1.2
DEFINITION**

Linear Combination

Let $\{v_1, v_2, \ldots, v_m\}$ be a set of vectors in F^n. Then $\alpha_1 v_1 + \alpha_2 v_2 + \cdots + \alpha_m v_m$ is a vector in F^n, where $\alpha_1, \alpha_2, \ldots, \alpha_m$ are scalars in F. The sum $\alpha_1 v_1 + \alpha_2 v_2 + \cdots + \alpha_m v_m$ is called a *linear combination* of $\{v_1, v_2, \ldots, v_m\}$.

3.1.3
EXAMPLES

a. A linear combination of a single vector, say $v = [1 \ \ 2]$, is simply $\alpha[1 \ \ 2]$, where α is any scalar.

Thus if **OP** represents the vector v and Q is any point on the line passing through O and P, then the coordinates of Q are $[\alpha \ \ \ 2\alpha]$ for some α.

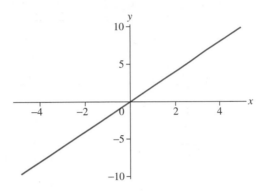

Thus $\mathbf{OQ} = [\alpha \ \ \ 2\alpha] = \alpha[1 \ \ \ 2] = \alpha v$.

In other words, the collection of all linear combinations of the vector $v = [1 \ \ \ 2]$ represents a line passing through the origin and the point $[1 \ \ \ 2]$.

b. Consider linear combinations of two vectors $v_1 = [1 \ \ \ 0]$ and $v_2 = [0 \ \ \ 1]$. Any linear combination of v_1 and v_2 is $\alpha_1[1 \ \ \ 0] + \alpha_2[0 \ \ \ 1] = [\alpha_1 \ \ \ \alpha_2]$, where α_1 and α_2 are scalars. Since α_1 and α_2 are arbitrary numbers, it follows that any point P on the plane with $\mathbf{OP} = [\alpha_1 \ \ \ \alpha_2]$, $\mathbf{OP}$ is a linear combination of v_1 and v_2.

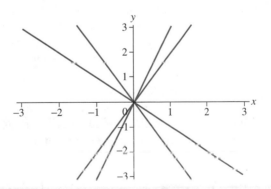

Therefore the set of all linear combinations of v_1 and v_2 represents the entire plane.

It may be remarked that, in general, if v_1 and v_2 are any two nonparallel vectors in a plane, then the set of all linear combinations of v_1 and v_2 represents the whole plane.

c. Suppose Tom makes three kinds of soup using the following ingredients measured in ounces:

	Soup 1	Soup 2	Soup 3
Tomatoes	4	9	11
Carrots	10	6	11
Mushrooms	10	6	11
Lentils	10	0	5

Show that Tom need not prepare Soup 3 separately because he can make Soup 3 as a combination of Soups 1 and 2. In other words, the vector $\begin{bmatrix} 11 \\ 11 \\ 11 \\ 5 \end{bmatrix}$ is a linear combination of the vectors $\begin{bmatrix} 4 \\ 10 \\ 10 \\ 10 \end{bmatrix}$ and $\begin{bmatrix} 9 \\ 6 \\ 6 \\ 0 \end{bmatrix}$.

To see this, we write

$$\begin{bmatrix} 11 \\ 11 \\ 11 \\ 5 \end{bmatrix} = a \begin{bmatrix} 4 \\ 10 \\ 10 \\ 10 \end{bmatrix} + b \begin{bmatrix} 9 \\ 6 \\ 6 \\ 0 \end{bmatrix}$$

and solve for a and b. By equating corresponding entries, we get the following equations:

$$4a + 9b = 11, \qquad 10a + 6b = 11, \qquad 10a + 6b = 11, \qquad \text{and} \qquad 10a = 5.$$

The last equation gives $a = \frac{1}{2}$. By substituting the value of a in the first equation, we obtain $b = 1$. Now we must verify whether these values of a and b satisfy the rest of the equations. This is indeed true. Hence the Soup 3 vector is a linear combination of the vectors representing Soups 1 and 2. We remark that if the values of a and b obtained above do not satisfy all equations, then the linear system is inconsistent, and the conclusion would be that the Soup 3 vector cannot be expressed as a linear combination of the Soup 1 vector and the Soup 2 vector. We also remark that if the Soup 3 vector is a linear combination of the Soup 1 vector and the Soup 2 vector with a or b negative, then we cannot prepare Soup 3 using Soups 1 and 2, although the Soup 3 vector is a linear combination of the vectors representing Soup 1 and Soup 2.

d.

$$2 \begin{bmatrix} 1 \\ 2 \\ 3 \end{bmatrix} - 5 \begin{bmatrix} 6 \\ 7 \\ 8 \end{bmatrix} + 10 \begin{bmatrix} 1 \\ 2 \\ 0 \end{bmatrix} = \begin{bmatrix} -18 \\ -11 \\ -34 \end{bmatrix} = - \begin{bmatrix} 18 \\ 11 \\ 34 \end{bmatrix}$$

is a linear combination of the columns of the matrix

$$\begin{bmatrix} 1 & 6 & 1 \\ 2 & 7 & 2 \\ 3 & 8 & 0 \end{bmatrix}.$$

e. Let

$$A = \begin{bmatrix} 1 & 1 & 0 & 6 \\ 2 & 3 & 5 & 0 \\ 6 & 0 & 1 & 1 \end{bmatrix}.$$

We claim that no linear combination of the first two rows of the matrix A can equal the third row. If we write the third row as a linear combination of the first two rows, that is,

$$[\,6 \quad 0 \quad 1 \quad 1\,] = a[\,1 \quad 1 \quad 0 \quad 6\,] + b[\,2 \quad 3 \quad 5 \quad 0\,],$$

then by equating corresponding components on both sides, we obtain $6 = a + 2b$, $0 = a + 3b$, $1 = 3b$, and $1 = 6a$. The last two equations give $b = \frac{1}{3}$ and $a = \frac{1}{6}$. But these values of a and b do not satisfy the first two equations. Thus we cannot express the third row as a linear combination of the first two rows. ∎

3.1 Exercises

1. (Drill 3.1) Let

$$A = \begin{bmatrix} -1 & 2 & 3 & 4 \\ 5 & 0 & -1 & -1 \\ 8 & -6 & -10 & -13 \end{bmatrix}.$$

a. Show that the last row of A is a linear combination of the first two rows.

b. Suppose $A^{(k)}$ denotes the kth column of A. Find scalars x_1, x_2, and x_4 such that

$$x_1 A^{(1)} + x_2 A^{(2)} + x_4 A^{(4)} = A^{(3)}.$$

c. Find scalars α_2, α_3, and α_4, not all 0, such that

$$\alpha_2 A^{(2)} + \alpha_3 A^{(3)} + \alpha_4 A^{(4)} = \begin{bmatrix} 0 \\ 0 \\ 0 \end{bmatrix}.$$

2. Let A be the diagonal matrix $\begin{bmatrix} 1 & 0 & 0 \\ 0 & 2 & 0 \\ 0 & 0 & 3 \end{bmatrix}$.

a. Show that no column of A is a linear combination of the other two columns.

b. Show that no row of A is a linear combination of the other rows of A.

c. For the diagonal matrix $\begin{bmatrix} 1 & 0 & 0 \\ 0 & 2 & 0 \\ 0 & 0 & 0 \end{bmatrix}$, check whether statements (a) and (b) hold.

3. For each of the coefficient matrices A, solve the linear system of equations $Ax = b$. If

$$x = \begin{bmatrix} \alpha \\ \beta \\ \gamma \\ \vdots \end{bmatrix}$$

is a solution vector, verify that

$$b = \alpha A^{(1)} + \beta A^{(2)} + \gamma A^{(3)} + \cdots.$$

In other words, b is a linear combination of the columns $A^{(1)}$, $A^{(2)}$, $A^{(3)}$, ... of the coefficient matrix A.

a. $2x_1 + 3x_2 + x_3 + 5x_4 = 2,$
$3x_1 + 2x_2 + 4x_3 + 2x_4 = 3,$
$x_1 + x_2 + 2x_3 + 4x_4 = 1.$

b. $x_1 + 2x_2 + x_3 + 2x_4 = 9,$
$2x_1 + x_2 + 3x_3 + x_4 = 5,$
$x_1 + 3x_2 + 2x_3 + 4x_4 = 15,$
$3x_1 + 2x_2 + x_3 + 5x_4 = 22.$

4. Let $AX = b$ be a linear system of m equations in n variables x_1, x_2, ..., x_n. Show that b is a linear combination of the columns $A^{(1)}$, ..., $A^{(n)}$ of A if and only if the linear system is consistent.

3.2 Vector Subspaces

Let S be any set, and let x be a member of S. We generally say that x belongs to S and write it as $x \in S$. To describe a set mathematically, we use the symbol $|$ to mean "such that." For example, if we want to write the set F^n of all $n \times 1$ matrices in the set theoretic language, we write

$$F^n = \{x \mid x \text{ is an } n \times 1 \text{ matrix}\}$$

and we read it "x such that x is an $n \times 1$ matrix." To give another example, suppose we want to describe the set of all points (a, b) in the xy plane, denoted by $\mathbf{R}^2$, that lie on the line $x + y = 0$; then we write

$$\{(a, b) \in \mathbf{R}^2 \mid a + b = 0\}$$

and read it "(a, b) in $\mathbf{R}^2$ such that $a + b = 0$."

3.2.1 Property   An important property of the set $W = \{\alpha_1 v_1 + \cdots + \alpha_m v_m \mid \alpha_i \in \mathbf{R}\}$ of all linear combinations of n-dimensional vectors $v_1, v_2, \ldots, v_m$ is that for any two vectors in W, their sum and difference belong to W. Also, for any number

β and for any vector w of W, βw is also in W. This can be seen by rearranging the terms in the sum and using commutativity of addition and other properties of vectors as sketched below:

$$(\alpha_1 v_1 + \cdots + \alpha_m v_m) + (\beta_1 v_1 + \cdots + \beta_m v_m) = (\alpha_1 + \beta_1)v_1 + \cdots + (\alpha_m + \beta_m)v_m,$$

and

$$\beta(\alpha_1 v_1 + \cdots + \alpha_m v_m) = (\beta\alpha_1)v_1 + \cdots + (\beta\alpha_m)v_m.$$

Since for any vector $w \in W$ and any number β, βw is in W, it follows that $0 \in W$ and $-w \in W$.

Recall, as was stated earlier, that F^n will denote the set of all $n \times 1$ or $1 \times n$ matrices whose entries are real or complex numbers. The context will make it clear whether F^n denotes the set of all $n \times 1$ matrices or $1 \times n$ matrices. The symbol **R** will denote the set of all real numbers.

The above remarks lead to the definition of a vector subspace in F^n. Members of F are usually referred to as *scalars*.

3.2.2 DEFINITION

Subspace

A nonempty subset W of F^n is called a *vector subspace* of F^n, or simply a *subspace*, if the following axioms hold:

1. Whenever $x \in W$ and $y \in W$, then $x + y \in W$.

2. Whenever $x \in W$ and $\alpha \in F$, then $\alpha x \in W$.

3.2.3 Remark

Axiom 1 can be replaced by $x - y \in W$ whenever $x \in W$ and $y \in W$. If $y \in W$, then by axiom 2, $-y = (-1)y \in W$, and so by axiom 1, $x - y = x + (-1)y \in W$. In particular, if W is a subspace, then $0 \in W$.

Note that $W_1 = \{0, 1, -1\}$ and $W_2 = \{\ldots, -2, -1, 1, 2, 3, \ldots\}$ are not subspaces. W_1 is not a subspace because $1 - (-1) = 2$ is not in W_1. W_2 is not a subspace because 0 is not in W_2.

Trivially, the set consisting of zero vector alone satisfies the axioms of a vector subspace. This subspace is denoted by (0). Furthermore, the set F^n of all $1 \times n$ (or $n \times 1$) matrices over the set of real or complex numbers F also satisfies the above axioms, and so F^n is a vector subspace.

The word "vector" preceding subspace is omitted, and we refer to vector subspace as *subspace*.

For $F = \mathbf{R}$, the set of real numbers, the subspace $\mathbf{R}^2$ is the usual xy plane in plane geometry, and the subspace $\mathbf{R}^3$ is the usual three-dimensional space in solid geometry. The general definition of vector space whose members are not necessarily row or column vectors will be given in *Chapter 9*.

3.2.4 EXAMPLES

a. Let $W = \left\{ \begin{bmatrix} x_1 \\ x_2 \end{bmatrix} \middle| x_1 + x_2 = 0 \right\}$. Then W is a subspace of $\mathbf{R}^2$, as we will now show. (Geometrically, W represents the straight line $y = -x$ in $\mathbf{R}^2$.)

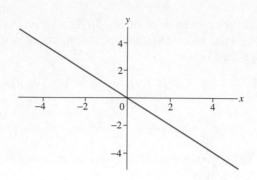

Let $\begin{bmatrix} x_1 \\ x_2 \end{bmatrix}$ and $\begin{bmatrix} x_1' \\ x_2' \end{bmatrix} \in W$. Then $x_1 + x_2 = 0$ and $x_1' + x_2' = 0$. This gives $(x_1 + x_1') + (x_2 + x_2') = 0$. So

$$\begin{bmatrix} x_1 \\ x_2 \end{bmatrix} + \begin{bmatrix} x_1' \\ x_2' \end{bmatrix} = \begin{bmatrix} x_1 + x_1' \\ x_2 + x_2' \end{bmatrix} \in W.$$

Thus axiom 1 is satisfied.

If $\alpha \in \mathbf{R}$, then

$$\alpha \begin{bmatrix} x_1 \\ x_2 \end{bmatrix} = \begin{bmatrix} \alpha x_1 \\ \alpha x_2 \end{bmatrix}$$

also belongs to W, since $\alpha x_1 + \alpha x_2 = \alpha(x_1 + x_2) = 0$. Thus axiom 2 is satisfied as well, and W is indeed a subspace of F^2.

b. Let

$$W = \left\{ \begin{bmatrix} x_1 \\ x_2 \\ x_3 \end{bmatrix} \in F^3 \Big| \ x_1 - x_2 - x_3 = 0 \right\}.$$

For $F = \mathbf{R}$, this represents, geometrically, the plane $x = y + z$ in the three-dimensional space $\mathbf{R}^3$.

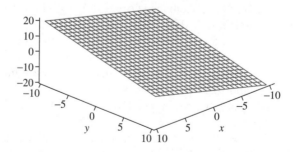

We show that W is a subspace of F^3. We first note that

$$\begin{bmatrix} x_1 \\ x_2 \\ x_3 \end{bmatrix} \in W$$

if and only if $x_3 = x_1 - x_2$.

Let

$$\begin{bmatrix} a_1 \\ a_2 \\ a_1 - a_2 \end{bmatrix}, \quad \begin{bmatrix} b_1 \\ b_2 \\ b_1 - b_2 \end{bmatrix}$$

be any two vectors in W. Then by the definition of sum of vectors,

$$\begin{bmatrix} a_1 \\ a_2 \\ a_1 - a_2 \end{bmatrix} + \begin{bmatrix} b_1 \\ b_2 \\ b_1 - b_2 \end{bmatrix} = \begin{bmatrix} a_1 + b_1 \\ a_2 + b_2 \\ a_1 - a_2 + b_1 - b_2 \end{bmatrix} = \begin{bmatrix} a_1 + b_1 \\ a_2 + b_2 \\ (a_1 + b_1) - (a_2 + b_2) \end{bmatrix},$$

which, by the definition of W, is an element of W. Let α be any scalar. Then

$$\alpha \begin{bmatrix} a_1 \\ a_2 \\ a_1 - a_2 \end{bmatrix} = \begin{bmatrix} \alpha a_1 \\ \alpha a_2 \\ \alpha(a_1 - a_2) \end{bmatrix} = \begin{bmatrix} \alpha a_1 \\ \alpha a_2 \\ \alpha a_1 - \alpha a_2 \end{bmatrix} \in W$$

by using the properties of scalar multiplication of a matrix. Hence W is a subspace.

c. Let W be the set of solutions

$$x = \begin{bmatrix} x_1 \\ x_2 \\ x_3 \end{bmatrix}$$

of the homogeneous linear system

$$2x_1 + 3x_2 + 4x_3 = 0,$$
$$x_2 - x_3 = 0.$$

Then W is a subspace.

We can rewrite the given LS as $Ax = 0$, where

$$A = \begin{bmatrix} 2 & 3 & 4 \\ 0 & 1 & -1 \end{bmatrix} \quad \text{and} \quad x = \begin{bmatrix} x_1 \\ x_2 \\ x_3 \end{bmatrix}.$$

Let $u, v \in W$. Then $Au = 0$, and $Av = 0$. So $A(u + v) = Au + Av = 0$. Thus $u + v \in W$. Similarly, if $\alpha \in F$, then $\alpha u \in W$. Thus W is a subspace.

The reader will notice that the above solution does not explicitly make use of a particular matrix and a particular vector x. The same solution holds for any homogeneous linear system of m equations in n unknowns. In other words, the set of solutions of any homogeneous linear system is a subspace.

Recall that the set of all linear combinations of vectors $v_1, v_2, \ldots, v_m$ satisfies the axioms 1 and 2 for the subspaces as explained in Definition 3.2.2. ∎

3.2.5
DEFINITION

Subspace Generated by a Set
The set of all linear combinations of vectors $v_1, v_2, \ldots, v_m$ is called the *subspace spanned* (or *generated*) by $v_1, v_2, \ldots, v_m$.

The subspace spanned by $v_1, v_2, \ldots, v_m$ is denoted by $\langle v_1, v_2, \ldots, v_m \rangle$ and the vectors $v_1, v_2, \ldots, v_m$ are called *generators*.

The subspace $\mathbf{R}^3$ is generated by $e_1 = [1 \quad 0 \quad 0]$, $e_2 = [0 \quad 1 \quad 0]$, and $e_3 = [0 \quad 0 \quad 1]$. If $[a_1 \quad a_2 \quad a_3]$ is any vector in $\mathbf{R}^3$, then

$$[a_1 \quad a_2 \quad a_3] = a_1[1 \quad 0 \quad 0] + a_2[0 \quad 1 \quad 0] + a_3[0 \quad 0 \quad 1].$$

The generators are not unique. For example, $\mathbf{R}^3$ is also generated by $[1 \quad 1 \quad 0]$, $[0 \quad 1 \quad 1]$, and $[1 \quad 0 \quad 0]$. Note we could represent vectors as column vectors instead of row vectors. Using the column vector notation, one would say in the latter case that $\mathbf{R}^3$ is also generated by $\begin{bmatrix} 1 \\ 1 \\ 0 \end{bmatrix}$, $\begin{bmatrix} 0 \\ 1 \\ 1 \end{bmatrix}$, and $\begin{bmatrix} 1 \\ 0 \\ 0 \end{bmatrix}$. Under this notation, let $\begin{bmatrix} a \\ b \\ c \end{bmatrix} \in \mathbf{R}^3$. Then we find $\alpha, \beta, \gamma \in \mathbf{R}$ such that

$$\begin{bmatrix} a \\ b \\ c \end{bmatrix} = \alpha \begin{bmatrix} 1 \\ 1 \\ 0 \end{bmatrix} + \beta \begin{bmatrix} 0 \\ 1 \\ 1 \end{bmatrix} + \gamma \begin{bmatrix} 1 \\ 0 \\ 0 \end{bmatrix}$$

as follows. By equating corresponding entries, we have

$$a = \alpha + \gamma,$$
$$b = \alpha + \beta,$$
$$c = \beta.$$

So $\beta = c$, $\alpha = b - c$, and $\gamma = a - b + c$.

It is important to note that the set of generators of a subspace is not unique. Trivially, the set of all vectors of a subspace generate it. Indeed, it is an important question in linear algebra to find a minimal set of generators of a subspace.

3.2.6 DEFINITION

Row Space

The subspace spanned by the rows of a matrix A is called the *row space* of A.

3.2.7 EXAMPLES

a. Let $A = \begin{bmatrix} 1 & 2 & 7 \\ 2 & 5 & 2 \\ -1 & 3 & 3 \end{bmatrix}$. The rows of A are vectors in F^3. Does the vector $[1 \quad -1 \quad 2]$ belong to the row space of A? We proceed to look for scalars $x_1, x_2,$ and x_3 such that

$$x_1[1 \quad 2 \quad 7] + x_2[2 \quad 5 \quad 2] + x_3[-1 \quad 3 \quad 3] = [1 \quad -1 \quad 2],$$

that is,

$$[x_1 + 2x_2 - x_3 \quad 2x_1 + 5x_2 + 3x_3 \quad 7x_1 + 2x_2 + 3x_3] = [1 \quad -1 \quad 2].$$

Equating the corresponding components of the vector on the left and the vector on the right, we obtain

$$x_1 + 2x_2 - x_3 = 1,$$
$$2x_1 + 5x_2 + 3x_3 = -1,$$
$$7x_1 + 2x_2 + 3x_3 = 2.$$

This LS is equivalent to

$$x_1 + 2x_2 - x_3 = 1,$$
$$x_2 + 5x_3 = -3,$$
$$-12x_2 + 10x_3 = -5,$$

which in turn is equivalent to

$$x_1 + 2x_2 - x_3 = 1,$$
$$x_2 + 5x_3 = -3,$$
$$70x_3 = -41.$$

Solving backward, we obtain $x_3 = -\frac{41}{70}$, and so on. Having found the scalars x_1, x_2, and x_3 that we sought, we conclude that the vector $[\, 1 \quad -1 \quad 2 \,]$, indeed belongs to the row space of A.

b. Let

$$A = \begin{bmatrix} -1 & 1 & 2 & 7 \\ 2 & 0 & 0 & 1 \\ 1 & 2 & 1 & 0 \end{bmatrix}.$$

Let $R(A)$ be the subspace spanned by the columns of A. Does the vector $\begin{bmatrix} 1 \\ 2 \\ 1 \end{bmatrix}$ belong to $R(A)$? It will be so if we can find scalars x_1, x_2, x_3, and x_4 such that

$$x_1 \begin{bmatrix} -1 \\ 2 \\ 1 \end{bmatrix} + x_2 \begin{bmatrix} 1 \\ 0 \\ 2 \end{bmatrix} + x_3 \begin{bmatrix} 2 \\ 0 \\ 1 \end{bmatrix} + x_4 \begin{bmatrix} 7 \\ 1 \\ 0 \end{bmatrix} = \begin{bmatrix} 1 \\ 2 \\ 1 \end{bmatrix},$$

that is,

$$\begin{bmatrix} -x_1 \\ 2x_1 \\ x_1 \end{bmatrix} + \begin{bmatrix} x_2 \\ 0 \\ 2x_2 \end{bmatrix} + \begin{bmatrix} 2x_3 \\ 0 \\ x_3 \end{bmatrix} + \begin{bmatrix} 7x_4 \\ x_4 \\ 0 \end{bmatrix} = \begin{bmatrix} 1 \\ 2 \\ 1 \end{bmatrix},$$

that is,

$$\begin{bmatrix} -x_1 & +x_2 & +2x_3 & +7x_4 \\ 2x_1 & & & +x_4 \\ x_1 & +2x_2 & +x_3 & \end{bmatrix} = \begin{bmatrix} 1 \\ 2 \\ 1 \end{bmatrix},$$

that is,

$$-x_1 + x_2 + 2x_3 + 7x_4 = 1,$$
$$2x_1 + x_4 = 2,$$
$$x_1 + 2x_2 + x_3 = 1.$$

We solve for x_1, x_2, x_3, x_4 by reducing the LS to echelon form:

$$-x_1 + x_2 + 2x_3 + 7x_4 = 1,$$
$$2x_2 + 4x_3 + 15x_4 = 4,$$
$$3x_2 + 3x_3 + 7x_4 = 2,$$

that is,

$$-x_1 + x_2 + 2x_3 + 7x_4 = 1,$$
$$x_2 + 2x_3 + \frac{15}{2}x_4 = 2,$$
$$-3x_3 - \frac{31}{2}x_4 = -4.$$

Solving backward, we get more than one set of values for x_1, x_2, x_3, and x_4. So the answer is yes.

We note that if $A^{(1)}, A^{(2)}, A^{(3)}$, and $A^{(4)}$ denote the columns of the matrix A in Example (b), then

$$R(A) = \{x_1 A^{(1)} + x_2 A^{(2)} + x_3 A^{(3)} + x_4 A^{(4)} \mid x_1, x_2, x_3, x_4 \in F\},$$

that is,

$$R(A) = \left\{ x_1 \begin{bmatrix} -1 \\ 2 \\ 1 \end{bmatrix} + x_2 \begin{bmatrix} 1 \\ 0 \\ 2 \end{bmatrix} + x_3 \begin{bmatrix} 2 \\ 0 \\ 1 \end{bmatrix} + x_4 \begin{bmatrix} 7 \\ 1 \\ 0 \end{bmatrix} \;\middle|\; x_1, x_2, x_3, x_4 \in F \right\},$$

that is,

$$R(A) = \left\{ \begin{bmatrix} -1 & 1 & 2 & 7 \\ 2 & 0 & 0 & 1 \\ 1 & 2 & 1 & 0 \end{bmatrix} \begin{bmatrix} x_1 \\ x_2 \\ x_3 \\ x_4 \end{bmatrix} \;\middle|\; x_1, x_2, x_3, x_4 \in F \right\}.$$

■

3.2.8 DEFINITION

Range of a Matrix

If A is an $m \times n$ matrix, then the *range* of A is the set $\{Ax \mid x \in F^n\}$, denoted by $R(A)$.

Note that given an $m \times n$ matrix A and a vector $x \in F^n$, Ax is a linear combination of the columns of A. For this reason, the range of A is also called the *column space* of A (see *Chapter 4*).

3.2 Exercises

In Exercises 1–6, prove or disprove that W is a subspace of some F^n.

1. $W = \{[\ a\quad b\quad 0\ ]\mid a,b \in F\}$.

2. $W = \left\{ \begin{bmatrix} a \\ b \\ c \end{bmatrix} \,\middle|\, a+b+c=0 \right\}$.

3. $W = \left\{ \begin{bmatrix} x_1 \\ x_2 \\ x_3 \end{bmatrix} \,\middle|\, x_3 = 2x_1 - x_2 \right\}$.

4. $W = \left\{ \begin{bmatrix} x_1 \\ x_2 \end{bmatrix} \,\middle|\, x_2 = 1 \right\}$.

5. $W = \{[1\quad 1\quad 1], [1\quad 0\quad 0], [0\quad 1\quad 0]\}$.

6. $W = \{x \in F^3 \mid Ax = 0\}$, where A is a fixed 3×3 matrix.

7. Which of the following regions are subspaces in $\mathbf{R}^2$? Give a reason.

a. Elliptic disk: $2x^2 + 3y^2 \leqslant 6$.

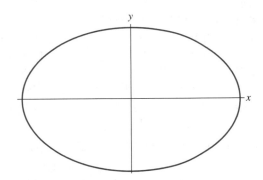

b. Triangular disk with vertices $(0, 0)$, $(1, 0)$, $(0, 1)$.

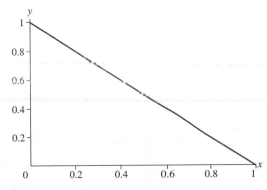

c. The whole x axis in the xy plane.

8. Show that the subset W of $\mathbf{R}^3$, where the elements of W are vectors with rational entries, cannot be a subspace of $\mathbf{R}^3$. (Hint: The product of a rational number and an irrational number is always an irrational number.)

9. Show that the intersection of two subspaces is again a subspace. (Hint: The set of vectors common to a family of subspaces is called the intersection of subspaces.)

10. Show algebraically and graphically that the set of solutions $x = \begin{bmatrix} x_1 \\ x_2 \\ x_3 \end{bmatrix}$ of the LS

$$2x_1 + x_2 + x_3 = 1,$$
$$x_2 - x_3 = 2,$$

is not a subspace.

11. Prove, in general, that the set of solutions

$$x = \begin{bmatrix} x_1 \\ x_2 \\ \vdots \\ x_3 \end{bmatrix}$$

of any nonhomogeneous LS $Ax = b$ is not a subspace. However, if the LS is homogeneous, then the set of solutions is a subspace.

12. Determine whether $[1 \quad 2]$ belongs to the subspace spanned by $[1 \quad 0], [1 \quad 1]$, and $[-1 \quad 0]$. Also, find whether $[1 \quad 2]$ belongs to the subspace spanned by $[1 \quad 0]$ and $[1 \quad 1]$. Is it possible that $[1 \quad 2]$ belongs to the subspace spanned by $[1 \quad 0]$?

 13. (Drill 3.2) Determine whether $[1 \quad 1 \quad 1]$ belongs to the subspace spanned by $[1 \quad 3 \quad 4], [4 \quad 0 \quad 1]$, and $[3 \quad 1 \quad 2]$.

14. Determine whether $[2 \ 0 \ 4 \ -2]$ belongs to the subspace spanned by $[0 \ 2 \ 1 \ -1]$, $[1 \ -1 \ 1 \ 0]$, and $[2 \ 1 \ 0 \ -2]$.

 15. (Drill 3.3) Let

$$A = \begin{bmatrix} 1 & 1 & 3 & 1 \\ 2 & 1 & 5 & 4 \\ 1 & 2 & 4 & -1 \end{bmatrix}.$$

a. Find two vectors other than the rows of A in the space spanned by the rows.

b. Find a vector in the space spanned by the columns of A not equal to any of the columns.

c. Show that the fourth column of A belongs to the subspace spanned by the first three columns of A.

d. Does the vector $[-1 \quad 1 \quad 0 \quad 1]$ belong to the space spanned by the rows of A?

e. Does the vector $\begin{bmatrix} 1 \\ 0 \\ -1 \end{bmatrix}$ or $\begin{bmatrix} 0 \\ 0 \\ 0 \end{bmatrix}$ belong to the subspace spanned by the columns of A?

f. Does the vector [1   0   0   0] belong to the subspace spanned by the rows of A^T?

3.3   Linear Dependence, Linear Independence, and Basis

3.3.1
DEFINITION

Linearly Dependent
A set of vectors $v_1, v_2, \ldots, v_m$ in F^n is said to be *linearly dependent* if at least one of the vectors can be expressed as a linear combination of the rest of the vectors.

In other words, if $v_1, v_2, \ldots, v_m$ are linearly dependent, then some $v_i = \alpha_1 v_1 + \cdots + \alpha_{i-1} v_{i-1} + \alpha_{i+1} v_{i+1} + \cdots \alpha_m v_m$, where $\alpha_1, \ldots, \alpha_{i-1}, \alpha_{i+1}, \ldots, \alpha_m$ are scalars. This implies that there exists a relation

$$\alpha_1 v_1 + \cdots + \alpha_m v_m = 0$$

among $v_1, v_2, \ldots, v_m$ with at least one of the scalars α_i equal to -1. Clearly, the converse also holds, that is, if there exists a relation

$$\alpha_1 v_1 + \cdots + \alpha_m v_m = 0$$

with some $\alpha_i = -1$, then the vector v_i is a linear combination of the rest of the vectors. Indeed, it is enough to require that some α_i is not zero. By dividing by α_i, we get

$$v_i = \left(-\frac{\alpha_1}{\alpha_i} \right) v_1 + \cdots + \left(-\frac{\alpha_m}{\alpha_i} \right) v_m,$$

that is, v_i is a linear combination of $v_1, \ldots, v_{i-1}, v_{i+1}, \ldots, v_m$.

3.3.2
DEFINITION

Linearly Independent
If a set of vectors in F^n is not linearly dependent, then it is called *linearly independent*.

Thus if $v_1, v_2, \ldots, v_m$ are linearly independent, then no vector can be expressed as a linear combination of the rest of the vectors. Equivalently,

$$\alpha_1 v_1 + \cdots + \alpha_m v_m = 0 \tag{1}$$

implies that $\alpha_1 = 0 = \alpha_2 = \cdots = \alpha_m$, that is, the only solution of the homogenous linear system of equations given by Equation (1) in the unknowns $\alpha_1, \alpha_2, \ldots, \alpha_m$ is the trivial solution. Otherwise, if some α_i is not zero, then, as was argued above, v_i is a linear combination of $v_1, \ldots, v_{i-1}, v_{i+1}, \ldots, v_m$ which is not true by the definition of linear independence of $v_1, v_2, \ldots, v_m$.

3.3.3
EXAMPLES

a. The set of vectors $[1 \quad 1 \quad 0], [1 \quad 2 \quad 3]$, and $[0 \quad 0 \quad 1]$ are linearly independent.

Let $\alpha_1[1 \quad 1 \quad 0] + \alpha_2[1 \quad 2 \quad 3] + \alpha_3[0 \quad 0 \quad 1] = 0$.

Then after adding the three vectors and equating each entry to 0, we get a homogeneous linear system of equations in α_1, α_2, and α_3 given by

$$\alpha_1 + \alpha_2 = 0, \quad \alpha_1 + 2\alpha_2 = 0, \quad \text{and} \quad 3\alpha_2 + \alpha_3 = 0.$$

This system has only a trivial solution, $\alpha_1 = 0, \alpha_2 = 0, \alpha_3 = 0$. Thus the given vectors are linearly independent.

b. The vectors $[1 \quad 2 \quad 3]$ and $[2 \quad 4 \quad 6]$ are linearly dependent.

For, let $\alpha_1[1 \quad 2 \quad 3] + \alpha_2[2 \quad 4 \quad 6] = 0$. Then by adding the two vectors and equating each entry to 0, we get $\alpha_1 + 2\alpha_2 = 0$, $2\alpha_1 + 4\alpha_2 = 0$, and $3\alpha_1 + 6\alpha_2 = 0$. But these equations are all the same, and so we have only one equation, say, $\alpha_1 + 2\alpha_2 = 0$. There are infinitely many solutions satisfying this equation $\alpha_1 + 2\alpha_2 = 0$, and one such solution is $\alpha_1 = -2, \alpha_2 = 1$, giving us a relation $(-2)[1 \quad 2 \quad 3] + (1)[2 \quad 4 \quad 6] = 0$. Thus the given vectors are linearly dependent.

c. Any set of vectors $\{v_1, v_2, \ldots, v_m\}$ in F^n that contains the zero vector is linearly dependent. For suppose $v_2 = 0$. Then

$$0v_1 + \alpha v_2 + 0v_3 + \cdots + 0v_m = 0,$$

where α is any nonzero scalar. So the set is linearly dependent.

d. Let $v = [x_1 \quad x_2 \ldots x_n]$ be a nonzero vector, so at least one of the numbers in the list $x_1, x_2, \ldots, x_n$ is not zero. We show that the set $\{v\}$ with one nonzero vector is linearly independent. Let $\alpha v = 0$. Then $\alpha v = [\alpha x_1 \quad \alpha x_2 \ldots \alpha x_n] = [0 \quad 0 \ldots 0]$ would give $\alpha x_i = 0$ for each i, that is, $\alpha = 0$, since the product of two numbers can be 0 only if at least one of them is 0. Thus $\alpha v = 0$ implies $\alpha = 0$, since $v \neq 0$. This shows that $\{v\}$ is linearly independent.

e. The set of nonzero rows of an echelon matrix forms a linearly independent set of row vectors. For example, consider

$$A = \begin{bmatrix} 1 & -2 & 3 & 4 & -6 \\ 0 & 2 & 0 & 6 & 1 \\ 0 & 0 & 5 & -1 & 3 \\ 0 & 0 & 0 & 0 & 0 \end{bmatrix},$$

which is in row echelon form.

Let

$$x_1[1 \quad -2 \quad 3 \quad 4 \quad -6] + x_2[0 \quad 2 \quad 0 \quad 6 \quad 1] + x_3[0 \quad 0 \quad 5 \quad -1 \quad 3]$$
$$= [0 \quad 0 \quad 0 \quad 0 \quad 0].$$

The left-hand side is the vector

$$[x_1 \quad -2x_1 + 2x_2 \quad 3x_1 + 5x_3 \quad 4x_1 + 6x_2 - x_3 \quad -6x_1 + x_2 + 3x_3],$$

which is equal to $[0 \quad 0 \quad 0 \quad 0 \quad 0]$. Thus

$$x_1 = 0,$$
$$-2x_1 + 2x_2 = 0,$$
$$3x_1 + 5x_3 = 0,$$
$$4x_1 + 6x_2 - x_3 = 0,$$
$$-6x_1 + x_2 + 3x_3 = 0.$$

The first equation gives $x_1 = 0$. Inserting this value into the second equation, we get $x_2 = 0$. Similarly, $x_3 = 0$. Hence the nonzero rows of A are linearly independent.

f. The set of vectors

$$e_1 = \begin{bmatrix} 1 \\ 0 \\ 0 \end{bmatrix}, \qquad e_2 = \begin{bmatrix} 0 \\ 1 \\ 0 \end{bmatrix}, \qquad e_3 = \begin{bmatrix} 0 \\ 0 \\ 1 \end{bmatrix}$$

in F^3 is linearly independent. Consider the equation

$$\alpha_1 \begin{bmatrix} 1 \\ 0 \\ 0 \end{bmatrix} + \alpha_2 \begin{bmatrix} 0 \\ 1 \\ 0 \end{bmatrix} + \alpha_3 \begin{bmatrix} 0 \\ 0 \\ 1 \end{bmatrix} = \begin{bmatrix} 0 \\ 0 \\ 0 \end{bmatrix}.$$

The left-hand side is

$$\begin{bmatrix} \alpha_1 \\ 0 \\ 0 \end{bmatrix} + \begin{bmatrix} 0 \\ \alpha_2 \\ 0 \end{bmatrix} + \begin{bmatrix} 0 \\ 0 \\ \alpha_3 \end{bmatrix} = \begin{bmatrix} \alpha_1 \\ \alpha_2 \\ \alpha_3 \end{bmatrix},$$

and this can be equal to $\begin{bmatrix} 0 \\ 0 \\ 0 \end{bmatrix}$ only if $\alpha_1 = 0$, $\alpha_2 = 0$, and $\alpha_3 = 0$. So the given set is linearly independent. Further, let

$$\begin{bmatrix} x_1 \\ x_2 \\ x_3 \end{bmatrix} \in F^3.$$

Then

$$x_1 \begin{bmatrix} 1 \\ 0 \\ 0 \end{bmatrix} + x_2 \begin{bmatrix} 0 \\ 1 \\ 0 \end{bmatrix} + x_3 \begin{bmatrix} 0 \\ 0 \\ 1 \end{bmatrix} = \begin{bmatrix} x_1 \\ x_2 \\ x_3 \end{bmatrix}.$$

This means that

$$F^3 = \left\langle \begin{bmatrix} 1 \\ 0 \\ 0 \end{bmatrix}, \begin{bmatrix} 0 \\ 1 \\ 0 \end{bmatrix}, \begin{bmatrix} 0 \\ 0 \\ 1 \end{bmatrix} \right\rangle.$$

g. Consider the subspace

$$W = \left\{ \begin{bmatrix} x_1 \\ x_2 \\ x_3 \end{bmatrix} \in F^3 \ \middle| \ x_3 = x_1 - x_2 \right\}.$$

Any vector $\begin{bmatrix} x_1 \\ x_2 \\ x_3 \end{bmatrix}$ in W can be written as $\begin{bmatrix} x_1 \\ x_2 \\ x_1 - x_2 \end{bmatrix}$, and this can be rewritten as

$$\begin{bmatrix} x_1 \\ 0 \\ x_1 \end{bmatrix} + \begin{bmatrix} 0 \\ x_2 \\ -x_2 \end{bmatrix} = x_1 \begin{bmatrix} 1 \\ 0 \\ 1 \end{bmatrix} + x_2 \begin{bmatrix} 0 \\ 1 \\ -1 \end{bmatrix}.$$

Thus every vector of W is a linear combination of the vectors $\begin{bmatrix} 1 \\ 0 \\ 1 \end{bmatrix}$ and $\begin{bmatrix} 0 \\ 1 \\ -1 \end{bmatrix}$.
So

$$W = \left\langle \begin{bmatrix} 1 \\ 0 \\ 1 \end{bmatrix}, \begin{bmatrix} 0 \\ 1 \\ -1 \end{bmatrix} \right\rangle.$$

Further, the set $\left\{ \begin{bmatrix} 1 \\ 0 \\ 1 \end{bmatrix}, \begin{bmatrix} 0 \\ 1 \\ -1 \end{bmatrix} \right\}$ is linearly independent. Consider the equation

$$\alpha \begin{bmatrix} 1 \\ 0 \\ 1 \end{bmatrix} + \beta \begin{bmatrix} 0 \\ 1 \\ -1 \end{bmatrix} = \begin{bmatrix} 0 \\ 0 \\ 0 \end{bmatrix}.$$

The left-hand side is equal to

$$\begin{bmatrix} \alpha \\ 0 \\ \alpha \end{bmatrix} + \begin{bmatrix} 0 \\ \beta \\ -\beta \end{bmatrix} = \begin{bmatrix} \alpha \\ \beta \\ \alpha - \beta \end{bmatrix},$$

which is equal to $\begin{bmatrix} 0 \\ 0 \\ 0 \end{bmatrix}$ only if $\alpha = 0, \beta = 0$.

Caution. The coefficient matrix of a homogeneous linear system in $a_1, a_2, \ldots, a_m$ given by a relation $a_1 v_1 + \cdots + a_m v_m = 0$, is $[v_1^T \ \ v_2^T \ \ \cdots \ \ v_m^T]$ or $[v_1 \ v_2 \ \cdots \ v_m]$ according to whether the vectors $v_1, v_2, \ldots, v_m$ are row vectors or column vectors.

Students tend to make the common mistake of assuming the matrix $\begin{bmatrix} v_1 \\ v_2 \\ \vdots \\ v_m \end{bmatrix}$ to be the coefficient matrix while solving for $a_1, a_2, \ldots, a_m$ when $v_1, v_2, \ldots, v_m$ are row vectors. ∎

> **3.3.4**
> **DEFINITION**
>
> **Basis**
>
> Let W be a subspace of F^n. Then a set of vectors $\{v_1, v_2, \ldots, v_m\}$ in F^n is called a *basis* of W if
>
> (i) $W = \langle v_1, \ldots, v_m \rangle$, that is, W is spanned by the vectors $v_1, v_2, \ldots, v_m$, and
>
> (ii) $\{v_1, \ldots, v_m\}$ is a linearly independent set.

It is a deep result in linear algebra that every subspace has a basis. The zero subspace has no linearly independent vectors, and its basis, by convention, is defined to be the empty set.

With reference to the preceding examples, we note that:

1. In Example (e), the nonzero rows of A form a basis of the subspace spanned by the rows.

2. In Example (f), the vectors e_1, e_2, and e_3 form a basis of F^3, called the *standard basis* of F^3. This follows from the fact that every vector $\begin{bmatrix} \alpha \\ \beta \\ \gamma \end{bmatrix}$ in F^3 can be expressed as a linear combination,

$$\alpha \begin{bmatrix} 1 \\ 0 \\ 1 \end{bmatrix} + \beta \begin{bmatrix} 0 \\ 1 \\ 0 \end{bmatrix} + \gamma \begin{bmatrix} 0 \\ 0 \\ 1 \end{bmatrix}$$

of

$$e_1 = \begin{bmatrix} 1 \\ 0 \\ 0 \end{bmatrix}, \qquad e_2 = \begin{bmatrix} 0 \\ 1 \\ 0 \end{bmatrix}, \qquad e_3 = \begin{bmatrix} 0 \\ 0 \\ 1 \end{bmatrix},$$

and e_1, e_2, and e_3 are linearly independent.

If $F = \mathbf{R}$, the field of real numbers, then e_1, e_2, and e_3 are the unit vectors in $\mathbf{R}^3$ (vectors of length 1 in the three-dimensional space) along the x axis, the y axis, and the z axis, respectively.

In general, the vectors

$$e_1 = \begin{bmatrix} 1 \\ 0 \\ \vdots \\ 0 \end{bmatrix}, \qquad e_2 = \begin{bmatrix} 0 \\ 1 \\ \vdots \\ 0 \end{bmatrix}, \qquad \ldots, \qquad e_n = \begin{bmatrix} 0 \\ \vdots \\ 0 \\ 1 \end{bmatrix}$$

form a basis of F^n, called the *standard basis*.

3. In Example (g), the vectors $\begin{bmatrix} 1 \\ 0 \\ 1 \end{bmatrix}$ and $\begin{bmatrix} 0 \\ 1 \\ -1 \end{bmatrix}$ form a basis of W.

3.3.5   A basis of a subspace is not unique. However, *Theorem 5* in *Section 3.4* shows that any
Remark   two bases of a subspace must have the same number of elements.

3.3.6
DEFINITION

Dimension of a Subspace
The number of vectors in any basis of a subspace is called the *dimension* of the
subspace.

The dimension of subspace W is denoted by dim W. By convention, the dimension
of the zero subspace is defined to be 0.

We now give a method of finding a basis of the subspace spanned by a given set of
vectors $v_1, v_2, \ldots, v_s$.

Step 1   Delete every zero vector, if any, from the collection $\{v_1, \ldots, v_s\}$ because
any set containing a zero vector is linearly dependent (see Example (c)).
So without any loss of generality, we may assume that the vectors $v_1, \ldots,$
v_s are nonzero.

Step 2   Test for linear independence of $v_1, \ldots, v_s$. If these are linearly independent, then $v_1, \ldots, v_s$ form a basis. Otherwise, find a dependence relation.
Suppose

$$\alpha_1 v_1 + \cdots + \alpha_s v_s = 0 \qquad (1)$$

is a dependence relation where not all α_i are zero.
Choose $\alpha_i \neq 0$, and rewrite Equation (1) as

$$v_i = -\frac{\alpha_1}{\alpha_i} v_1 - \cdots - \frac{\alpha_s}{\alpha_i} v_s.$$

So v_i can be expressed as a linear combination of the rest of the
vectors, and therefore any linear combination of the vectors $v_1, \ldots, v_s$ can
be expressed as a linear combination of all but v_i. Thus

$$\langle v_1, \ldots, v_{i-1}, v_{i+1}, \ldots, v_s \rangle = \langle v_1, \ldots, v_s \rangle.$$

In other words, we may delete v_i and begin with the question of finding
a basis of the subspace spanned by the remaining vectors $v_1, \ldots, v_{i-1},$
$v_{i+1}, \ldots, v_s$.

Step 3   Repeat Step 2 to delete vectors until arriving at a linearly independent
subset. The linearly independent subset that is so obtained is then a basis.

3.3.7
EXAMPLE

Determine whether the vectors

$$v_1 = [1 \quad 2 \quad 3], \quad v_2 = [1 \quad 2 \quad -1], \quad v_3 = [3 \quad -1 \quad 0], \quad \text{and} \quad v_4 = [2 \quad 1 \quad 2]$$

form a basis of F^3. If not, choose, if possible, a basis of F^3 consisting of vectors out
of the given set of vectors.

Let

$$\alpha_1[\,1 \quad 2 \quad 3\,] + \alpha_2[\,1 \quad 2 \quad -1\,] + \alpha_3[\,3 \quad -1 \quad 0\,] + \alpha_4[\,2 \quad 1 \quad 2\,] = 0.$$

This gives rise to the homogeneous LS

$$\alpha_1 + \alpha_2 + 3\alpha_3 + 2\alpha_4 = 0,$$
$$2\alpha_1 + 2\alpha_2 - \alpha_3 + \alpha_4 = 0,$$
$$3\alpha_1 - \alpha_2 + 0\alpha_3 + 2\alpha_4 = 0,$$

whose coefficient matrix is

$$\begin{bmatrix} 1 & 1 & 3 & 2 \\ 2 & 2 & -1 & 1 \\ 3 & -1 & 0 & 2 \end{bmatrix} \begin{array}{c} R_2 + (-2)R_1 \\ \longrightarrow \\ R_3 + (-3)R_1 \end{array} \begin{bmatrix} 1 & 1 & 3 & 2 \\ 0 & 0 & -7 & -3 \\ 0 & -4 & -9 & -4 \end{bmatrix}$$

$$\begin{array}{c} R_2 \longleftrightarrow R_3 \\ \longrightarrow \end{array} \begin{bmatrix} 1 & 1 & 3 & 2 \\ 0 & -4 & -9 & -4 \\ 0 & 0 & -7 & -3 \end{bmatrix}.$$

The transformed LS is

$$\alpha_1 + \alpha_2 + 3\alpha_3 + 2\alpha_4 = 0,$$
$$-4\alpha_2 - 9\alpha_3 - 4\alpha_4 = 0,$$
$$-7\alpha_3 - 3\alpha_4 = 0.$$

Since the last equation has two unknowns, we can choose an arbitrary value of one of the variables, say $\alpha_4 = t$. Here, we choose a special value of t (whatever you like, but not zero). Choose $t = 7$. Then

$$\alpha_4 = 7,$$
$$\alpha_3 = -3,$$
$$\alpha_1 + \alpha_2 = -3\alpha_3 - 2\alpha_4 = 9 - 14 = -5.$$

Choose $\alpha_2 = 1$. Then $\alpha_1 = -6$. So we have the relation

$$(-6)v_1 + v_2 + (-3)v_3 + 7v_4 = 0.$$

Hence

$$v_4 = \frac{1}{7}(6v_1 - v_2 + 3v_3),$$

which shows that v_4 can be expressed as a linear combination of v_1, v_2, and v_3, and so the given set of vectors is linearly dependent. However, it can be shown that v_1, v_2, and v_3 are linearly independent, and therefore this set forms a basis of F^3, because any set of three linearly independent vectors in F^3 forms a basis of F^3 (*Theorem 3, Section 3.4*). ■

3.3 Exercises

1. Is the set $\left\{ \begin{bmatrix} 2 \\ 6 \\ -2 \end{bmatrix}, \begin{bmatrix} 3 \\ 1 \\ 2 \end{bmatrix}, \begin{bmatrix} 8 \\ 16 \\ -3 \end{bmatrix} \right\}$ linearly independent?

2. Find the value of a such that $\begin{bmatrix} 4 \\ 5 \\ 1 \end{bmatrix}, \begin{bmatrix} 3 \\ 0 \\ 2 \end{bmatrix}, \begin{bmatrix} a \\ 10 \\ 9 \end{bmatrix}$ are linearly dependent.

3. Find a such that $[2 \quad 1 \quad 1 \quad 1], [3 \quad -2 \quad 1 \quad 0], [a \quad -1 \quad 2 \quad 0]$ are linearly independent.

4. Find a, b such that $\begin{bmatrix} 1 \\ 2 \\ 3 \end{bmatrix}, \begin{bmatrix} a \\ 0 \\ 0 \end{bmatrix}, \begin{bmatrix} 0 \\ b \\ 1 \end{bmatrix}$ are linearly independent.

5. Determine whether or not $\begin{bmatrix} 1 \\ 1 \end{bmatrix}$ is a basis of $\mathbf{R}^2$.

6. Determine whether $\begin{bmatrix} 1 \\ 1 \end{bmatrix}$ and $\begin{bmatrix} 2 \\ 3 \end{bmatrix}$ form a basis of $\mathbf{R}^2$.

7. Determine whether $\begin{bmatrix} 1 \\ 1 \end{bmatrix}, \begin{bmatrix} 2 \\ 3 \end{bmatrix}$ and $\begin{bmatrix} 1 \\ 0 \end{bmatrix}$ form a basis of $\mathbf{R}^2$.

8. Determine whether $\begin{bmatrix} 1 \\ 0 \end{bmatrix}, \begin{bmatrix} 0 \\ 1 \end{bmatrix}$ and $\begin{bmatrix} 2 \\ 4 \end{bmatrix}$ form a basis of $\mathbf{R}^2$. If not, find a subset that forms a basis.

9. Determine whether the vectors $[1 \quad 1 \quad 1], [1 \quad 2 \quad 3], [2 \quad -1 \quad 1]$ form a basis of the subspace spanned by them. If not, find a subset that forms a basis of this subspace.

10. (Drill 3.4) Find a basis of the subspace W of $\mathbf{R}^4$ spanned by $[1 \quad 4 \quad -1 \quad 3]$, $[2 \quad 1 \quad -3 \quad 1], [0 \quad 2 \quad 1 \quad -5]$, which is a subset of the set consisting of the given vectors.

11. (Drill 3.5) Let $v_1 = [3 \quad -2 \quad 2 \quad -1], v_2 = [2 \quad -6 \quad 4 \quad 0], v_3 = [4 \quad 8 \quad -4 \quad -3]$, $v_4 = [1 \quad 10 \quad -6 \quad -2], v_5 = [1 \quad -1 \quad 8 \quad 5], v_6 = [6 \quad -2 \quad 4 \quad 8]$. Let $S = \langle v_1, v_2, v_3, v_4 \rangle$, the space generated by the vectors $\{v_1, v_2, v_3, v_4\}$ and $T = \langle v_4, v_5, v_6 \rangle$, the space generated by the vectors $\{v_4, v_5, v_6\}$. Find bases of S, T, and $S + T$. ($S + T$ is defined to be the subspace generated by the union of the bases for S and T. Every element of $S + T$ is of the form $s + t$, where s is in S and t is in T.)

12. Find the dimension of each of the subspaces:

   a. $\mathbf{R}$

   b. $\left\{ \begin{bmatrix} x \\ y \end{bmatrix} \mid y = x \right\}$.

c. $\left\{ \begin{bmatrix} x \\ y \end{bmatrix} \mid y = 3x \right\}.$

d. $\left\{ \begin{bmatrix} x \\ y \\ z \end{bmatrix} \mid z = x + y \right\}.$

13. Find the dimension of the subspace consisting of solutions of each of the following linear systems $Ax = 0$, where

a. $A = \begin{bmatrix} 1 & 2 \\ 2 & 3 \end{bmatrix}, x = \begin{bmatrix} x_1 \\ x_2 \end{bmatrix}.$

b. $A = \begin{bmatrix} 1 & 2 \\ 0 & 1 \end{bmatrix}, x = \begin{bmatrix} x_1 \\ x_2 \end{bmatrix}.$

c. $A = \begin{bmatrix} 1 & 2 & 3 \\ 0 & 1 & 2 \\ 0 & 1 & 1 \end{bmatrix}, x = \begin{bmatrix} x_1 \\ x_2 \\ x_3 \end{bmatrix}.$

d. $A = \begin{bmatrix} 1 & 2 \\ 2 & 4 \end{bmatrix}, x = \begin{bmatrix} x_1 \\ x_2 \end{bmatrix}.$

e. $A = \begin{bmatrix} 1 & 2 & 3 \\ 0 & 1 & 2 \\ 1 & 3 & 5 \end{bmatrix}, x = \begin{bmatrix} x_1 \\ x_2 \\ x_3 \end{bmatrix}.$

f. $A = \begin{bmatrix} 0 & 0 \\ 0 & 0 \end{bmatrix}, x = \begin{bmatrix} x_1 \\ x_2 \end{bmatrix}.$

g. $A = \begin{bmatrix} 0 & 1 & -1 \\ 1 & 0 & 1 \\ 2 & 1 & 1 \end{bmatrix}, x = \begin{bmatrix} x_1 \\ x_2 \\ x_3 \end{bmatrix}.$

14. For the matrix

$$A = \begin{bmatrix} 3 & 3 & 3 \\ 4 & 5 & 6 \\ 7 & 8 & 9 \end{bmatrix},$$

find the dimension of

a. the subspace of $\mathbf{R}^3$ spanned by the rows of A,
b. the subspace of $\mathbf{R}^3$ spanned by the columns of A.

15. For the matrix

$$A = \begin{bmatrix} 0 & 1 \\ 1 & 1 \\ 2 & 3 \end{bmatrix},$$

find the dimension of

a. the subspace of $\mathbf{R}^2$ spanned by the rows of A,
b. the subspace of $\mathbf{R}^3$ spanned by the columns of A.

16. (Drill 3.6) Consider the matrix

$$A = \begin{bmatrix} -1 & 2 & 3 & 1 & 11 \\ 0 & 2 & 3 & 4 & 8 \\ 1 & -2 & 3 & -1 & 1 \\ 1 & -2 & 9 & -1 & 13 \\ -1 & 4 & 6 & 5 & 19 \end{bmatrix}.$$

   a. Find a maximal linearly independent set of vectors from the columns of A.

   b. Reduce the matrix to its row echelon form and find a basis for the row space of A.

17. (Drill 3.7) For the matrices

$$\begin{bmatrix} 1 & 2 & 3 \\ 4 & 5 & 6 \\ 7 & 8 & 9 \end{bmatrix}, \qquad \begin{bmatrix} 1 & 2 & 1 & 2 & 2 & 1 \\ 2 & 4 & 2 & 4 & -1 & 0 \\ 1 & 2 & 1 & 2 & -1 & 0 \\ 2 & 4 & 1 & 2 & 0 & -1 \end{bmatrix},$$

   a. find the dimension of the subspace spanned by the rows and the dimension of the subspace spanned by the columns.

   b. Let A be a random matrix of any size (not necessarily square). Find the dimensions of the subspaces spanned by the rows and the columns of A. What do you conjecture?

18. (Drill 3.8) Consider the two planes in the three-dimensional space $\mathbf{R}^3$ given by

$$S = \{[x, y, z] \mid x - 2y + z = 0\} \quad \text{and} \quad T = \{[x, y, z] \mid -x + 2y + z = 0\}.$$

Find bases for S, T, $S \cap T$, and $S + T$. Also, graph the planes S and T.

3.4 Proofs of Facts

In this section, we give proofs of key results about bases of subspaces. Although their proofs may be optional, familiarity with these results would help the reader to appreciate the concepts of basis and dimension of subspaces.

THEOREM 1

If $\{v_1, \ldots, v_p\}$ is a basis of a subspace W, then any vector w in W can be expressed as a linear combination of $v_1, \ldots, v_p$ in only one way.

Proof Suppose

$$w = \alpha_1 v_1 + \cdots + \alpha_p v_p, \qquad \text{and also} \qquad w = \alpha_1' v_1 + \cdots + \alpha_p' v_p.$$

We now show that $\alpha_1 = \alpha_1', \ldots, \alpha_p = \alpha_p'$. Subtracting one equation from the other, we get $(\alpha_1 - \alpha_1')v_1 + \cdots + (\alpha_p - \alpha_p')v_p = 0$. Since $v_1, \ldots, v_p$ are linearly independent, the last equation gives $\alpha_1 = \alpha_1', \ldots, \alpha_p = \alpha_p'$. ∎

The next theorem is one of the basic theorems of linear algebra and leads to the concept of dimension of subspace.

THEOREM 2

Any set of p vectors in F^m is linearly dependent if $p > m$.

Proof Let

$$v_1 = \begin{bmatrix} a_{11} \\ \vdots \\ a_{m1} \end{bmatrix}, \qquad \ldots, \qquad v_p = \begin{bmatrix} a_{1p} \\ \vdots \\ a_{mp} \end{bmatrix} \in F^m.$$

Then the homogeneous $m \times p$ LS

$$a_{11}x_1 + \cdots + a_{1p}x_p = 0,$$
$$\vdots$$
$$a_{m1}x_1 + \cdots + a_{mp}x_p = 0,$$

has a nontrivial solution, say $\begin{bmatrix} \alpha_1 \\ \vdots \\ \alpha_p \end{bmatrix}$, since $p > m$ (*Theorem 1* of *Section 1.6*).

Therefore

$$a_{11}\alpha_1 + \cdots + a_{1p}\alpha_p = 0,$$
$$\vdots$$
$$a_{m1}\alpha_1 + \cdots + a_{mp}\alpha_p = 0,$$

and so

$$\begin{bmatrix} a_{11} \\ \vdots \\ a_{m1} \end{bmatrix} \alpha_1 + \cdots + \begin{bmatrix} a_{1p} \\ \vdots \\ a_{mp} \end{bmatrix} \alpha_p = 0.$$

This yields $v_1\alpha_1 + \cdots + v_p\alpha_p = 0$, where not all α_i are zero. Hence $v_1, \ldots, v_p$ are linearly dependent. ∎

THEOREM 3

Any linearly independent set of n vectors in F^n is a basis of F^n.

Proof Let $\{v_1, \ldots, v_n\}$ be a linearly independent set of n vectors in F^n. Suppose there exists $v \in F^n$ such that v is not a linear combination of $v_1, \ldots, v_n$. Then v must be distinct from $v_1, \ldots, v_n$, and therefore, by the foregoing theorem, the set $\{v_1, \ldots, v_n, v\}$ is linearly dependent. Hence there exist $\alpha_1, \ldots, \alpha_n, \beta \in F$ (not all zero) such that $\alpha_1 v_1 + \cdots + \alpha_n v_n + \beta v = 0$. Now, β cannot be zero, since otherwise the set $\{v_1, \ldots, v_n\}$ would be linearly dependent. So we get

$$v = \left[-\frac{\alpha_1}{\beta} \right] v_1 + \cdots + \left[-\frac{\alpha_n}{\beta} \right] v_n,$$

which contradicts the hypothesis that v is not a linear combination of $v_1, \ldots, v_n$. Hence we conclude that every element in F^n is a linear combination of $v_1, \ldots, v_n$. Therefore the set $\{v_1, \ldots, v_n\}$ is a basis of F^n. ∎

The following is another version of Theorem 3. The vector space F^m and any subspace of dimension m of F^n ($n > m$) are regarded as the "same" abstractly (called *isomorphic spaces*) (*Section 9.4, Property 9.4.14*).

THEOREM 4

If a subspace W of F^n has a basis consisting of m vectors, then any set of p vectors, where $p > m$, is linearly dependent.

Proof Let $\{v_1, \ldots, v_m\}$ be a basis of W, and let $\{w_1, \ldots, w_p\}$ be a subset of W, where $p > m$. Then we can express each w_i as a unique linear combination of the basis vectors:

$$\begin{aligned}
w_1 &= \alpha_{11}v_1 + \alpha_{21}v_2 + \cdots + \alpha_{m1}v_m, \\
w_2 &= \alpha_{12}v_1 + \alpha_{22}v_2 + \cdots + \alpha_{m2}v_m, \\
&\vdots \\
w_p &= \alpha_{1p}v_1 + \alpha_{2p}v_2 + \cdots + \alpha_{mp}v_m \,.
\end{aligned} \tag{1}$$

We now show that there exist scalars $x_1, \ldots, x_p$, not all 0, such that

$$x_1 w_1 + x_2 w_2 + \cdots + x_p w_p = 0. \tag{2}$$

By replacing $w_1, w_2, \ldots, w_p$ in Equation (2) by their corresponding linear combinations in Equation (1), we get

$$\begin{aligned}
x_1(\alpha_{11}v_1 + \alpha_{21}v_2 + \cdots + \alpha_{m1}v_m) \\
+ x_2(\alpha_{12}v_1 + \alpha_{22}v_2 + \cdots + \alpha_{m2}v_m) \\
\vdots \\
+ x_p(\alpha_{1p}v_1 + \alpha_{2p}v_2 + \cdots + \alpha_{mp}v_m) = 0.
\end{aligned}$$

This equation can now be rewritten as

$$\begin{aligned}
(\alpha_{11}x_1 + \alpha_{12}x_2 + \cdots + \alpha_{1p}x_p)v_1 \\
+ (\alpha_{21}x_1 + \alpha_{22}x_2 + \cdots + \alpha_{2p}x_p)v_1 \\
\vdots \\
+ (\alpha_{m1}x_1 + \alpha_{m2}x_2 + \cdots + \alpha_{mp}x_p)v_m = 0.
\end{aligned}$$

Since $v_1, v_2, \ldots, v_m$ are linearly independent vectors, this equation gives the following linear system of homogeneous equations:

$$\begin{aligned}
a_{11}x_1 + \alpha_{12}x_2 + \cdots + \alpha_{1p}x_p &= 0, \\
a_{21}x_1 + \alpha_{22}x_2 + \cdots + \alpha_{2p}x_p &= 0, \\
&\vdots \\
a_{m1}x_1 + \alpha_{m2}x_2 + \cdots + \alpha_{mp}x_p &= 0.
\end{aligned}$$

This is an $m \times p$ LS, $p > m$. So by *Theorem 1* in *Section 1.6*, it has a nontrivial solution.

This proves our assertion that there exist scalars $x_1, \ldots, x_p$ (not all zero) such that Equation (2) holds. ∎

The next theorem is fundamental.

THEOREM 5

If a subspace W of F^n has a basis consisting of p vectors, then every basis of W consists of p vectors.

Proof Suppose B is a basis of W consisting of p vectors. Let B' be another basis of W. If the number of vectors in B' is m, then $m \leqslant p$. Otherwise, by the above theorem, B' would be a linearly dependent set, which is not the case. By a similar argument, $p \leqslant m$. Hence $p = m$. ∎

THEOREM 6

Let A be an $n \times n$ matrix. If the columns of A are linearly independent, then the LS

$$Ax = b$$

has a unique solution.

Proof Since F^n has the standard basis of n vectors, any set of n linearly independent vectors in F^n is a basis of F^n. Thus the n columns of F^n also form a basis of F^n. By Theorem 1, the vector $b \in F^n$ is a unique linear combination of the columns $A^{(1)}, \ldots, A^{(n)}$ of A. Thus there exist unique numbers $\alpha_1, \ldots, \alpha_n$ such that

$$\alpha_1 A^{(1)} + \alpha_2 A^{(2)} + \cdots + \alpha_n A^{(n)} = b.$$

Now, if $\begin{bmatrix} x_1 \\ x_2 \\ \vdots \\ x_n \end{bmatrix}$ is a solution of $Ax = b$, then

$$x_1 A^{(1)} + x_2 A^{(2)} + \cdots + x_n A^{(n)} = b.$$

Therefore, $x_i = \alpha_i$ for all i. Hence $Ax = b$ has a unique solution. ∎

3.5 Chapter Review Questions and Project

1. Let $u, v, w, \ldots$ be a set of vectors. What is meant by a linear combination of $u, v, w, \ldots$? (Section 3.1)

2. Interpret geometrically the set of all linear combinations of a single vector $[1 \quad 0]$ in xy plane. (Section 3.1)

3. Show that the set of all linear combinations of any two nonparallel vectors in $\mathbf{R}^2$ generates the whole plane $\mathbf{R}^2$. (Section 3.2)

4. **a.** Define carefully the concepts of linearly independent and linearly dependent vectors. (Section 3.3)

   b. Show that $u = [1 \quad 1 \quad 1]$, $v = [1 \quad 1 \quad 0]$, $w = [1 \quad 0 \quad 0]$ are linearly independent. Also show that the subsets $\{u\}$ and $\{u, v\}$ are linearly independent. (Section 3.3)

5. Show that $u = [1 \quad 2 \quad 0]$, $v = [-1 \quad 1 \quad 1]$, $w = [0 \quad 3 \quad 1]$ are linearly dependent vectors. (Section 3.3)

   Suppose $x = [a \quad b \quad c]$ is any other vector. Show that the superset $\{u, v, w, x\}$ is also a linearly dependent set.

   Can you generalize the above result to any set of linearly dependent vectors? (Section 3.3)

6. Show that the vectors $u = [1 \quad 2 \quad 0]$, $v = [2 \quad 4 \quad 0]$, and $w = [0 \quad 0 \quad 1]$ are linearly dependent. (Section 3.3)

   Show also that u can be expressed as a linear combination of v and w, and show that v can be expressed as a linear combination of u and w but w cannot be expressed as a linear combination of u and v. (Section 3.3)

7. What do we mean by a subspace? (Section 3.2)

   a. Is $\{0, 1, 2, 3, \ldots\}$ a subspace?

   b. Is $\{\ldots, -3, -2, -1, 1, 2, 3, \ldots\}$ a subspace?

8. **a.** Let

$$2x_1 + 3x_2 - x_3 = 0,$$
$$x_2 + x_3 = 0,$$
$$5x_1 + x_2 + 6x_3 = 0$$

be a homogeneous system of linear equations. Show that the solution set is a subspace. (Section 3.2)

9. **a.** Show that the subspace $\mathbf{R}^2$ can be spanned by vectors $[1 \quad 0]$ and $[0 \quad 1]$. (Section 3.3)

   b. Show that the subspace $\mathbf{R}^3$ can be spanned by vectors $[1 \quad 0 \quad 0]$, $[0 \quad 1 \quad 0]$, and $[0 \quad 0 \quad 1]$.

10. **a.** Find the subspace spanned by

    (i) the rows of $A = \begin{bmatrix} 1 & 2 \\ 0 & 1 \end{bmatrix}$,

    (ii) the columns of $A = \begin{bmatrix} 1 & 2 \\ 0 & 1 \end{bmatrix}$. (Section 3.3)

    b. Show every vector in $\mathbf{R}^2$ is in the subspace given by (i) as well as in the subspace given by (ii). (Section 3.3)

11. What do we mean by a basis of a subspace? Can a zero vector be a member of a basis of a subspace? Explain carefully. Find a basis and the dimension of the

subspaces spanned by

a. $[2 \quad 1 \quad 0], [-1 \quad 3 \quad 1], [0 \quad 7 \quad 3]$.
b. $[0 \quad 0 \quad 1 \quad 1], [-1 \quad -1 \quad 2 \quad 3], [1 \quad 0 \quad -2 \quad 1], [1 \quad -1 \quad -1 \quad 6]$.
(Section 3.3)

12. Do you know that a subspace can have infinitely many bases? Give an example of at least two bases of (i) $\mathbf{R}^2$, (ii) $\mathbf{R}^3$. What is a common feature of two different bases? (Section 3.3)

Chapter Project

Spice Island, Inc., a leading spice maker, makes several kinds of spice mixes using nine kinds of ingredients. The following table gives in ounces the amount of ingredients used for Mix A, Mix B, Mix C, Mix D, Mix E, and Mix F to make one packet of each mix:

	Mix A	Mix B	Mix C	Mix D	Mix E	Mix F
Paprika	3	1.5	4.5	7.5	9	4.5
Turmeric	2	4	0	8	1	6
Dried chili peppers	1	2	0	4	2	3
Cumin	1	2	0	4	1	3
Oregano	1	2	0	4	1	3
Cayenne	0.5	1	0	2	2	1.5
Garlic powder	0.5	1	0	2	2	1.5
Salt	0.5	1	0	2	2	1.5
Ground cloves	0.25	0.5	0	2	1	0.75

a. One customer thinks that instead of buying all six mixes, he can purchase a few of them and get all other mixes made from these. What is the least number of mixes he can buy to create every other mix from these? Write down your set of fewest mixes.
b. Is the minimal set of mixes you obtained in part (a) unique? Can you find another such minimal set of mixes having the same number of mixes?
c. Using the minimum number of mixes you discovered in part (a), make a new mix that has the following amounts of ingredients:

Paprika	18
Turmeric	18
Dried chili peppers	9
Cumin	9
Oregano	9
Cayenne	4.5
Garlic powder	4.5
Salt	4.5
Ground cloves	3.25

Write down the number of packets you need of each kind.
d. The prices per packet in US dollars for the mixes are as below:

Mix	A	B	C	D	E	F
Price	2.30	1.15	1.00	3.20	2.50	3.00

Compute the price of making the new mix defined in part (c), using the minimal sets obtained in parts (a) and (b).

e. Another customer wants to create a mix that has the following composition:

Paprika	12
Turmeric	14
Dried Chili Peppers	7
Cumin	7
Oregano	7
Cayenne	35
Garlic Powder	35
Salt	35
Ground Cloves	175

What is the minimum number of mixes he can buy to create this mix?

f. Several concepts have been invoked in this project. List all those concepts.

Key Words

Row Vector	Row Space
Column Vector	Range of a Matrix
Linear Combination of Vectors	Column Space
Subspace	Linearly Independent
Subspace Spanned or Generated by a Set	Linearly Dependent
	Basis of a Subspace
Generators of a Subspace	Dimension of a Subspace

Key Phrases

■ Any $n + 1$ or more vectors in an n-dimensional subspace are linearly dependent.

4 | Rank

Introduction

The solutions to a homogeneous linear system of equations $Ax = 0$ form a subspace called the null space of A. Many properties of the solution(s) of $Ax = b$ are inherently embedded in the solution space of the $Ax = 0$. We study conditions regarding consistency of the system $Ax = b$ in terms of the concept of rank defined in the chapter. A highly useful fact giving the relationship between the rank of a matrix A and the dimension of the null space is also given.

4.1 Elementary Operations and Rank

Suppose $v_1, v_2, \ldots, v_p$ is a list of vectors in F^m. Then the following operations are called *elementary operations* on this set. These are of three types:

> **1.** Interchange two vectors, say v_i and v_j, in the list.
> **2.** Multiply a vector v_i in the list by a nonzero scalar α.
> **3.** Multiply v_i by a scalar β and add it to v_j.

Let $v_1, v_2, \ldots, v_p$ be a list of vectors in F^m. Let $W = \langle v_1, v_2, \ldots, v_p \rangle$ be the subspace spanned by these vectors. Then $\{v_1, \ldots, v_p\}$ is called a *spanning set* of the subspace W.

If $\langle v_1, \ldots, v_p \rangle = \langle w_1, \ldots, w_k \rangle$, then the sets $\{v_1, \ldots, v_p\}$ and $\{w_1, \ldots, w_k\}$ will be called *equivalent spanning sets*.

The following result regarding the effect of performing elementary operations on a spanning set has a number of applications throughout this chapter.

4.1.1 THEOREM	**(Theorem 1, Section 4.4)** Elementary operations on a spanning set of a subspace transform it into an equivalent spanning set.

We are now in a position to discuss row space, column space, row rank, column rank, and related concepts.

If $A = (a_{ij})$ is an $m \times n$ matrix, then each row of A can be considered as a $1 \times n$ matrix, and so it is an n-dimensional row vector. Similarly, any column of A is an m-dimensional column vector. We will write the ith row of A as A_i, $i = 1, \ldots, m$, and the jth column of A as $A^{(j)}$, $j = 1, \ldots, n$. We can write the matrix A as

$$A = \begin{bmatrix} A_1 \\ A_2 \\ \vdots \\ A_m \end{bmatrix}$$

in terms of the rows or as

$$A = [A^{(1)} \ A^{(2)} \ \cdots \ A^{(n)}]$$

in terms of the columns. For example, if

$$A = \begin{bmatrix} -1 & 2 & 3 \\ 5 & 2 & 9 \end{bmatrix},$$

then we can write

$$A = \begin{bmatrix} A_1 \\ A_2 \end{bmatrix},$$

where $A_1 = [-1 \ 2 \ 3]$ and $A_2 = [5 \ 2 \ 9]$.

We recall the definitions of row space and column space (Section 3.3).

4.1.2
DEFINITION

Row Space and Column Space

Let $A = (a_{ij})_{m \times n}$. Then the *row space* of A is the subspace of F^n (the set of all n-dimensional row vectors) spanned by the rows of A. The *column space* of A is the subspace of F^m (the set of m-dimensional column vectors) spanned by the columns of A. The column space of A is also called the *range* of A.

4.1.3
DEFINITION

Row Rank and Column Rank

The dimension of the row space of A is called the *row rank* of A, and the dimension of the column space of A is called the *column rank* of A.

According to our convention mentioned in *Chapter 3* that the dimension of the zero space is defined to be 0, the row (column) rank of a zero matrix is 0.

In the following examples, we illustrate a method of finding a basis of the row (column) space and the row (column) rank of a matrix A. This method will also provide an alternative method to check whether a set of row (or column) vectors is linearly independent or dependent. We do this by the Gauss elimination process. We reduce the matrix A to a row echelon matrix by elementary row operations of Types I, II, and III. We note that the subspace spanned by the rows of A is the same as the subspace spanned by the rows of the echelon matrix or, equivalently, by the nonzero rows of the echelon matrix (see *Theorem 4.1.1*).

Recall that the nonzero rows of a row echelon matrix are always linearly independent.

Similar remarks hold for column operations and the column space of A.

It is to be noted that a column operation on A is the same as a row operation on A^T, the transpose of A. This allows us to focus only on the row operations on matrices.

4.1.4 EXAMPLES

a. Let

$$A = \begin{bmatrix} 1 & 2 & 3 & -1 \\ 3 & 5 & 8 & -2 \\ 1 & 1 & 2 & 0 \end{bmatrix}.$$

The rows of A are four-dimensional row vectors, that is, vectors in F^4. The row space of A is the subspace $\langle A_1, A_2, A_3 \rangle$ spanned by the three rows A_1, A_2, A_3 of A. We now perform elementary row operations on A:

$$A = \begin{bmatrix} 1 & 2 & 3 & -1 \\ 3 & 5 & 8 & -2 \\ 1 & 1 & 2 & 0 \end{bmatrix} \xrightarrow[R_3+(-1)R_1]{R_2+(-3)R_1} \begin{bmatrix} 1 & 2 & 3 & -1 \\ 0 & -1 & -1 & 1 \\ 0 & -1 & -1 & 1 \end{bmatrix}$$

$$\xrightarrow[R_3+(-1)R_2]{} \begin{bmatrix} 1 & 2 & 3 & -1 \\ 0 & -1 & -1 & 1 \\ 0 & 0 & 0 & 0 \end{bmatrix}.$$

The row space of A is $\langle [1\ 2\ 3\ -1], [0\ -1\ -1\ 1] \rangle$. The spanning set is linearly independent. So the dimension of the row space of A is 2. Thus the row rank of A is 2.

Next we find a basis of the column space of A. We write

$$A^T = \begin{bmatrix} 1 & 3 & 1 \\ 2 & 5 & 1 \\ 3 & 8 & 2 \\ -1 & -2 & 0 \end{bmatrix}$$

and perform elementary row operations on A^T; these operations are the same as elementary column operations on A, since the columns of A are the rows of A^T:

$$A^T = \begin{bmatrix} 1 & 3 & 1 \\ 2 & 5 & 1 \\ 3 & 8 & 2 \\ -1 & -2 & 0 \end{bmatrix} \xrightarrow[\substack{R_3+(-3)R_1 \\ R_4+R_1}]{R_2+(-2)R_1} \begin{bmatrix} 1 & 3 & 1 \\ 0 & -1 & -1 \\ 0 & 1 & 1 \\ 0 & 1 & 1 \end{bmatrix} \xrightarrow[R_4+R_2]{R_3+(-1)R_2} \begin{bmatrix} 1 & 3 & 1 \\ 0 & -1 & -1 \\ 0 & 0 & 0 \\ 0 & 0 & 0 \end{bmatrix}$$

The matrix A^T has been reduced to row echelon form. The transpose of the row echelon matrix is the matrix

$$\begin{bmatrix} 1 & 0 & 0 & 0 \\ 3 & -1 & 0 & 0 \\ 1 & -1 & 0 & 0 \end{bmatrix},$$

which is a column echelon form of A.

So the column space of A is equal to $\left\langle \begin{bmatrix} 1 \\ 3 \\ 1 \end{bmatrix}, \begin{bmatrix} 0 \\ -1 \\ -1 \end{bmatrix} \right\rangle$. The two column

vectors $\begin{bmatrix} 1 \\ 3 \\ 1 \end{bmatrix}$ and $\begin{bmatrix} 0 \\ -1 \\ -1 \end{bmatrix}$ are linearly independent, and so they form a basis of the column space. Also, the dimension of the column space is 2, the number of nonzero columns in the column echelon matrix. Therefore the column rank of A is 2.

b. Find a basis of the subspace spanned by $\begin{bmatrix} 1 \\ 2 \\ 0 \end{bmatrix}, \begin{bmatrix} 1 \\ 1 \\ 1 \end{bmatrix}$, and $\begin{bmatrix} -1 \\ 0 \\ -2 \end{bmatrix}$.

We write the matrix

$$A = \begin{bmatrix} 1 & 1 & -1 \\ 2 & 1 & 0 \\ 0 & 1 & -2 \end{bmatrix}$$

whose columns are the given column vectors. A basis of the subspace spanned by the given vectors is the same as a basis of the column space of A. To do this, we write

$$A^T = \begin{bmatrix} 1 & 2 & 0 \\ 1 & 1 & 1 \\ -1 & 0 & -2 \end{bmatrix}$$

and reduce it to row echelon form:

$$\begin{bmatrix} 1 & 2 & 0 \\ 1 & 1 & 1 \\ -1 & 0 & -2 \end{bmatrix} \xrightarrow[R_3 + R_1]{R_2 + (-1)R_1} \begin{bmatrix} 1 & 2 & 0 \\ 0 & -1 & 1 \\ 0 & 2 & -2 \end{bmatrix} \xrightarrow{R_3 + 2R_2} \begin{bmatrix} 1 & 2 & 0 \\ 0 & -1 & 1 \\ 0 & 0 & 0 \end{bmatrix}.$$

The transpose of this row echelon matrix is $\begin{bmatrix} 1 & 0 & 0 \\ 2 & -1 & 0 \\ 0 & 1 & 0 \end{bmatrix}$. So a basis of the

column space of A is $\left\{ \begin{bmatrix} 1 \\ 2 \\ 0 \end{bmatrix}, \begin{bmatrix} 0 \\ -1 \\ 1 \end{bmatrix} \right\}.$ ∎

4.1.5
Remarks

i. The method in Example (b) gives another method to find a basis of a subspace spanned by a set of row or column vectors. Form a matrix with the given row vectors as its rows (or given column vectors as its columns). The set of nonzero vectors in the row echelon form of A gives a basis of the subspace spanned by the given set of row vectors. The transpose of each of the nonzero vectors in the row echelon form of A^T gives a basis of the subspace spanned by the given set of column vectors.

ii. It is known that for any matrix A, row rank $A =$ column rank A.

The reader who is interested in the proof may see *Theorem 3* in *Section 4.4*.

4.1.6 **DEFINITION**	**Rank** The *rank* of a matrix A is defined to be the row rank or the column rank of A.

The notation rank A or rank(A) is used interchangeably to denote the rank of A.

4.1.7 An Application to Linear Systems

A linear system of equations $Ax = b$ is consistent if and only if the rank of the coefficient matrix A is equal to the rank of the augmented matrix $[A \ b]$.

Let us illustrate this by considering the linear system

$$a_1 x_1 + a_2 x_2 + a_3 x_3 = b_1, \qquad a_1' x_1 + a_2' x_2 + a_3' x_3 = b_2.$$

Suppose this system is consistent and $x_1 = \alpha_1$, $x_2 = \alpha_2$, $x_3 = \alpha_3$ is a solution. Then, by substituting the values of the variables in the linear system, we get

$$a_1 \alpha_1 + a_2 \alpha_2 + a_3 \alpha_3 = b_1, \qquad a_1' \alpha_1 + a_2' \alpha_2 + a_3' \alpha_3 = b_2.$$

We rewrite the above equations into one matrix equation as

$$\begin{bmatrix} a_1 \alpha_1 + a_2 \alpha_2 + a_3 \alpha_3 \\ a_1' \alpha_1 + a_2' \alpha_2 + a_3' \alpha_3 \end{bmatrix} = \begin{bmatrix} b_1 \\ b_2 \end{bmatrix}$$

and further rewrite this as

$$\begin{bmatrix} a_1 \\ \alpha_1' \end{bmatrix} \alpha_1 + \begin{bmatrix} a_2 \\ a_2' \end{bmatrix} \alpha_2 + \begin{bmatrix} a_3 \\ a_3' \end{bmatrix} \alpha_3 = \begin{bmatrix} b_1 \\ b_2 \end{bmatrix}.$$

This shows that the vector

$$b = \begin{bmatrix} b_1 \\ b_2 \end{bmatrix}$$

is a linear combination of the column vectors of the coefficient matrix A. Thus rank $A = $ rank$[A \ b]$. This is equivalent to saying that the subspace spanned by the columns of A is equal to the subspace spanned by the columns of A and the column vector b. The converse follows by retracing steps backwards.

As an example, consider the system of equations $2x + 3y + 4z = 1$ and $4x + 6y + 8z = 35$.

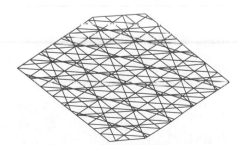

These equations represent two parallel planes, and so the system is inconsistent. By actual computation, we find that the rank of the coefficient matrix $\begin{bmatrix} 2 & 3 & 4 \\ 4 & 6 & 8 \end{bmatrix}$ is 1 but the rank of the augmented matrix $\begin{bmatrix} 2 & 3 & 4 & 1 \\ 4 & 6 & 8 & 35 \end{bmatrix}$ is 2.

4.1 Exercises

1. Find the rank and bases of both the row and the column spaces for each of the following matrices:

   a. $A = \begin{bmatrix} 1 & 2 \\ 3 & 4 \end{bmatrix}$.

   b. $A = \begin{bmatrix} 1 & 0 & 1 \\ 3 & 2 & 1 \end{bmatrix}$.

   c. $A = \begin{bmatrix} 2 & 3 \\ 4 & 5 \\ 6 & 8 \end{bmatrix}$.

   d. $A = \begin{bmatrix} 1 & 2 & 0 \\ -1 & 3 & 4 \\ 0 & 4 & 3 \end{bmatrix}$.

   e. $A = \begin{bmatrix} 5 & 6 & 7 \\ 0 & 5 & 6 \\ 0 & 0 & 5 \end{bmatrix}$.

2. (Drill 4.1) Find the rank and bases of both the row and column spaces for each of the following matrices:

   a. $A = \begin{bmatrix} 2 & 3 & 4 & 5 \\ 0 & 1 & 5 & 6 \\ 0 & 0 & 7 & 8 \\ 0 & 0 & 5 & 3 \end{bmatrix}$.

   b. $B = \begin{bmatrix} 1 & 2 & 3 & 4 \\ 2 & 4 & 6 & 8 \\ 3 & 5 & 7 & 9 \\ 4 & 6 & 8 & 10 \end{bmatrix}$.

   c. $C = [A \ B]$, where A and B are as above.

3. Find a basis for each of the subspaces spanned by the given vectors:

   a. $[1 \ 1 \ 1], [2 \ 2 \ 2], [0 \ 0 \ 0]$.

   b. $[3 \ 2 \ 1], [4 \ 3 \ 2], [1 \ 1 \ 1]$.

   c. $\begin{bmatrix} 2 \\ 7 \\ 0 \end{bmatrix}, \begin{bmatrix} 3 \\ 5 \\ 1 \end{bmatrix}, \begin{bmatrix} 1 \\ 0 \\ 0 \end{bmatrix}$.

4. Find a basis of the subspace consisting of the solutions of the following homogeneous LS:

a. $x_1 + x_2 - x_3 = 0.$

b. $2x + y + z = 0,$
$$x - z = 0.$$

c. $x_1 + 2x_2 + 3x_3 = 0,$
$$4x_1 + 5x_2 + 6x_3 = 0,$$
$$x_1 - x_2 + x_3 = 0.$$

 d. (Drill 4.2) $x_1 + 2x_2 + 3x_4 = 0,$
$$2x_1 + 5x_2 + 2x_3 + x_4 = 0,$$
$$x_1 + 3x_2 + x_3 - x_4 = 0.$$

 5. (Drill 4.3) Find a basis of the subspace consisting of the solutions of the following homogeneous LS, and graph the solution set:

$$x_1 + 2x_2 = 0,$$
$$2x_1 - x_2 + 3x_3 = 0.$$

6. Find a basis of the subspace of the solutions of $Ax = 0$, where

a. $A = \begin{bmatrix} 1 & 5 \\ 2 & 6 \end{bmatrix}.$ **b.** $A = \begin{bmatrix} 2 & 3 & 5 \\ 1 & 0 & 1 \end{bmatrix}.$

c. $A = \begin{bmatrix} 0 & 1 & 2 \\ 3 & 0 & 1 \\ 0 & 0 & 1 \end{bmatrix}.$ **d.** $A = [1 \ 2 \ 3 \ 4].$

4.2 Null Space and Nullity of a Matrix

Let A be an $m \times n$ matrix, and let $W = \{x \in F^n \mid Ax = 0\}$ be the set of solutions of the homogeneous linear system $Ax = 0$. Recall that W is a subspace.

4.2.1
DEFINITION

Null Space of a Matrix and Nullity
Let A be an $m \times n$ matrix. Then $W = \{x \in F^n \mid Ax = 0\}$ is called the *null space or solution space* of A. The dimension of the null space of A is called the *nullity* of A.

The nullity of A indeed gives the number of unknowns in the linear system $Ax = b$ whose values in the solution set can be chosen arbitrarily as demonstrated in the following Examples (a)–(e). The following is an important property of the rank and the nullity of a matrix, known as the *rank-nullity theorem*.

4.2.2
THEOREM

Rank-Nullity Theorem (Theorem 5, Section 4.4)
Let A be an $m \times n$ matrix. Then, rank A + nullity of $A = n$.

The proof of the above result is included in *Section 4.4 (Theorem 5)*.

4.2.3 EXAMPLES

a. Consider the linear system of equations given by a single equation $x_1 + x_2 + x_3 = 1$. The coefficient matrix is $A = \begin{bmatrix} 1 & 1 & 1 \end{bmatrix}$ and rank $A = 1$. Since there are three unknowns and one equation, we can only solve it for one unknown, and the remaining two unknowns can be chosen arbitrarily. Also, by the rank-nullity theorem, the nullity is $3 - 1 = 2$. This verifies the fact that the nullity gives the number of unknowns that can be chosen arbitrarily in the solutions of $Ax = b$.

b. Find a basis of the null space of

$$A = \begin{bmatrix} 2 & 3 & 1 \\ 5 & 6 & 7 \end{bmatrix}.$$

We first reduce A to row echelon form:

$$\begin{bmatrix} 2 & 3 & 1 \\ 5 & 6 & 7 \end{bmatrix} \xrightarrow{\frac{1}{2}R_1} \begin{bmatrix} 1 & \frac{3}{2} & \frac{1}{2} \\ 5 & 6 & 7 \end{bmatrix} \xrightarrow{R_2 + (-5)R_1} \begin{bmatrix} 1 & \frac{3}{2} & \frac{1}{2} \\ 0 & -\frac{3}{2} & \frac{9}{2} \end{bmatrix}$$

$$\xrightarrow{-\frac{2}{3}R_2} \begin{bmatrix} 1 & \frac{3}{2} & \frac{1}{2} \\ 0 & 1 & -3 \end{bmatrix}.$$

The LS $Ax = 0$ is then transformed into

$$x_1 + \frac{3}{2}x_2 + \frac{1}{2}x_3 = 0,$$
$$x_2 - 3x_3 = 0.$$

We now solve backwards. Since the last equation has two unknowns, we put one of the variables, say x_3, as $x_3 = t$, where t is arbitrary. Then $x_2 = 3t$, and $x_1 = -\frac{9}{2}t - \frac{1}{2}t = -5t$. Thus the solutions

$$x = \begin{bmatrix} x_1 \\ x_2 \\ x_3 \end{bmatrix}$$

of $Ax = 0$ are given by

$$x = \begin{bmatrix} -5t \\ 3t \\ t \end{bmatrix} = t \begin{bmatrix} -5 \\ 3 \\ 1 \end{bmatrix}.$$

This shows that the set of solutions of $Ax = 0$ is spanned by a single vector $\begin{bmatrix} -5 \\ 3 \\ 1 \end{bmatrix}$.

This yields that the nullity of A is 1.

c. Find the dimension of the solution space of the LS

$$x_1 + x_2 - x_3 + x_4 = 0,$$
$$x_2 - 2x_3 - x_4 = 0.$$

Here

$$A = \begin{bmatrix} 1 & 1 & -1 & 1 \\ 0 & 1 & -2 & -1 \end{bmatrix}$$

is already in row echelon form. So we have to solve

$$x_1 + x_2 - x_3 + x_4 = 0,$$
$$x_2 - 2x_3 - x_4 = 0.$$

Since the last equation has three unknowns, we assign arbitrary values to any two of them. So let $x_4 = t_1$, $x_3 = t_2$. This gives $x_2 = 2t_2 + t_1$. Then

$$x_1 = -x_2 + x_3 - x_4,$$
$$x_1 = -2t_2 - t_1 + t_2 - t_1 = -t_2 - 2t_1.$$

Therefore the set of solutions is

$$x = \begin{bmatrix} x_1 \\ x_2 \\ x_3 \\ x_4 \end{bmatrix} = \begin{bmatrix} -t_2 - 2t_1 \\ 2t_2 + t_1 \\ t_2 \\ t_1 \end{bmatrix}$$

$$= \begin{bmatrix} -t_2 \\ 2t_2 \\ t_2 \\ 0 \end{bmatrix} + \begin{bmatrix} -2t_1 \\ t_1 \\ 0 \\ t_1 \end{bmatrix}$$

$$= t_2 \begin{bmatrix} -1 \\ 2 \\ 1 \\ 0 \end{bmatrix} + t_1 \begin{bmatrix} -2 \\ 1 \\ 0 \\ 1 \end{bmatrix}.$$

Therefore the solution space is spanned by $\begin{bmatrix} -1 \\ 2 \\ 1 \\ 0 \end{bmatrix}$ and $\begin{bmatrix} -2 \\ 1 \\ 0 \\ 1 \end{bmatrix}$.

Is this a basis of the solution space too? It is so if the vectors are linearly independent.

Indeed, one can show that $\begin{bmatrix} -1 \\ 2 \\ 1 \\ 0 \end{bmatrix}$ and $\begin{bmatrix} -2 \\ 1 \\ 0 \\ 1 \end{bmatrix}$ are linearly independent. Thus

the dimension of the solution space is 2.

d. Find a basis of the subspace formed by the set of solutions of the homogeneous LS

$$x_1 + 2x_2 + x_3 + 3x_4 = 0,$$
$$2x_1 + 5x_2 + 2x_3 + x_4 = 0,$$
$$x_1 + 3x_2 + x_3 - x_4 = 0.$$

We reduce the matrix of the LS to echelon form by elementary row operations:

$$\begin{bmatrix} 1 & 2 & 1 & 3 \\ 2 & 5 & 2 & 1 \\ 1 & 3 & 1 & -1 \end{bmatrix} \xrightarrow[R_3 - R_1]{R_2 - 2R_1} \begin{bmatrix} 1 & 2 & 1 & 3 \\ 0 & 1 & 0 & -5 \\ 0 & 1 & 0 & -4 \end{bmatrix} \xrightarrow{R_3 - R_2} \begin{bmatrix} 1 & 2 & 1 & 3 \\ 0 & 1 & 0 & -5 \\ 0 & 0 & 0 & 1 \end{bmatrix}.$$

So the given LS is equivalent to

$$x_1 + 2x_2 + x_3 + 3x_4 = 0,$$
$$x_2 - 5x_4 = 0,$$
$$x_4 = 0.$$

Solving backwards, we get

$$x_4 = 0, \qquad x_2 = 0, \qquad x_3 = t, \qquad x_1 = -t,$$

that is, the solution space is the set of vectors

$$\begin{bmatrix} -t \\ 0 \\ t \\ 0 \end{bmatrix} = t \begin{bmatrix} -1 \\ 0 \\ 1 \\ 0 \end{bmatrix},$$

where t is a parameter, that is, an arbitrary number. Thus $\begin{bmatrix} -1 \\ 0 \\ 1 \\ 0 \end{bmatrix}$ is a basis of the

subspace of solutions, and so its dimension is 1.

e. Find a basis of the subspace formed by the solution of the homogeneous LS whose coefficient matrix is

$$\begin{bmatrix} 1 & 2 & 1 & 0 \\ 2 & 5 & 3 & -1 \\ 2 & 2 & 0 & 2 \\ 0 & 1 & 1 & -1 \end{bmatrix}.$$

We reduce the matrix to echelon form:

$$\begin{bmatrix} 1 & 2 & 1 & 0 \\ 0 & 1 & 1 & -1 \\ 0 & -2 & -2 & 2 \\ 0 & 1 & 1 & -1 \end{bmatrix} \xrightarrow[R_3-2R_1]{R_2-2R_1} \begin{bmatrix} 1 & 2 & 1 & 0 \\ 0 & 1 & 1 & -1 \\ 0 & 0 & 0 & 0 \\ 0 & 1 & 1 & -1 \end{bmatrix} \xrightarrow[R_4-R_2]{R_3+2R_2} \begin{bmatrix} 1 & 2 & 1 & 0 \\ 0 & 1 & 1 & -1 \\ 0 & 0 & 0 & 0 \\ 0 & 0 & 0 & 0 \end{bmatrix}.$$

So the homogeneous LS is equivalent to the LS

$$x_1 + 2x_2 + x_3 = 0,$$
$$x_2 + x_3 - x_4 = 0.$$

Solving backwards, we get the general solution

$$x_4 = t, \qquad x_3 = u, \qquad x_2 = t - u, \qquad x_1 = -2(t - u) - u = -2t + u,$$

that is, the general solution is

$$\begin{bmatrix} -2t + u \\ t - u \\ u \\ t \end{bmatrix},$$

where t and u are arbitrary numbers. So the subspace formed by the solutions has

$$\text{basis} \left\{ \begin{bmatrix} -2 \\ 1 \\ 0 \\ 1 \end{bmatrix}, \begin{bmatrix} 1 \\ -1 \\ 1 \\ 0 \end{bmatrix} \right\}. \qquad \blacksquare$$

4.2.4 Remarks

1. The arbitrary numbers $t, t_1, t_2,$ and u assigned in Examples (b), (c), and (d) are called parameters.

2. The number of parameters in the solution set of a linear system $Ax = b$ is the dimension of the solution space of the homogeneous linear system $Ax = 0$ (nullity). It may be noted that any solution of $Ax = b$ is of the form $x_0 + z$, where x_0 is a fixed solution (also called a *particular solution*) of $Ax = b$ and z is any solution of $Ax = 0$.

4.2 Exercises

1. Find a basis and the dimension of the null space for each of the following matrices.

   a. $[1 \ 1 \ 1]$ 　　　**b.** $\begin{bmatrix} 1 & 2 \\ 0 & 1 \end{bmatrix}$ 　　　**c.** $\begin{bmatrix} 0 & 1 & 1 \\ 1 & 0 & 0 \end{bmatrix}$

   d. $\begin{bmatrix} 2 & 5 \\ 3 & 6 \\ 4 & 7 \end{bmatrix}$ 　　　**e.** $\begin{bmatrix} 0 & -1 & 1 \\ 2 & 1 & 2 \\ 3 & 0 & -1 \end{bmatrix}$

2. Find the dimension of the solution space for the homogeneous LS

$$x_1 + x_2 - 2x_3 + x_4 = 0,$$
$$2x_1 - x_2 + x_4 = 0.$$

3. Without actually solving them, determine whether there is a nontrivial solution of the following linear systems:

   a. $2x_1 + x_2 - 3x_3 + x_4 = 0,$
   $3x_1 - x_2 + x_3 - x_4 = 0,$
   $5x_1 + x_2 + x_3 + 9x_4 = 0,$
   $10x_1 + x_2 - x_3 + 9x_4 = 0.$

   b. $x + y + z = 0,$
   $2y - z = 0,$
   $2x + 4y + z = 0.$

   c. $x_1 + x_2 + x_3 + x_4 = 0.$

4. Find a basis of the null space of each of the following matrices:

   a. $\begin{bmatrix} 1 & -1 & 2 & 0 \\ 2 & 1 & 3 & 7 \\ 1 & -1 & 3 & 2 \end{bmatrix}$

b. (Drill 4.4)

$$\begin{bmatrix} 1 & 1 & 1 & 1 & 1 \\ 1 & 2 & 1 & 2 & 1 \\ 1 & 3 & 3 & 1 & 1 \\ 1 & 4 & 4 & 1 & 1 \end{bmatrix}$$

c.
$$\begin{bmatrix} 1 & 2 & 3 & 4 & 5 \\ 2 & 3 & 4 & 5 & 1 \\ 1 & 1 & 1 & 1 & 2 \\ 2 & 3 & 4 & 5 & 4 \end{bmatrix}$$

5. For each of the following matrices, find the row rank and the column rank, and show that they are equal. Also find the nullity by using the rank-nullity theorem.

a.
$$\begin{bmatrix} 1 & 2 & 3 & -1 \\ 3 & 5 & 8 & -1 \\ 1 & 1 & 2 & 0 \end{bmatrix}$$

b.
$$\begin{bmatrix} 2 & 4 & 8 & 0 \\ 1 & 3 & -1 & 2 \\ 1 & 2 & 4 & 0 \\ 1 & 2 & 2 & 4 \end{bmatrix}$$

c.
$$\begin{bmatrix} 1 & 2 & 3 & 1 \\ 2 & 5 & 7 & 1 \\ 1 & 0 & 1 & 3 \end{bmatrix}$$

6. Verify the rank-nullity theorem for each of the following matrices:

a.
$$\begin{bmatrix} 1 & 2 & 3 & -1 \\ 3 & 5 & 8 & -1 \\ 1 & 1 & 2 & 0 \end{bmatrix}$$

b.
$$\begin{bmatrix} 1 & 2 & 1 \\ 1 & 2 & 1 \\ 5 & 3 & 2 \end{bmatrix}.$$

7. (Drill 4.5) Verify the rank-nullity theorem for each of the following matrices:

a. $A = \begin{bmatrix} 1 & 2 & 1 & 0 \\ 2 & 5 & 3 & -1 \\ 2 & 2 & 0 & 2 \\ 0 & 1 & 1 & -1 \end{bmatrix}.$

b. B is a 5×7 matrix with random integer entries.

8. By using Matlab to generate nonzero random vectors, say $x = 1 : 5$ (that is, $x = [1\ 2\ 3\ 4\ 5]$) and $y = 0 : 2 : 8$ (that is, $y = [0\ 2\ 4\ 6\ 8]$), show that $x^T y$ is of rank 1.

9. (Drill 4.6) Use random matrices to verify the following statements:

    a. $\text{rank}(AB) \leqslant \min(\text{rank}(A), \text{rank}(B))$.
    b. $\text{rank}(A + B) \leqslant \text{rank}(A) + \text{rank}(B)$.
    c. $\text{rank}(AA^T) = \text{rank}(A) = \text{rank}(A^T A)$.

10. A square matrix is called an *idempotent matrix* if $A = A^2$. Choose idempotent matrices A of various sizes to compute $\text{rank}(A) + \text{rank}(I - A)$. What conclusion does this suggest? Can you prove this conjecture for any idempotent matrix? (Hint: Nullity of $A = \text{rank}(I - A)$.)

4.3 Elementary Matrices

Analogous to the three elementary row operations defined in Section 1.6, we can define elementary column operations as follows:

Type I: $C_i \longleftrightarrow C_j$: Interchange of column i and column j.

Type II: αC_i: Multiply the ith column by a nonzero scalar α.

Type III: $C_j + \alpha C_i$: Add α times the ith column to the jth column $(i \neq j)$.

**4.3.1
DEFINITION**

Elementary Matrix
Let I_n be the $n \times n$ identity matrix. Then any matrix that is obtained by performing an elementary row or column operation on I_n is called an *elementary matrix* of order n.
    The elementary matrices are of three types:

$E_{(i,j)} = $ the matrix obtained by interchanging the ith and jth rows (or equivalently the ith and jth columns) of I.

$E_{\lambda(i)} = $ the matrix obtained by multiplying the ith row (or equivalently the ith column) of I by $\lambda \neq 0$.

$E_{(i)+\lambda(j)} = $ the matrix obtained by adding λ times the jth row of I to the ith row of I $(i \neq j)$ (or equivalently by adding λ times the ith column of I to the jth column of I).

**4.3.2
EXAMPLE**

Let

$$I - \begin{bmatrix} 1 & 0 & 0 \\ 0 & 1 & 0 \\ 0 & 0 & 1 \end{bmatrix}.$$

Then the elementary matrices of order 3 are

$$E_{(1,2)} = \begin{bmatrix} 0 & 1 & 0 \\ 1 & 0 & 0 \\ 0 & 0 & 1 \end{bmatrix}, \quad E_{(1,3)} = \begin{bmatrix} 0 & 0 & 1 \\ 0 & 1 & 0 \\ 1 & 0 & 0 \end{bmatrix}, \quad E_{(2,3)} = \begin{bmatrix} 1 & 0 & 0 \\ 0 & 0 & 1 \\ 0 & 1 & 0 \end{bmatrix},$$

$$E_{\lambda(1)} = \begin{bmatrix} \lambda & 0 & 0 \\ 0 & 1 & 0 \\ 0 & 0 & 1 \end{bmatrix}, \qquad E_{\lambda(2)} = \begin{bmatrix} 1 & 0 & 0 \\ 0 & \lambda & 0 \\ 0 & 0 & 1 \end{bmatrix}, \qquad E_{\lambda(3)} = \begin{bmatrix} 1 & 0 & 0 \\ 0 & 1 & 0 \\ 0 & 0 & \lambda \end{bmatrix},$$

$$E_{(1)+\lambda(2)} = \begin{bmatrix} 1 & \lambda & 0 \\ 0 & 1 & 0 \\ 0 & 0 & 1 \end{bmatrix}, \quad E_{(1)+\lambda(3)} = \begin{bmatrix} 1 & 0 & \lambda \\ 0 & 1 & 0 \\ 0 & 0 & 1 \end{bmatrix}, \quad E_{(2)+\lambda(1)} = \begin{bmatrix} 1 & 0 & 0 \\ \lambda & 1 & 0 \\ 0 & 0 & 1 \end{bmatrix},$$

$$E_{(2)+\lambda(3)} = \begin{bmatrix} 1 & 0 & 0 \\ 0 & 1 & \lambda \\ 0 & 0 & 1 \end{bmatrix}, \quad E_{(3)+\lambda(1)} = \begin{bmatrix} 1 & 0 & 0 \\ 0 & 1 & 0 \\ \lambda & 0 & 1 \end{bmatrix}, \quad E_{(3)+\lambda(2)} = \begin{bmatrix} 1 & 0 & 0 \\ 0 & 1 & 0 \\ 0 & \lambda & 1 \end{bmatrix}.$$

We recall that $E_{(i)+\lambda(j)}$ is the matrix that is obtainable by multiplying column i of I by λ and adding it to column j of I.

Elementary matrices are useful tools for proving a number of properties in the theory of matrices.

By premultiplying a matrix A by a matrix B, we mean multiplying A on the left by B, that is, forming the product BA. Likewise, postmultiplying means multiplying on the right.

■

4.3.3 THEOREM

(Theorems 6 and 7, Section 4.4)

Let B be the matrix obtained from the matrix A by performing an elementary row (column) operation. Let I be the identity matrix of the same order as that of A. Suppose E is the matrix obtained from I by performing the same elementary row (column) operation. Then $B = EA(AE)$, that is, B can be obtained by premultiplying (postmultiplying) A by E, and furthermore, rank $A = \text{rank}(EA) = \text{rank}(AE)$.

The above results about elementary matrices will be used frequently in Chapter 5.

4.3.4 EXAMPLES

a. Let $A = \begin{bmatrix} 2 & 3 & 1 \\ 4 & 1 & 2 \end{bmatrix}$ be a 2×3 matrix.

Interchanging row 1 and row 2 of A gives the matrix $\begin{bmatrix} 4 & 1 & 2 \\ 2 & 3 & 1 \end{bmatrix}$. Interchanging row 1 and row 2 of $\begin{bmatrix} 1 & 0 \\ 0 & 1 \end{bmatrix}$ gives the matrix $E = \begin{bmatrix} 0 & 1 \\ 1 & 0 \end{bmatrix}$.

If we premultiply A by $\begin{bmatrix} 0 & 1 \\ 1 & 0 \end{bmatrix}$, we get

$$\begin{bmatrix} 0 & 1 \\ 1 & 0 \end{bmatrix} \begin{bmatrix} 2 & 3 & 1 \\ 4 & 1 & 2 \end{bmatrix} = \begin{bmatrix} 4 & 1 & 2 \\ 2 & 3 & 1 \end{bmatrix}.$$

Suppose we multiply row 1 of A by -2 and add it to row 2. We get the matrix

$$B = \begin{bmatrix} 2 & 3 & 1 \\ 0 & -5 & 0 \end{bmatrix}.$$

The same row operation on $\begin{bmatrix} 1 & 0 \\ 0 & 1 \end{bmatrix}$ gives the matrix $\begin{bmatrix} 1 & 0 \\ -2 & 1 \end{bmatrix}$, and

$$\begin{bmatrix} 1 & 0 \\ -2 & 1 \end{bmatrix} \begin{bmatrix} 2 & 3 & 1 \\ 4 & 1 & 2 \end{bmatrix} = \begin{bmatrix} 2 & 3 & 1 \\ 0 & -5 & 0 \end{bmatrix} = B.$$

These confirm the above theorem for row operations. Next, suppose we add -2 times column 3 of A to column 1 of A. We get the matrix

$$C = \begin{bmatrix} 0 & 3 & 1 \\ 0 & 1 & 2 \end{bmatrix}.$$

The same column operation on $\begin{bmatrix} 1 & 0 & 0 \\ 0 & 1 & 0 \\ 0 & 0 & 1 \end{bmatrix}$ gives the matrix $\begin{bmatrix} 1 & 0 & 0 \\ 0 & 1 & 0 \\ -2 & 0 & 1 \end{bmatrix}$,

and

$$\begin{bmatrix} 2 & 3 & 1 \\ 4 & 1 & 2 \end{bmatrix} \begin{bmatrix} 1 & 0 & 0 \\ 0 & 1 & 0 \\ -2 & 0 & 1 \end{bmatrix} = \begin{bmatrix} 0 & 3 & 1 \\ 0 & 1 & 2 \end{bmatrix} = C.$$

b. Let

$$A = \begin{bmatrix} 1 & 2 & -1 & 4 \\ 2 & 1 & 3 & 0 \\ 1 & 3 & 3 & -1 \end{bmatrix}.$$

If we perform successive row operations $R_2 - 2R_1$ and $R_3 - R_1$, on A, we get the matrix

$$B = \begin{bmatrix} 1 & 2 & -1 & 4 \\ 0 & -3 & 5 & -8 \\ 0 & 1 & 4 & -5 \end{bmatrix}.$$

If we do the same successive operations on the 3×3 identity matrix $\begin{bmatrix} 1 & 0 & 0 \\ 0 & 1 & 0 \\ 0 & 0 & 1 \end{bmatrix}$,

we get the matrix

$$C = \begin{bmatrix} 1 & 0 & 0 \\ -2 & 1 & 0 \\ -1 & 0 & 1 \end{bmatrix}.$$

Note that if we premultiply A by C, we get B.

c. Let

$$A = \begin{bmatrix} 1 & 2 & 3 & -1 \\ 3 & 5 & 8 & -2 \\ 1 & 1 & 2 & 0 \end{bmatrix}.$$

Find a matrix P such that the product PA is a row echelon matrix. Express P as a product of elementary matrices.

Let us perform elementary row operations on A to reduce it to echelon form:

$$\begin{bmatrix} 1 & 2 & 3 & -1 \\ 3 & 5 & 8 & -2 \\ 1 & 1 & 2 & 0 \end{bmatrix} \xrightarrow[R_3-R_1]{R_2-3R_1} \begin{bmatrix} 1 & 2 & 3 & -1 \\ 0 & -1 & -1 & 1 \\ 0 & -1 & -1 & 1 \end{bmatrix} \xrightarrow{R_3-R_2} \begin{bmatrix} 1 & 2 & 3 & -1 \\ 0 & -1 & -1 & 1 \\ 0 & 0 & 0 & 0 \end{bmatrix},$$

which is in row echelon form.

So P is the product of the elementary matrices corresponding to these elementary operations, that is, $P = E_{(3)-(2)} \cdot E_{(3)-(1)} \cdot E_{(2)-3(1)}$.

Let

$$I = \begin{bmatrix} 1 & 0 & 0 \\ 0 & 1 & 0 \\ 0 & 0 & 1 \end{bmatrix}.$$

Then

$$P = \begin{bmatrix} 1 & 0 & 0 \\ 0 & 1 & 0 \\ 0 & -1 & 1 \end{bmatrix} \begin{bmatrix} 1 & 0 & 0 \\ 0 & 1 & 0 \\ -1 & 0 & 1 \end{bmatrix} \begin{bmatrix} 1 & 0 & 0 \\ -3 & 1 & 0 \\ 0 & 0 & 1 \end{bmatrix}.$$

To verify, we compute PA:

$$PA = \begin{bmatrix} 1 & 0 & 0 \\ 0 & 1 & 0 \\ 0 & -1 & 1 \end{bmatrix} \begin{bmatrix} 1 & 0 & 0 \\ 0 & 1 & 0 \\ -1 & 0 & 1 \end{bmatrix} \begin{bmatrix} 1 & 0 & 0 \\ -3 & 1 & 0 \\ 0 & 0 & 1 \end{bmatrix} \begin{bmatrix} 1 & 2 & 3 & -1 \\ 3 & 5 & 8 & -2 \\ 1 & 1 & 2 & 0 \end{bmatrix}$$

$$= \begin{bmatrix} 1 & 0 & 0 \\ 0 & 1 & 0 \\ 0 & -1 & 1 \end{bmatrix} \begin{bmatrix} 1 & 0 & 0 \\ 0 & 1 & 0 \\ -1 & 0 & 1 \end{bmatrix} \begin{bmatrix} 1 & 2 & 3 & -1 \\ 0 & -1 & -1 & 1 \\ 1 & 1 & 2 & 0 \end{bmatrix}$$

$$= \begin{bmatrix} 1 & 0 & 0 \\ 0 & 1 & 0 \\ 0 & -1 & 1 \end{bmatrix} \begin{bmatrix} 1 & 2 & 3 & -1 \\ 0 & -1 & -1 & 1 \\ 0 & -1 & -1 & 1 \end{bmatrix} = \begin{bmatrix} 1 & 2 & 3 & -1 \\ 0 & -1 & -1 & 1 \\ 0 & 0 & 0 & 0 \end{bmatrix}.$$

■

4.3.5
Remark The matrix P in Example (c) can be obtained from the identity matrix I by performing successively the operations $R_2 - 3R_1$, $R_3 - R_1$, and $R_3 - R_2$ in this order, as shown below:

$$\begin{bmatrix} 1 & 0 & 0 \\ 0 & 1 & 0 \\ 0 & 0 & 1 \end{bmatrix} \xrightarrow{R_2-3R_1} \begin{bmatrix} 1 & 0 & 0 \\ -3 & 1 & 0 \\ 0 & 0 & 1 \end{bmatrix} \xrightarrow{R_3-R_1} \begin{bmatrix} 1 & 0 & 0 \\ -3 & 1 & 0 \\ -1 & 0 & 1 \end{bmatrix}$$

$$\xrightarrow{R_3-R_2} \begin{bmatrix} 1 & 0 & 0 \\ -3 & 1 & 0 \\ 2 & -1 & 1 \end{bmatrix}.$$

So

$$P = \begin{bmatrix} 1 & 0 & 0 \\ -3 & 1 & 0 \\ 2 & -1 & 1 \end{bmatrix}.$$

4.3 Exercises

1. Let
$$A = \begin{bmatrix} 5 & 6 & 7 \\ 1 & 1 & 1 \\ 0 & 1 & 1 \end{bmatrix}.$$

Verify Theorem 4.3.3 for each of the following elementary row operations on A:

   a. $2R_1$ **b.** $R_3 + 2R_2$ **c.** $R_2 \longleftrightarrow R_3$

2. Let
$$A = \begin{bmatrix} 2 & 3 & 7 \\ 1 & 0 & 0 \end{bmatrix}.$$

Verify Theorem 4.3.3 for each of the following elementary column operations on A:

   a. $10C_2$ **b.** $C_3 + 2C_1$ **c.** $C_1 \longleftrightarrow C_2$

 3. a. (Drills 4.7) For the matrix

$$\begin{bmatrix} 2 & 3 & 7 & 6 \\ 5 & 4 & 2 & 0 \\ 1 & -1 & 1 & -1 \end{bmatrix},$$

verify Theorem 4.3.3 for each of the following elementary operations:

   i. $\frac{1}{2}R_1$
   ii. $(-3)C_2 + C_3$

b. For the matrix
$$\begin{bmatrix} 1 & 2 \\ 3 & 4 \\ 0 & 1 \end{bmatrix},$$

verify Theorem 4.3.3 for each of the following elementary operations:

   i. $(-1)R_2$
   ii. $R_3 + 2R_1$
   iii. $R_1 + 5R_3$
   iv. $R_2 \longleftrightarrow R_3$
   v. $(-1)C_2$
   vi. $5C_1$
   vii. $C_2 + 3C_1$

4. For each of the following matrices A, find a matrix P, expressed as a product of elementary matrices, such that PA is a row echelon matrix:

   a. (Drill 4.8) $A = \begin{bmatrix} 1 & 1 & 1 & 1 \\ 2 & 3 & 1 & 2 \\ 1 & -1 & 3 & 2 \end{bmatrix}.$

b. (Drill 4.9) $A = \begin{bmatrix} 1 & 1 & 2 & 6 \\ 3 & 4 & -1 & 5 \\ -1 & 1 & 1 & 2 \end{bmatrix}$.

c. $A = \begin{bmatrix} 1 & 3 & -1 & 1 \\ 2 & 5 & 1 & 5 \\ 1 & 1 & 1 & 3 \end{bmatrix}$.

5. For each matrix A in Exercise 4, find a matrix Q, expressed as a product of elementary matrices, such that AQ is a column echelon matrix, that is, $(AQ)^T$ is a row echelon matrix.

4.4 Proofs of Facts

THEOREM 1

Elementary operations on a spanning set of a subspace transform it into an equivalent spanning set.

Proof We shall give the proof for the simple case $W = \langle v_1, v_2 \rangle$.

i. $\langle v_1, v_2 \rangle = \langle v_2, v_1 \rangle$ is obvious.

ii. We show that $\langle v_1, v_2 \rangle = \langle \alpha v_1, v_2 \rangle$, $\alpha \neq 0$. Any vector in $\langle v_1, v_2 \rangle$ is of the form $a_1 v_1 + a_2 v_2$, where a_1, a_2 are scalars. Then $a_1 v_1 + a_2 v_2 = \frac{a_1}{\alpha}(\alpha v_1) + a_2(v_2)$, which is in $\langle \alpha v_1, v_2 \rangle$. Also, any vector in $\langle \alpha v_1, v_2 \rangle$ is of the form $c_1(\alpha v_1) + c_2(v_2)$, where c_1 and c_2 are scalars. Then $c_1(\alpha v_1) + c_2(v_2) = (c_1 \alpha)v_1 + c_2(v_2)$, which is in $\langle v_1, v_2 \rangle$. So $\langle v_1, v_2 \rangle = \langle \alpha v_1, v_2 \rangle$.

iii. Next we show that $\langle v_1, v_2 \rangle = \langle v_1, \alpha v_1 + v_2 \rangle$. Any vector in $\langle v_1, v_2 \rangle$ is of the form $\alpha_1 v_1 + \alpha_2 v_2 = (\alpha_1 - \alpha \alpha_2)v_1 + \alpha_2(\alpha v_1 + v_2)$, which is in $\langle v_1, \alpha v_1 + v_2 \rangle$. Similarly, any vector in $\langle v_1, \alpha v_1 + v_2 \rangle$ is of the form $c_1 v_1 + c_2(\alpha v_1 + v_2)$, where c_1, c_2 are scalars. Then $c_1 v_1 + c_2(\alpha v_1 + v_2) = (c_1 + c_2 \alpha)v_1 + c_2 v_2$, which is an element of $\langle v_1, v_2 \rangle$. Thus

$$\langle v_1, v_2 \rangle = \langle v_1, \alpha v_1 + v_2 \rangle.$$

■

THEOREM 2

If A is an $m \times n$ matrix of row rank r, then the dimension of the solution space of $Ax = 0$ is $n - r$.

Proof Recall that the row rank of A is equal to the number of nonzero rows in row echelon form. If the row rank of A is r, then the echelon form of the LS $Ax = 0$ has precisely r nontrivial equations. (By a trivial equation we mean $0 = 0$.) Therefore, while solving backwards for the variables $x_1, x_2, \ldots, x_n$, we need to assign arbitrary values to $n - r$ variables. This proves the theorem. ■

We shall now prove the important result that the row rank of a matrix is equal to its column rank. Recall that the common value is called the rank of the matrix.

THEOREM 3

For any matrix A,

$$\text{row rank } A = \text{column rank } A.$$

First we prove the following lemma:

LEMMA 4

Let A be an $m \times n$ matrix, and let A' be an $m \times k$ matrix consisting of any k columns of A $(k \leqslant n)$. Then

$$\text{row rank } A' \leqslant \text{row rank } A.$$

Proof We may assume without loss of generality that A' consists of the first k columns of A, that is,

$$A = \begin{bmatrix} a_{11} & a_{12} & \cdots & a_{1n} \\ \vdots & \vdots & & \vdots \\ a_{m1} & a_{m2} & \cdots & a_{mn} \end{bmatrix}, \quad A' = \begin{bmatrix} a_{11} & a_{12} & \cdots & a_{1k} \\ \vdots & \vdots & & \vdots \\ a_{m1} & a_{m2} & \cdots & a_{mk} \end{bmatrix}.$$

Let row rank $A' = s$, so A' has s linearly independent rows. Then the corresponding rows of A are also linearly independent. Let the first s rows of A' be linearly independent, and suppose that $x_1 A_1 + x_2 A_2 + \cdots + x_s A_s = 0$, where $x_1, x_2, \ldots, x_s \in F$. That is,

$$x_1[a_{11} \cdots a_{1n}] + \cdots + x_s[a_{s1} \cdots a_{sn}] = [0 \cdots 0].$$

But this implies that

$$x_1[a_{11} \cdots a_{1k}] + \cdots + x_s[a_{s1} \cdots a_{sk}] = [0 \cdots 0],$$

that is, $x_1 A'_1 + \cdots + x_s A'_s = 0$. Since $A'_1, \ldots, A'_s$ are linearly independent rows of A', it follows that $x_1 = \cdots = x_s = 0$. This proves that $A_1, \ldots, A_s$ are linearly independent. So A has at least s linearly independent rows, and therefore row rank $A \geqslant s$. This proves that row rank $A' \leqslant \text{row rank } A$. ∎

We now give the proof of Theorem 3:

Proof Let A be an $m \times n$ matrix. Let row rank $A = r$ and column rank $A = s$. Then by the definition of column rank, A has s linearly independent columns. Let A' by the $m \times s$ matrix consisting of these s linearly independent columns. Then by the foregoing lemma,

$$\text{row rank } A' \leqslant r. \tag{1}$$

Consider now the LS $A'x = 0$. Since the columns of A' are linearly independent, by Theorem 1, the nullity of $A' = 0$. Therefore by Theorem 2, row rank $A' = s$. Hence by Equation (1), $s \leqslant r$. By considering the matrix A^T, we similarly obtain $r \leqslant s$. Thus $r = s$. ∎

THEOREM 5

Let A be an $m \times n$ matrix. Then rank $A+$ nullity of $A = n$.

Proof Nullity of $A =$ dimension of the null space of $A = n-$ row rank A, by Theorem 2. But since row rank $A =$ column rank $A =$ rank A, we obtain

$$\text{nullity } A + \text{row rank } A = n.$$

■

THEOREM 6

Let $A = (a_{ij})$ be an $m \times n$ matrix. Then any elementary row operation on A is equivalent to premultiplying A by E, where E is the elementary matrix obtained by the same elementary row operation on the $m \times m$ identity matrix.

Proof Consider the elementary row operation $R_p + \lambda R_q$ on A. This gives the matrix

$$B = \begin{bmatrix} a_{11} & \cdots & a_{1n} \\ \vdots & & \vdots \\ a_{p1} + \lambda a_{q1} & \cdots & a_{pn} + \lambda a_{qn} \\ \vdots & & \vdots \\ a_{m1} & \cdots & a_{mn} \end{bmatrix} \rightarrow p\text{th row.}$$

The same row operation on the $m \times m$ identity matrix gives

$$E_{(p)+\lambda(q)} = \begin{matrix} \\ \\ p\text{th} \rightarrow \\ \\ q\text{th} \rightarrow \\ \\ \\ \end{matrix} \begin{bmatrix} 1 & \cdots & 0 & \cdots & \overset{\overset{p\text{th}}{\downarrow}}{0} & \overset{\overset{q\text{th}}{\downarrow}}{0} & \cdots & 0 \\ \vdots & & \vdots & & \vdots & \vdots & & \vdots \\ 0 & \cdots & 1 & \cdots & \lambda & 0 & \cdots & 0 \\ \vdots & & \vdots & & \vdots & \vdots & & \vdots \\ 0 & \cdots & 0 & \cdots & 1 & 0 & \cdots & 0 \\ \vdots & & \vdots & & \vdots & \vdots & & \vdots \\ 0 & \cdots & 0 & \cdots & 0 & 0 & \cdots & 1 \end{bmatrix}.$$

By directly multiplying, it is quite easy to verify that $EA = B$.

■

We can similarly verify the other cases of elementary operations.

THEOREM 7

Any elementary column operation on a matrix A is equivalent to postmultiplying the matrix A by the corresponding elementary matrix.

Proof The proof of this theorem is similar to the proof of Theorem 6.

■

4.5 Chapter Review Questions and Project

1. **a.** What do we mean by the row rank of a matrix? Find the row rank of each of the following matrices: (Section 4.1)

   (i) $[2 \ 3 \ 4 \ 5]$

   (ii) $\begin{bmatrix} 1 & -2 & 1 \\ 2 & 1 & -1 \\ 2 & 0 & 1 \end{bmatrix}$

   (iii) $\begin{bmatrix} 1 & 2 & -4 & 3 \\ 2 & -1 & -3 & 5 \\ -1 & 8 & -6 & -1 \end{bmatrix}$

   b. Calculate the column rank of each of the matrices given above.

   c. What conclusion can you draw about the row rank and the column rank of a matrix?

2. Find the nullity of each of the matrices given in Question 1. (Section 4.2)

3. State the rank-nullity theorem, and verify the theorem for the following matrices: (Section 4.2)

   a. $[1 \ 0 \ -1 \ 1]$

   b. $\begin{bmatrix} 0 & 0 & 1 \\ 0 & 0 & 0 \\ -1 & 0 & 0 \end{bmatrix}$

   c. $\begin{bmatrix} 1 & 2 & 3 & 4 \\ 5 & 6 & 7 & 8 \\ 9 & 10 & 11 & 12 \end{bmatrix}$

4. Let

$$A = \begin{bmatrix} 0 & 1 & -1 & 1 \\ 2 & 3 & 4 & 5 \\ 0 & 1 & 0 & 1 \end{bmatrix}.$$

   a. Let A_1 and E_1 be the matrices obtained from A and I_3, respectively, by performing $R_1 \leftrightarrow R_2$. Show that $E_1 A = A_1$. (Section 4.3)

   b. Let A_2 and E_2 be the matrices obtained from A_1 and I_3, respectively, by performing $\frac{1}{2}R_2$. Show that $E_2 A_1 = A_2$. (Section 4.3)

   c. Show that $E_2 E_1$ is the matrix obtained by performing $R_1 \leftrightarrow R_2$ and $\frac{1}{2}R_2$, in succession, on I_3. (Section 4.3)

   d. State a general fact that gives the relation between the matrix obtained by performing elementary row operations on a given matrix A and the matrix obtained by performing the same elementary row operations on I. (Section 4.3)

5. Change row operations into column operations in Question 4, and show the corresponding results (a)–(d). (Section 4.3)

Chapter Project

Eight students are asked to help in a computer lab in connection with six different projects for ten days. They are to be paid different hourly rates depending on their experience, out of a fixed daily budget assigned for each project. The following matrix gives the number of hours that each student can put in on each of the six projects. The (i, j) entry of the matrix is the number of hours by the jth student on the ith day.

Student

Day 1	1	0.5	2	1	0.5	1.5	3	2.5
Day 2	2	1.5	3	1	2	1.5	1	3
Day 3	1.5	2	3	1	2	1	1	3
Day 4	1.5	0	1.5	2	1	1	2.5	1.5
Day 5	2	1	2	1.5	1	2	1.5	3
Day 6	0	1	1	2	1.5	1	3	1.5
Day 7	1	1.5	2	3	1	1.5	1	1
Day 8	2	1	1.5	1	0	2	3	2
Day 9	1	1	2	2	1	1	2.5	1.5
Day 10	0	1.5	1	2	1	1	2.5	2

Let A be the 10×8 matrix given above.

a. The daily budgets for the six projects are given by the following vectors:

$$
\begin{bmatrix} 109 \\ 132 \\ 128 \\ 98 \\ 124 \\ 96 \\ 103 \\ 110 \\ 106 \\ 97 \end{bmatrix} ,
\begin{bmatrix} 110 \\ 125 \\ 127 \\ 90 \\ 120 \\ 100 \\ 100 \\ 110 \\ 110 \\ 100 \end{bmatrix} ,
\begin{bmatrix} 108 \\ 132 \\ 127 \\ 100 \\ 122 \\ 99 \\ 99 \\ 111 \\ 106 \\ 96 \end{bmatrix} ,
\begin{bmatrix} 115 \\ 110 \\ 120 \\ 100 \\ 110 \\ 120 \\ 110 \\ 100 \\ 90 \\ 100 \end{bmatrix} ,
\begin{bmatrix} 90 \\ 110 \\ 120 \\ 100 \\ 130 \\ 150 \\ 125 \\ 115 \\ 112 \\ 90 \end{bmatrix} ,
\begin{bmatrix} 88 \\ 113 \\ 111 \\ 79 \\ 104 \\ 84 \\ 88 \\ 90 \\ 88 \\ 85 \end{bmatrix} .
$$

Find the ranks of A and augmented matrices $[A \ \ b]$, where b is a daily budget vector.
b. Which budget vectors are linear combinations of the columns of A?
c. Use part (a) to find the maximal number of linearly independent rows and columns of $[A \ \ b]$.
d. Find the hourly wages of the students for each of the daily budget vectors given in part (a), whenever possible. Explain why this is not possible in some cases.
e. Find the nullity of A without actually computing it.

Key Words

Row Space of a Matrix
Column Space of a Matrix
Row Rank of a Matrix
Column Rank of a Matrix
Rank of a Matrix
Null Space of a Matrix

Nullity of a Matrix
Rank-Nullity Theorem
Elementary Matrices
Elementary Row Operations
Elementary Column Operations

Key Phrases

■ For any matrix, the row rank equals the column rank, called the rank.

■ Rank plus nullity equals the number of columns.

■ Performing an elementary row operation on a matrix A results in a matrix that is equal to EA, where E is the matrix obtained from the identity matrix of suitable size by performing the same elementary row operation.

Similar statement holds when an elementary column operation is performed, excepting for the difference that E is multiplied on the right of A.

5 Inverse, Rank Factorization, and LU-Decomposition

Introduction

For certain square matrices A, we can find a matrix B whose "behavior" is inverse to that of A in the sense of inverse problems in the physical sciences. For a linear system $Ax = b$, where A has an inverse, the unique solution to the system is given by $A^{-1}b$. This is analogous to the solution of $ax = b$, where a, x, and b are any numbers and a is not equal to zero. In this chapter, we invoke the fact that the effect of performing an elementary row operation on a matrix is equivalent to premultiplying by an elementary matrix. This fact is used for developing an effective technique for computing the inverse of a matrix. Two important factorizations of a matrix, namely, full-rank factorization, a factorization of an $m \times n$ matrix of rank r into a product of $m \times r$ and $r \times n$ matrices, and LU-decomposition, a factorization of certain matrices into a product of lower triangular and upper triangular matrices, are also given.

5.1 Inverse of a Matrix and Its Properties

5.1.1 DEFINITION

Inverse
Let A be an $n \times n$ matrix. If there exists an $n \times n$ matrix B such that

$$AB = I = BA,$$

where I is the $n \times n$ identity matrix, then B is called an *inverse* of A, and A is said to be *invertible*. An invertible matrix is also called *nonsingular*.

Clearly, if B is an inverse of A, then A is an inverse of B.

It is a known fact that $AB = I$ holds if and only if $BA = I$. Thus it is enough to check $AB = I$ to prove that B is an inverse of A.

Furthermore, it is important to note that not every nonzero matrix has an inverse. For example, let

$$A = \begin{bmatrix} 0 & 1 \\ 0 & 0 \end{bmatrix}.$$

If possible, let $B = \begin{bmatrix} a & b \\ c & d \end{bmatrix}$ be an inverse of A. Then $AB = I$. This gives, on multiplying A with B in this order, the following equality:

$$\begin{bmatrix} c & d \\ 0 & 0 \end{bmatrix} = \begin{bmatrix} 1 & 0 \\ 0 & 1 \end{bmatrix}.$$

On equating the $(2, 2)$ entry, we get $0 = 1$, which is absurd. Therefore A is not invertible. In case an inverse of a matrix exists, however, it is unique, as shown in Property 5.1.2.

5.1.2 Property The inverse of a matrix (if it exists) is unique.

Proof Suppose that B and C are inverses of A. Then

$$AB = I = BA,$$

and

$$AC = I = CA.$$

From these equations and the associative law, we have

$$C(AB) = (CA)B = IB = B.$$

Also,

$$C(AB) = CI = C.$$

Thus $B = C$, completing the proof. ■

Notation The inverse of a matrix A, whenever it exists, is denoted by A^{-1} (not by $1/A$).

5.1.3 Property Let A and B be invertible $n \times n$ matrices. Then

$$(AB)^{-1} = B^{-1}A^{-1}.$$

Proof
$$(AB)(B^{-1}A^{-1}) = A(BB^{-1})A^{-1} = I.$$

Also,

$$(B^{-1}A^{-1})(AB) = I.$$

Hence

$$(AB)^{-1} = B^{-1}A^{-1}.$$

This completes the proof of Property 5.1.3. ■

5.1.4 Property Let $A_1, A_2, \ldots, A_m$ be invertible $n \times n$ matrices. Then

$$(A_1 A_2 \cdots A_m)^{-1} = A_m^{-1} \cdots A_2^{-1} A_1^{-1}.$$

Its proof is similar to the proof of Property 5.1.3.

The following fact gives a condition for the existence of the inverse of a matrix. The interested reader may see its proof in Section 5.5.

5.1.5 THEOREM

(Theorem 1, Section 5.5)

If a row echelon form of an $n \times n$ matrix A has no zero rows, then A^{-1} exists. Indeed, the converse also holds.

We now proceed to develop a method to compute the inverse of a matrix, normally used for matrices with numerical entries. The idea is to perform elementary row operations on a matrix A so as to reduce it to reduced row echelon form, which will be the identity matrix if the row echelon form has no zero rows. This process is equivalent to premultiplying the matrix A with a suitable matrix E (*Theorem 6, Section 4.4*). The matrix E is the result of performing the same elementary row operations on I as on A. In other words, we have $EA = I$. This suggests a procedure to compute the inverse of an $n \times n$ matrix A as illustrated in Examples 5.1.7.

5.1.6 Procedure for Computing the Inverse

We start with

$$[I \mid A],$$

where I is the $n \times n$ identity matrix. We perform elementary row operations simultaneously on A and I such that A is reduced to reduced row echelon form. If the row echelon form has no zero rows, then the reduced echelon form of A will be the identity matrix I. In this process, I is transformed to A^{-1}.

5.1.7 EXAMPLES

a. Let $A = \begin{bmatrix} 1 & 3 \\ 2 & -4 \end{bmatrix}$.

Now,

$$\begin{bmatrix} 1 & 0 & | & 1 & 3 \\ 0 & 1 & | & 2 & -4 \end{bmatrix} \xrightarrow{R_2 + (-2)R_1} \begin{bmatrix} 1 & 0 & | & 1 & 3 \\ -2 & 1 & | & 0 & -10 \end{bmatrix} \xrightarrow{-\frac{1}{10}R_2}$$

$$\begin{bmatrix} 1 & 0 & | & 1 & 3 \\ \frac{1}{5} & -\frac{1}{10} & | & 0 & 1 \end{bmatrix} \xrightarrow{R_1 + (-3)R_2} \begin{bmatrix} \frac{2}{5} & \frac{3}{10} & | & 1 & 0 \\ \frac{1}{5} & -\frac{1}{10} & | & 0 & 1 \end{bmatrix}.$$

Thus

$$A^{-1} = \begin{bmatrix} \frac{2}{5} & \frac{3}{10} \\ \frac{1}{5} & -\frac{1}{10} \end{bmatrix}.$$

b. Let $A = \begin{bmatrix} 3 & -1 & 4 \\ 0 & 2 & 1 \\ 1 & -1 & -2 \end{bmatrix}$.

Performing the elementary row operations indicated on the left, we have

$$[I\,|\,A] = \begin{bmatrix} 1 & 0 & 0 & 3 & -1 & 4 \\ 0 & 1 & 0 & 0 & 2 & 1 \\ 0 & 0 & 1 & 1 & -1 & -2 \end{bmatrix}$$

$$\xrightarrow{R_1 \longleftrightarrow R_3} \begin{bmatrix} 0 & 0 & 1 & 1 & -1 & -2 \\ 0 & 1 & 0 & 0 & 2 & 1 \\ 1 & 0 & 0 & 3 & -1 & 4 \end{bmatrix}$$

$$\xrightarrow{R_3+(-3)R_1} \begin{bmatrix} 0 & 0 & 1 & 1 & -1 & -2 \\ 0 & 1 & 0 & 0 & 2 & 1 \\ 1 & 0 & -3 & 0 & 2 & 10 \end{bmatrix}$$

$$\xrightarrow{\frac{1}{2}R_2} \begin{bmatrix} 0 & 0 & 1 & 1 & -1 & -2 \\ 0 & \frac{1}{2} & 0 & 0 & 1 & \frac{1}{2} \\ 1 & 0 & -3 & 0 & 2 & 10 \end{bmatrix}$$

$$\xrightarrow{R_3+(-2)R_2} \begin{bmatrix} 0 & 0 & 1 & 1 & -1 & -2 \\ 0 & \frac{1}{2} & 0 & 0 & 1 & \frac{1}{2} \\ 1 & -1 & -3 & 0 & 0 & 9 \end{bmatrix}$$

$$\xrightarrow{\frac{1}{9}R_3} \begin{bmatrix} 0 & 0 & 1 & 1 & -1 & -2 \\ 0 & \frac{1}{2} & 0 & 0 & 1 & \frac{1}{2} \\ \frac{1}{9} & -\frac{1}{9} & -\frac{1}{3} & 0 & 0 & 1 \end{bmatrix}$$

$$\begin{array}{c} \xrightarrow{R_1+2R_3} \\ \xrightarrow{R_2+(-\frac{1}{2})R_3} \end{array} \begin{bmatrix} \frac{2}{9} & -\frac{2}{9} & \frac{1}{3} & 1 & -1 & 0 \\ -\frac{1}{18} & \frac{5}{9} & \frac{1}{6} & 0 & 1 & 0 \\ \frac{1}{9} & -\frac{1}{9} & -\frac{1}{3} & 0 & 0 & 1 \end{bmatrix}$$

$$\xrightarrow{R_1+R_2} \begin{bmatrix} \frac{1}{6} & \frac{1}{3} & \frac{1}{2} & 1 & 0 & 0 \\ -\frac{1}{18} & \frac{5}{9} & \frac{1}{6} & 0 & 1 & 0 \\ \frac{1}{9} & -\frac{1}{9} & -\frac{1}{3} & 0 & 0 & 1 \end{bmatrix}.$$

Hence

$$A^{-1} = \begin{bmatrix} \frac{1}{6} & \frac{1}{3} & \frac{1}{2} \\ -\frac{1}{18} & \frac{5}{9} & \frac{1}{6} \\ \frac{1}{9} & -\frac{1}{9} & -\frac{1}{3} \end{bmatrix} = \frac{1}{18} \begin{bmatrix} 3 & 6 & 9 \\ -1 & 10 & 3 \\ 2 & -2 & -6 \end{bmatrix}.$$

c. Solve the linear system

$$3x - y + 4z = 1,$$
$$2y + z = 2,$$
$$x - y - 2z = 3$$

by using the inverse of the coefficient matrix.

The coefficient matrix A of the linear system is $\begin{bmatrix} 3 & -1 & 4 \\ 0 & 2 & 1 \\ 1 & -1 & -2 \end{bmatrix}$. Write

$$b = \begin{bmatrix} 1 \\ 2 \\ 3 \end{bmatrix} \quad \text{and} \quad X = \begin{bmatrix} x \\ y \\ z \end{bmatrix}.$$

Then the linear system is $Ax = b$. Premultiplying both sides of the equation $AX = b$ by A^{-1}, we get $A^{-1}(AX) = A^{-1}b$, and so $X = A^{-1}b$. As shown in Example (b),

$$A^{-1} = \begin{bmatrix} \frac{1}{6} & \frac{1}{3} & \frac{1}{2} \\ -\frac{1}{18} & \frac{5}{9} & \frac{1}{6} \\ \frac{1}{9} & -\frac{1}{9} & -\frac{1}{3} \end{bmatrix}.$$

Thus

$$X = A^{-1}b = \begin{bmatrix} \frac{1}{6} & \frac{1}{3} & \frac{1}{2} \\ -\frac{1}{18} & \frac{5}{9} & \frac{1}{6} \\ \frac{1}{9} & -\frac{1}{9} & -\frac{1}{3} \end{bmatrix} \begin{bmatrix} 1 \\ 2 \\ 3 \end{bmatrix}$$

$$= \begin{bmatrix} \frac{7}{3} \\ \frac{14}{9} \\ -\frac{10}{9} \end{bmatrix}.$$

Thus $x = \frac{7}{3}$, $y = \frac{14}{9}$, $z = -\frac{10}{9}$.   ■

5.1 Exercises

1. For each of the following matrices A, determine whether A^{-1} exists by reducing it to row echelon form. If A^{-1} exists, find it. Also verify the fact that for each matrix A, the inverse of A exists if and only if the rank of A is equal to its order.

   a. $A = \begin{bmatrix} 1 & 1 \\ 2 & 2 \end{bmatrix}$.

   b. $A = \begin{bmatrix} -1 & 0 \\ 0 & 1 \end{bmatrix}$.

   c. $A = \begin{bmatrix} 2 & 3 \\ 1 & 1 \end{bmatrix}$.

   d. $A = \begin{bmatrix} 1 & 2 & 3 \\ 4 & 5 & 6 \\ 7 & 8 & 9 \end{bmatrix}$.

e. (Drill 5.1) $A = \begin{bmatrix} 1 & 0 & 1 \\ 0 & -1 & 1 \\ 2 & 3 & 4 \end{bmatrix}$.

f. $A = \begin{bmatrix} 0 & 1 & 2 \\ 0 & -1 & 1 \\ 1 & 5 & -4 \end{bmatrix}$.

2. Find the inverse of each of the following matrices by using elementary row operations. Also verify that the rank of each of the matrices is equal to its order.

a. $\begin{bmatrix} 3 & -2 & 1 \\ 1 & -1 & 2 \\ 1 & 0 & -2 \end{bmatrix}$.

b. (Drill 5.2) $\begin{bmatrix} 1 & 1 & 2 & 1 \\ 2 & -1 & 1 & 2 \\ -1 & 2 & 1 & -2 \\ 1 & -1 & 1 & -1 \end{bmatrix}$.

c. $\frac{1}{6}\begin{bmatrix} 3 & 3 & 3 & 3 \\ 3 & -5 & 1 & 1 \\ 3 & 1 & -5 & 1 \\ 3 & 1 & 1 & -5 \end{bmatrix}$.

3. Solve the following linear systems by using the inverse of the coefficient matrix if it exists:

a. $Ax = \begin{bmatrix} 1 \\ 1 \\ 1 \end{bmatrix}$, where $x = \begin{bmatrix} x_1 \\ x_2 \\ x_3 \end{bmatrix}$ and A is the matrix in Exercise 2(a).

b. $Ax = \begin{bmatrix} -1 \\ 1 \\ 0 \\ 1 \end{bmatrix}$, where $x = \begin{bmatrix} x_1 \\ x_2 \\ x_3 \\ x_4 \end{bmatrix}$ and A is the matrix given in Exercise 2(b) (or in Exercise 2(c)).

4. Solve the LS $Ax = b$, when

a. $A = \begin{bmatrix} 1 & 0 & 0 \\ 2 & 1 & 0 \\ 3 & 4 & 1 \end{bmatrix}$, $b = \begin{bmatrix} 1 \\ 0 \\ -1 \end{bmatrix}$.

b. $A = \begin{bmatrix} 1 & 3 & 5 \\ 0 & 1 & 4 \\ 0 & 2 & 7 \end{bmatrix}$, $b = \begin{bmatrix} 0 \\ 0 \\ 0 \end{bmatrix}$.

c. $A = \begin{bmatrix} 1 & 2 & 3 & 1 \\ 0 & 2 & 5 & 2 \\ 2 & 1 & -3 & 0 \\ -1 & 0 & 5 & 1 \end{bmatrix}$, $b = \begin{bmatrix} -2 \\ -1 \\ 0 \\ 2 \end{bmatrix}$.

5. Let A be an $n \times n$ matrix such that

$$A^3 - 2A^2 - I = 0,$$

where 0 denotes the $n \times n$ zero matrix. Show that A has an inverse.

6. Let A be an $n \times n$ invertible matrix. Can you find an $n \times n$ matrix $B \neq 0$ such that

$$AB = 0?$$

7. Let A, B, and C be $n \times n$ matrices such that A is invertible and $AB = AC$. Show that $B = C$.

8. Find all values of a such that the following matrices are invertible:

a. $A = \begin{bmatrix} 2 & a \\ 3 & 4 \end{bmatrix}$.

b. $A = \begin{bmatrix} 1 & a & 0 \\ -1 & 0 & 1 \\ 0 & 1 & 1 \end{bmatrix}$.

9. (Drill 5.3)

a. Show that the matrix

$$A = \begin{bmatrix} a & 1 & 1 & 1 \\ 1 & b & 2 & -1 \\ 2 & 2 & 3 & 1 \\ -1 & 1 & 1 & 1 \end{bmatrix}$$

is invertible if and only if a and b are any values except $a = -1$ and $b = \frac{1}{2}$.

b. Use part (a) to obtain a unique solution, if it exists, for the LS $Ax = c$, where $c = [1 \ 2 \ 3 \ 4]^T$.

5.2 Further Properties of Inverses

5.2.1 Property An $n \times n$ matrix A is invertible if and only if rank $A = n$.

The proof follows from the fact that the rank of A is the number of nonzero rows in row echelon form of A and the fact that A^{-1} exists if and only if the row echelon form of A has no zero rows.

Recall that (1) E_{ij} is the elementary matrix obtained from the identity matrix by interchanging the ith and jth rows, (2) $E_{\alpha(i)}$ is the elementary matrix obtained from the identity matrix by multiplying the ith row with α, and (3) $E_{(i)+\alpha(j)}$ is the elementary matrix obtained from the identity matrix by adding to the ith row α times the jth row. The property that follows describes the inverse of these matrices.

5.2.2 Property

a. $E_{ij}^{-1} = E_{ij}$.

b. $E^{-1}\alpha(i) = E_{(1/\alpha)(i)}$.

c. $E^{-1}_{(i)+\alpha(j)} = E_{(i)-\alpha(j)}$.

We illustrate by taking particular cases only.

$$E_{13} = \begin{bmatrix} 0 & 0 & 1 \\ 0 & 1 & 0 \\ 1 & 0 & 0 \end{bmatrix}.$$

Clearly, $E_{13}E_{13} = I$. So $E_{13}^{-1} = E_{13}$.

$$E_{2(2)} = \begin{bmatrix} 1 & 0 & 0 \\ 0 & 2 & 0 \\ 0 & 0 & 1 \end{bmatrix}.$$

Clearly, $E_{2(2)}E_{(1/2)(2)} = I$. So $E_{2(2)}^{-1} = E_{(1/2)(2)}$.

$$E_{(2)+4(3)} = \begin{bmatrix} 1 & 0 & 0 \\ 0 & 1 & 4 \\ 0 & 0 & 1 \end{bmatrix}, \quad \text{and} \quad E_{(2)-4(3)} = \begin{bmatrix} 1 & 0 & 0 \\ 0 & 1 & -4 \\ 0 & 0 & 1 \end{bmatrix}.$$

Again clearly,

$$\begin{bmatrix} 1 & 0 & 0 \\ 0 & 1 & 4 \\ 0 & 0 & 1 \end{bmatrix} \begin{bmatrix} 1 & 0 & 0 \\ 0 & 1 & -4 \\ 0 & 0 & 1 \end{bmatrix} = \begin{bmatrix} 1 & 0 & 0 \\ 0 & 1 & 0 \\ 0 & 0 & 1 \end{bmatrix}.$$

5.2.3 Property Every invertible matrix is a product of elementary matrices.

For example, let

$$A = \begin{bmatrix} 1 & 2 \\ 3 & 4 \end{bmatrix}.$$

By performing in succession the elementary row operations $R_2 - 3R_1$, $-\frac{1}{2}R_2$, and $R_1 - 2R_2$, we reduce A to the identity matrix $\begin{bmatrix} 1 & 0 \\ 0 & 1 \end{bmatrix}$. Then $A^{-1} = E_{(1)-2(2)} E_{-\frac{1}{2}(2)}$ $E_{(2)-3(1)}$ is the product of elementary matrices corresponding to the above elementary operations written in reverse order.

5.3 Full-Rank Factorization

5.3.1 DEFINITION

Full-Column (Row) Rank
If A is a matrix such that rank A = the number of columns in A, then A is said to have *full-column rank*. In this case, A is called a *full-column-rank* matrix. *Full-row rank* and a *full-row-rank* matrix are defined similarly.

For example, $A = [1\ 2\ 3]$ is a full-row-rank matrix because its rank is 1. The matrix

$$A = \begin{bmatrix} 1 & 0 \\ 1 & 0 \\ 1 & 1 \end{bmatrix}$$

is a full-column-rank matrix because its rank is 2.

An invertible matrix has both full-row rank and full-column rank because an $m \times m$ matrix is invertible if and only if its rank is m (Property 5.2.1).

5.3.2
DEFINITION

Full-Rank Factorization

Let A be an $m \times n$ matrix of rank r. If $A = FG$, where F is an $m \times r$ matrix of rank r (i.e., full-column rank) and G is an $r \times n$ matrix of rank r (i.e., full-row rank), then $A = FG$ is called a full-rank factorization of A.

Full-rank factorization of a matrix is used in *Chapter 8* to obtain the best approximate solution of a possibly inconsistent linear system.

5.3.3 Property Every matrix A has a full-rank factorization.

The proof given below shows the steps that yield full-rank factorization. The reader who is not interested in the proof may go directly to the steps summarized after the proof.

Let A be an $m \times n$ matrix of rank r. The matrix A can be reduced to row echelon form by performing elementary row operations. Since performing an elementary row operation on a matrix is equivalent to premultiplying it by an elementary matrix, we get $E_k \cdots E_1 A = B$, where $E_1, \ldots, E_k$ are elementary matrices and B is in row echelon form.

On writing $P = E_k \cdots E_1$, we get $PA = B$. Since each E_i is invertible, P is invertible and $P^{-1} = E_1^{-1} \ldots E_k^{-1}$. Write

$$B = \begin{bmatrix} G \\ 0 \end{bmatrix},$$

where G consists of the nonzero rows of B. So G must be of order $r \times n$, and $\text{rank}(G) = r$. Then

$$PA = \begin{bmatrix} G \\ 0 \end{bmatrix}.$$

This gives

$$A = P^{-1} \begin{bmatrix} G \\ 0 \end{bmatrix}.$$

Partition P^{-1} as $[F\ \ F']$, where F consists of the first r columns of P^{-1} and F' consists of the remaining $n - r$ columns.

It can be verified that

$$A = \begin{bmatrix} F & F' \end{bmatrix} \begin{bmatrix} G \\ 0 \end{bmatrix}$$

yields $A = FG$. We note that since the rank of P^{-1} is m (being an $m \times m$ invertible matrix), the columns of P^{-1} are linearly independent (since the rank is the maximal number of linearly independent columns). Thus the r columns of F are also linearly independent. Hence rank$(F) = r$. This completes the proof.

We now summarize the steps used to find a full-rank factorization of a matrix.

Step 1 Reduce A to row echelon form by performing elementary row operations, say $E_1, \ldots, E_m$:

$$A \xrightarrow{E_1} [\quad] \xrightarrow{E_2} \cdots \xrightarrow{E_m} \begin{bmatrix} G \\ 0 \end{bmatrix}.$$

Step 2 Obtain P^{-1} by performing the row operations $E_m^{-1}, \ldots, E_1^{-1}$ on I:

$$I \xrightarrow{E_m^{-1}} [\quad] \xrightarrow{E_{m-1}^{-1}} \cdots \xrightarrow{E_1^{-1}} [\quad] = P^{-1}.$$

Step 3 If r is the number of rows of G (of course, all of them are nonzero), then form the $m \times r$ matrix F whose columns are the first r columns of P^{-1} (in the same order).

Step 4 Write $A = FG$.

5.3.4 Note

In Step 2, the inverse of an elementary operation is obtained as follows:

The inverse of $R_i \leftrightarrow R_j$ is $R_i \leftrightarrow R_j$ itself,

the inverse of αR_i is $\frac{1}{\alpha} R_i$,

the inverse of $R_i + \alpha R_j$ is $R_i - \alpha R_j$.

(Refer to Property 5.2.2.)

5.3.5 EXAMPLES **a.** Find a full-rank factorization of

$$A = \begin{bmatrix} 1 & 1 \\ 0 & 0 \end{bmatrix}.$$

Here, A is already in row echelon form $\begin{bmatrix} G \\ 0 \end{bmatrix}$, where $G = [1\ \ 1]$.

To find F, we consider

$$I = \begin{bmatrix} 1 & 0 \\ 0 & 1 \end{bmatrix}$$

and take

$$F = \begin{bmatrix} 1 \\ 0 \end{bmatrix}$$

by taking the first column of I, since rank$(A) = 1$. Thus $A = \begin{bmatrix} 1 \\ 0 \end{bmatrix} [1\ \ 1]$ is a full-rank factorization of A.

b. Find a full-rank factorization of

$$A = \begin{bmatrix} 1 & 2 & 3 \\ 1 & 1 & 0 \\ 2 & 3 & 3 \end{bmatrix}.$$

First we reduce A to row echelon form as follows:

$$A = \begin{bmatrix} 1 & 2 & 3 \\ 1 & 1 & 0 \\ 2 & 3 & 3 \end{bmatrix} \xrightarrow[R_3+(-2)R_1]{R_2+(-1)R_1} \begin{bmatrix} 1 & 2 & 3 \\ 0 & -1 & -3 \\ 0 & -1 & -3 \end{bmatrix}$$

$$\xrightarrow{R_3+(-1)R_2} \begin{bmatrix} 1 & 2 & 3 \\ 0 & -1 & -3 \\ 0 & 0 & 0 \end{bmatrix} = \begin{bmatrix} G \\ 0 \end{bmatrix}.$$

To find F, we proceed as in Step 2:

$$I = \begin{bmatrix} 1 & 0 & 0 \\ 0 & 1 & 0 \\ 0 & 0 & 1 \end{bmatrix} \xrightarrow{R_3+1R_2} \begin{bmatrix} 1 & 0 & 0 \\ 0 & 1 & 0 \\ 0 & 1 & 1 \end{bmatrix} \xrightarrow[R_3+2R_1]{R_2+1R_1} \begin{bmatrix} 1 & 0 & 0 \\ 1 & 1 & 0 \\ 2 & 1 & 1 \end{bmatrix}.$$

So $F = \begin{bmatrix} 1 & 0 \\ 1 & 1 \\ 2 & 1 \end{bmatrix}$, since rank $A = 2$.

Therefore,

$$A = \begin{bmatrix} 1 & 0 \\ 1 & 1 \\ 2 & 1 \end{bmatrix} \begin{bmatrix} 1 & 2 & 3 \\ 0 & -1 & -3 \end{bmatrix}$$

is a full-rank factorization. ∎

5.3 Exercises

Find a full-rank factorization for each of the following matrices.

1. $[1 \ 2 \ 3]$

2. $\begin{bmatrix} 1 & 2 \\ 3 & 4 \end{bmatrix}$

 3. (Drill 5.4) $\begin{bmatrix} 1 & 2 & 4 \\ 3 & 0 & 1 \\ 1 & -4 & -7 \end{bmatrix}$

4. $\begin{bmatrix} 1 & 2 & 5 \\ 6 & 0 & 1 \end{bmatrix}$

5. (Drill 5.5)
$$\begin{bmatrix} 2 & 1 & -1 & 0 \\ 1 & 0 & 2 & 3 \\ -1 & 1 & 0 & 4 \\ 3 & 0 & 1 & 0 \\ 1 & 2 & 3 & 4 \end{bmatrix}$$

5.4 LU-Decomposition of a Matrix

To solve a linear system $Ax = b$ efficiently, especially using computers, we can factor A as PLU, where L is the lower triangular, U is the upper triangular, and P is a permutation matrix. (P is called a *permutation matrix* if each row and each column has exactly one nonzero entry, which is 1.) Since $P^{-1} = P^T$, solving $Ax = b$ amounts to solving $Ly = P^T b$ and $Ux = y$ for y and x, respectively. The latter equations are easy to solve because of the special form of the matrices L and U.

We state the following fact without proof.

5.4.1 Fact If no interchanges of rows are necessary in reducing A to row echelon form, then A can be expressed as $A = LU$.

In general (when interchanges of rows are necessary), A can be expressed as $A = PLU$, where P is a permutation matrix.

We will illustrate Fact 5.4.1 by examples. Indeed, the procedure is almost the same as that for full-rank factorization.

Step 1

$$A \xrightarrow{E_1} [\quad] \xrightarrow{E_2} [\quad] \cdots \xrightarrow{E_m} [\quad] = U.$$

Step 2

$$I \xrightarrow{E_m^{-1}} [\quad] \xrightarrow{E_{m-1}^{-1}} [\quad] \cdots \xrightarrow{E_1^{-1}} [\quad] = L.$$

Step 3

$$A = LU.$$

5.4.2
EXAMPLES
a. Find an LU-decomposition of

$$A = \begin{bmatrix} 1 & 2 & 3 \\ 4 & 5 & 6 \\ 0 & 0 & 1 \end{bmatrix}.$$

Step 1

$$A = \begin{bmatrix} 1 & 2 & 3 \\ 4 & 5 & 6 \\ 0 & 0 & 1 \end{bmatrix} \xrightarrow{R_2 + (-4)R_1} \begin{bmatrix} 1 & 2 & 3 \\ 0 & -3 & -6 \\ 0 & 0 & 1 \end{bmatrix}$$

$$\xrightarrow{-\frac{1}{3}R_2} \begin{bmatrix} 1 & 2 & 3 \\ 0 & 1 & 2 \\ 0 & 0 & 1 \end{bmatrix} = U.$$

Step 2

$$
\begin{bmatrix} 1 & 0 & 0 \\ 0 & 1 & 0 \\ 0 & 0 & 1 \end{bmatrix}
\xrightarrow{-3R_2}
\begin{bmatrix} 1 & 0 & 0 \\ 0 & -3 & 0 \\ 0 & 0 & 1 \end{bmatrix}
\xrightarrow{R_2+4R_1}
\begin{bmatrix} 1 & 0 & 0 \\ 4 & -3 & 0 \\ 0 & 0 & 1 \end{bmatrix} = L.
$$

Then $A = LU$.

b. Find an LU-decomposition of

$$
A = \begin{bmatrix} 1 & 3 & 0 \\ 2 & 1 & 0 \\ 3 & 4 & 1 \end{bmatrix}.
$$

$$
A = \begin{bmatrix} 1 & 3 & 0 \\ 2 & 1 & 0 \\ 3 & 4 & 1 \end{bmatrix}
\begin{array}{c} \xrightarrow{R_2+(-2)R_1} \\ \xrightarrow{R_3+(-3)R_1} \end{array}
\begin{bmatrix} 1 & 3 & 0 \\ 0 & -5 & 0 \\ 0 & -5 & 1 \end{bmatrix}
$$

$$
\xrightarrow{R_3+(-1)R_2}
\begin{bmatrix} 1 & 3 & 0 \\ 0 & -5 & 0 \\ 0 & 0 & 1 \end{bmatrix} = U,
$$

$$
I = \begin{bmatrix} 1 & 0 & 0 \\ 0 & 1 & 0 \\ 0 & 0 & 1 \end{bmatrix}
\xrightarrow{R_3+1R_2}
\begin{bmatrix} 1 & 0 & 0 \\ 0 & 1 & 0 \\ 0 & 1 & 1 \end{bmatrix}
$$

$$
\begin{array}{c} \xrightarrow{R_2+2R_1} \\ \xrightarrow{R_3+3R_1} \end{array}
\begin{bmatrix} 1 & 0 & 0 \\ 2 & 1 & 0 \\ 3 & 1 & 1 \end{bmatrix} = L.
$$

Thus $A = LU$, where L and U are as above.

c. Solve the linear system

$$
\begin{aligned}
x_1 + 3x_2 &= 1, \\
2x_1 + x_2 &= 2, \\
3x_1 + 4x_2 + x_3 &= 0
\end{aligned}
$$

by using an LU-decomposition of the coefficient matrix.
The coefficient matrix is

$$
A = \begin{bmatrix} 1 & 3 & 0 \\ 2 & 1 & 0 \\ 3 & 4 & 1 \end{bmatrix} = LU,
$$

where L and U are as in Example (b).
We want to solve $LUx = b$, where

$$
b = \begin{bmatrix} 1 \\ 2 \\ 0 \end{bmatrix}.
$$

Set $y = Ux$. Then $Ly = b$. From $Ly = b$, we obtain

$$y = \begin{bmatrix} 1 \\ 0 \\ -3 \end{bmatrix}.$$

Then $Ux = y$ gives

$$x = \begin{bmatrix} 1 \\ 0 \\ -3 \end{bmatrix}$$

as the solution of the linear system. ■

5.4 Exercises

1. Find an LU-decomposition of each of the following matrices:

a. $\begin{bmatrix} 1 & 2 \\ 2 & 3 \end{bmatrix}$

b. $\begin{bmatrix} 1 & 0 & 0 \\ 1 & 0 & 3 \\ 1 & 2 & 0 \end{bmatrix}$

c. $\begin{bmatrix} 2 & 3 & 4 \\ 0 & 1 & 3 \\ 1 & 1 & 1 \end{bmatrix}$

d. $\begin{bmatrix} 1 & 0 & 2 \\ 1 & 2 & 1 \\ 4 & 5 & 0 \end{bmatrix}$

e. $\begin{bmatrix} 0 & 1 & 1 \\ 1 & 0 & 0 \\ 1 & 1 & 1 \end{bmatrix}$

 f. (Drill 5.6) $\begin{bmatrix} 1 & 2 & 3 & 0 \\ 2 & 9 & 6 & 10 \\ 3 & 7 & 10 & 7 \end{bmatrix}$

 g. (Drill 5.7) $\begin{bmatrix} 1 & 1 & 1 & 7 \\ 6 & 12 & 18 & 0 \\ 3 & 9 & 6 & 10 \end{bmatrix}$

2. Use an LU-decomposition of the coefficient matrices to solve each of the following linear systems:

 a. (Drill 5.8)
$$\begin{aligned} x_1 + 2x_2 + 3x_3 &= 0, \\ 2x_1 + 9x_2 + 6x_3 &= 10, \\ 3x_1 + 2x_2 + 10x_3 &= 7. \end{aligned}$$

b. $x_1 + 2x_2 + x_3 = 4,$
$2x_1 - x_2 - x_3 = 1,$
$x_1 + x_2 + 3x_3 = 0.$

c. $x_1 - 2x_2 + x_3 = 2,$
$2x_1 + x_2 - x_3 = 1,$
$-3x_1 + x_2 - 2x_3 = -5.$

3. Consider the linear system given by

$$2x_1 + 4x_2 + 3x_3 + 2x_4 = 0,$$
$$3x_1 + 6x_2 + 5x_3 + 2x_4 = 10,$$
$$2x_1 + 5x_2 + 2x_3 - 3x_4 = 7,$$
$$4x_1 + 5x_2 + 14x_3 + 14x_4 = 2.$$

a. Find an LU-decomposition of the coefficient matrix, if possible.
b. If an LU-decomposition does not exist for the above matrix, can you suggest a permutation of rows that will give an LU-decomposition? (Hint: Try interchanging rows.)
c. Using the LU-decomposition obtained in part (b), solve the system $Ax = b$ by first solving the system $Ly = b$ for y and then solving the system $Ux = y$ for x.

5.5   Proofs of Facts

Recall that performing an elementary row operation on a matrix A is equivalent to premultiplying A by a suitable elementary matrix (*Theorem 6, Section 4.4*).

THEOREM 1

If a row echelon form of an $n \times n$ matrix A has no zero rows, then A^{-1} exists.

Proof By performing successively elementary row operations, say m times, we reduce A into row echelon form and obtain

$$E_m \cdots E_2 E_1 A = \begin{bmatrix} 1 & 0 & \cdots & 0 \\ 0 & 1 & \cdots & 0 \\ \vdots & \vdots & \vdots & \vdots \\ 0 & 0 & \cdots & 1 \end{bmatrix}, \tag{1}$$

where the diagonal consists of 1's alone and $E_1, E_2, \ldots, E_m$ are elementary matrices corresponding to the elementary row operations performed on A (cf. Theorem 6, Section 4.4).

Furthermore, we can reduce the right-hand side of Equation (1) to the identity matrix by performing successive elementary row operations of type III, starting from the last row.

This implies that

$$F_k \cdots F_1 E_m \cdots E_1 A = I.$$

Hence $BA = I$, where $B = F_k \cdots F_1 E_m \cdots E_1$. Therefore, B is the inverse of A. The converse follows by Property 5.2.1. ∎

5.6   Chapter Review Questions and Projects

1. a. What is meant by the inverse of a square matrix? (Section 5.1)

  b. Does every matrix possess an inverse? Give a reason to justify your answer.

  c. Can a matrix possess more than one inverse? Justify your answer.

2. a. Construct two invertible matrices A and B such that $A + B$ is not invertible. (Section 5.1)

  b. Construct two noninvertible matrices A and B such that $A + B$ is invertible.

3. Without actually attempting to compute the inverse, find which of the following row echelon matrices have inverses. Also give the rank of each matrix. (Section 5.2)

$$\begin{bmatrix} 2 & 3 & 5 \\ 0 & 4 & -1 \\ 0 & 0 & 0 \end{bmatrix}, \quad \begin{bmatrix} 1 & 0 & 0 \\ 0 & 1 & 2 \\ 0 & 0 & 3 \end{bmatrix}, \quad \begin{bmatrix} 1 & 1 & 2 & 2 \\ 0 & 1 & 3 & 3 \\ 0 & 0 & 1 & 4 \\ 0 & 0 & 0 & 1 \end{bmatrix}.$$

4. If A is an $n \times n$ matrix and rank $A = m$, what is the condition on m for A to be invertible? (Section 5.2)

5. Compute the inverse, if it exists, of the following matrices. (Section 5.1)

$$\begin{bmatrix} 1 & 1 & 1 \\ 1 & 2 & 2 \\ 1 & 2 & 3 \end{bmatrix}, \quad \begin{bmatrix} 1 & 2 & 3 \\ 4 & 5 & 6 \\ 7 & 8 & 9 \end{bmatrix}, \quad \begin{bmatrix} 3 & 3 & 0 & 0 \\ 4 & 3 & 0 & 0 \\ 0 & 0 & 6 & 5 \\ 0 & 0 & 7 & 6 \end{bmatrix}.$$

6. Find which of the following matrices have (i) full-row rank, (ii) full-column rank. Also find a full-rank factorization of each of the matrices. (Section 5.3)

  a. $[1 \ 2 \ 3]$

  b. $\begin{bmatrix} 1 & -2 & 3 \\ 2 & -5 & 1 \\ 1 & -4 & -7 \end{bmatrix}$

  c. $\begin{bmatrix} 1 & 2 & -1 & 1 \\ 2 & 4 & -3 & 0 \\ 3 & 6 & -4 & 1 \end{bmatrix}$

  d. $\begin{bmatrix} 1 & -2 & 1 & 1 & 2 \\ -1 & 3 & 0 & 2 & -2 \\ 0 & 1 & 1 & 3 & 4 \\ 1 & 2 & 5 & 0 & 5 \end{bmatrix}$

7. Find an LU-decomposition (if it exists) of the following matrices: (Section 5.4)

a.
$$\begin{bmatrix} -2 & 1 & 0 \\ 1 & -2 & 1 \\ 0 & 1 & -2 \end{bmatrix}$$

b.
$$\begin{bmatrix} 1 & 4 & 0 \\ 4 & 12 & 4 \\ 0 & 4 & 0 \end{bmatrix}$$

c.
$$\begin{bmatrix} 1 & 2 & 0 & 0 \\ 2 & 3 & 1 & 0 \\ 0 & 1 & 2 & 3 \\ 0 & 0 & 3 & 4 \end{bmatrix}$$

By choosing diagonal entries of matrices L and U as 1, rewrite the decomposition as LDU, where D is a diagonal matrix. Do you observe any relation between the newly obtained L and U for the symmetric matrix (b)?

8. Using an LU-decomposition of A, solve the LS $Ax = b$, where A is the matrix in Question 7(b), by taking

$$b = \begin{bmatrix} 1 \\ 1 \\ 1 \end{bmatrix} \quad \text{and} \quad b = \begin{bmatrix} 1 \\ 0 \\ 1 \end{bmatrix}$$

in succession, using an LU-decomposition of A. (Section 5.4)

Chapter Projects

1. a. Compute L and U for the symmetric matrix

$$A = \begin{bmatrix} a & a & a & a & a \\ a & b & b & b & b \\ a & b & c & c & c \\ a & b & c & d & d \\ a & b & c & d & e \end{bmatrix},$$

and show that $A = LDL^T$, where the diagonal entries of L are 1, and D is some diagonal matrix.

b. Write down 7×7 and 10×10 symmetric matrices of the type given in part (a) having numerical entries, by choosing $a = 1, b = 2, c = 3, d = 4, e = 5$, and so on, and obtain L in each case. Do you see any pattern?

2. a. Invertible matrices are used to send secret messages by assigning a numerical value to each letter of the alphabet. For example, if we assign $a = 1, b = 2, c = 3$, and so on, try to send the following message:

GO NORTHEAST NOW

You may first form a 4×4 matrix

$$A = \begin{bmatrix} 7 & 15 & 14 & 15 \\ 18 & 20 & 8 & 5 \\ 1 & 19 & 20 & 14 \\ 15 & 23 & 0 & 0 \end{bmatrix}$$

of the actual message and then premultiply A by a suitable invertible matrix, say

$$P = \begin{bmatrix} 1 & 0 & 0 & 0 \\ 0 & 0 & 0 & 1 \\ 0 & 0 & 1 & 0 \\ 0 & 1 & 0 & 0 \end{bmatrix},$$

to obtain the matrix R containing scrambled message. The person at the other end who knows the matrix P can decode the message. Describe the scrambled message, and explain the process of decoding.

b. Make up your own values of the alphabet, send a coded message using a suitable permutation matrix, and verify that it can be decoded by using the inverse of the matrix. (An $n \times n$ matrix is called a permutation matrix if each row and column contains only one nonzero entry that is 1.)

Key Words

Inverse of a Matrix
Invertible (Nonsingular) Matrix
LU-Decomposition of a Matrix
Full-Column-Rank Matrix
Full-Row-Rank Matrix

Full-Rank Factorization
Elementary Row Operations
Reduced Row Echelon Form
Lower and Upper Triangular
   Matrices

Key Phrases

■ The inverse of a matrix exists if and only if its row echelon form has no zero rows.

6 Determinants

Introduction

In this chapter, we consider an important property of square matrices. With every square matrix, we associate a number, called its *determinant*, in a manner to be explained below. We will state without proof the important properties of the determinant and give practical methods for working with the determinant. The determinant of a matrix provides quite useful information about the matrix. For example, if the determinant of a matrix is not zero, then the matrix is invertible, and conversely if a matrix is invertible, then its determinant is not zero. A rigorous development of the subject is given in Chapter 10.

6.1 Determinant

6.1.1 DEFINITION

Determinant of 1×1 and 2×2 Matrices

1. The *determinant* of a 1×1 matrix $[a]$ is a.

2. The *determinant* of a 2×2 matrix $\begin{bmatrix} a & b \\ c & d \end{bmatrix}$ is $ad - bc$.

Notation The determinant of a matrix A will be denoted by det A. So

$$\det [3] = 3, \qquad \det \begin{bmatrix} 1 & 2 \\ 3 & 4 \end{bmatrix} = 4 - 6 = -2.$$

For quick practice, compute the following:

a. $\det \begin{bmatrix} 1 & 0 \\ -1 & 0 \end{bmatrix}$

b. $\det \begin{bmatrix} -1 & -1 \\ -1 & -1 \end{bmatrix}$

c. $\det \begin{bmatrix} 1 & 2 \\ 6 & -3 \end{bmatrix}$

d. $\det [-3]$

e. $\det \begin{bmatrix} \sin \phi & \cos \phi \\ -\cos \phi & \sin \phi \end{bmatrix}$

f. $\det \begin{bmatrix} e^x & \sin x \\ x & e^x \end{bmatrix}$

Answers

a. 0
b. 0
c. -15
d. -3
e. 1
f. $e^{2x} - x \sin x$

**6.1.2
DEFINITION**

Determinant of a 3×3 Matrix

If
$$A = \begin{bmatrix} a_{11} & a_{12} & a_{13} \\ a_{21} & a_{22} & a_{23} \\ a_{31} & a_{32} & a_{33} \end{bmatrix},$$

then

$$\det A = a_{11}(-1)^{1+1} \det \quad \text{(the } 2 \times 2 \text{ matrix obtained by deleting row 1 and column 1 of } A)$$
$$+ a_{12}(-1)^{1+2} \det \quad \text{(the } 2 \times 2 \text{ matrix obtained by deleting row 1 and column 2 of } A)$$
$$+ a_{13}(-1)^{1+3} \det \quad \text{(the } 2 \times 2 \text{ matrix obtained by deleting row 1 and column 3 of } A),$$

that is,

$$\det A = a_{11}(-1)^2 \det \begin{bmatrix} a_{22} & a_{23} \\ a_{32} & a_{33} \end{bmatrix} + a_{12}(-1)^3 \det \begin{bmatrix} a_{21} & a_{23} \\ a_{31} & a_{33} \end{bmatrix}$$
$$+ a_{13}(-1)^4 \det \begin{bmatrix} a_{21} & a_{22} \\ a_{31} & a_{32} \end{bmatrix}.$$

Since we know, by Definition 6.1.1, how to write the determinant of a 2×2 matrix, the above definition tells us how to write the determinant of a 3×3 matrix.

6.1.3
EXAMPLE

Compute

$$\det \begin{bmatrix} 1 & 2 & -1 \\ -2 & 0 & 7 \\ 3 & 0 & 7 \end{bmatrix}.$$

By definition,

$$\det \begin{bmatrix} 1 & 2 & -1 \\ -2 & 0 & 7 \\ 3 & 0 & 7 \end{bmatrix} = 1(-1)^{1+1} \det \begin{bmatrix} 0 & 7 \\ 0 & 7 \end{bmatrix} + 2(-1)^{1+2} \det \begin{bmatrix} -2 & 7 \\ 3 & 7 \end{bmatrix}$$

$$+ (-1)(-1)^{1+3} \det \begin{bmatrix} -2 & 0 \\ 3 & 0 \end{bmatrix}$$

$$= 1(0 - 0) - 2(-14 - 21) - (0 - 0) = -70.$$

For practice, compute the determinant of the following matrices:

a. $\begin{bmatrix} 1 & 2 & 0 \\ 6 & 7 & 3 \\ -1 & 0 & 0 \end{bmatrix}$

b. $\begin{bmatrix} 2 & 3 & 7 \\ -1 & 5 & 0 \\ 0 & 1 & -1 \end{bmatrix}$

c. $\begin{bmatrix} 3 & 5 & 0 \\ -1 & 2 & 1 \\ 3 & -6 & 4 \end{bmatrix}$

d. $\begin{bmatrix} 1 & 2 \\ 3 & 0 \end{bmatrix}$

e. $\begin{bmatrix} 1 & 3 & 2 \\ -1 & 4 & 1 \\ 5 & 3 & 8 \end{bmatrix}$

f. $\begin{bmatrix} x & x^2 + 1 & -1 \\ 0 & -x & e^x \\ 1 & 0 & 0 \end{bmatrix}$

Answers
a. -6
b. -20
c. 77
d. -6
e. 22
f. $e^x(x^2 + 1) - x$

<table>
<tr><td>**6.1.4**
DEFINITION</td><td>**Determinant of a 4×4 Matrix**
If

$$A = \begin{bmatrix} a_{11} & a_{12} & a_{13} & a_{14} \\ a_{21} & a_{22} & a_{23} & a_{24} \\ a_{31} & a_{32} & a_{33} & a_{34} \\ a_{41} & a_{42} & a_{43} & a_{44} \end{bmatrix},$$

then

$$\begin{aligned} \det A = \; & a_{11}(-1)^{1+1} \det \quad \text{(the } 3 \times 3 \text{ matrix obtained by deleting row 1} \\ & \qquad\qquad\qquad\qquad \text{and column 1 of } A) \\ + \; & a_{12}(-1)^{1+2} \det \quad \text{(the } 3 \times 3 \text{ matrix obtained by deleting row 1} \\ & \qquad\qquad\qquad\qquad \text{and column 2 of } A) \\ + \; & a_{13}(-1)^{1+3} \det \quad \text{(the } 3 \times 3 \text{ matrix obtained by deleting row 1} \\ & \qquad\qquad\qquad\qquad \text{and column 3 of } A) \\ + \; & a_{14}(-1)^{1+4} \det \quad \text{(the } 3 \times 3 \text{ matrix obtained by deleting row 1} \\ & \qquad\qquad\qquad\qquad \text{and column 4 of } A). \end{aligned}$$
</td></tr>
</table>

Since we know how to compute the determinant of a 3×3 matrix, the above definition gives us a method of computing a 4×4 matrix. ∎

6.1.5
EXAMPLE

Compute

$$\det \begin{bmatrix} 1 & 0 & 5 & 2 \\ -1 & 4 & 1 & 0 \\ 3 & 0 & 4 & 1 \\ -2 & 1 & 1 & 3 \end{bmatrix}.$$

$$\det \begin{bmatrix} 1 & 0 & 5 & 2 \\ -1 & 4 & 1 & 0 \\ 3 & 0 & 4 & 1 \\ -2 & 1 & 1 & 3 \end{bmatrix} = (-1)^{1+1} \det \begin{bmatrix} 4 & 1 & 0 \\ 0 & 4 & 1 \\ 1 & 1 & 3 \end{bmatrix} + 0(-1)^{1+2} \det \begin{bmatrix} -1 & 1 & 0 \\ 3 & 4 & 1 \\ -2 & 1 & 3 \end{bmatrix}$$

$$+ 5(-1)^{1+3} \det \begin{bmatrix} -1 & 4 & 0 \\ 3 & 0 & 1 \\ -2 & 1 & 3 \end{bmatrix} + 2(-1)^{1+4} \det \begin{bmatrix} -1 & 4 & 1 \\ 3 & 0 & 4 \\ -2 & 1 & 1 \end{bmatrix}.$$

Let us first compute

$$\det \begin{bmatrix} 4 & 1 & 0 \\ 0 & 4 & 1 \\ 1 & 1 & 3 \end{bmatrix} = 4(-1)^{1+1} \det \begin{bmatrix} 4 & 1 \\ 1 & 3 \end{bmatrix}$$

$$+ 1(-1)^{1+2} \det \begin{bmatrix} 0 & 1 \\ 1 & 3 \end{bmatrix} + 0(-1)^{1+3} \det \begin{bmatrix} * & * \\ * & * \end{bmatrix}$$

$$= 4(12 - 1) = (0 - 1) = 44 + 1 = 45.$$

(Note: $\begin{bmatrix} * & * \\ * & * \end{bmatrix}$ denotes an appropriate matrix that need not be computed because it is being multiplied by zero.)

Next, compute

$$\det \begin{bmatrix} -1 & 4 & 0 \\ 3 & 0 & 1 \\ -2 & 1 & 3 \end{bmatrix} = -1(-1)^{1+1} \det \begin{bmatrix} 0 & 1 \\ 1 & 3 \end{bmatrix} + 4(-1)^{1+2} \det \begin{bmatrix} 3 & 1 \\ -1 & 3 \end{bmatrix}$$

$$+ 0(-1)^{1+3} \text{ (determinant of a matrix that we do not need}$$
$$\text{to worry about, since it is being multiplied by 0)}$$
$$= -1(0-1) - 4(9+2) = 1 - 44 = -43.$$

Finally, compute

$$\det \begin{bmatrix} -1 & 4 & 1 \\ 3 & 0 & 4 \\ -2 & 1 & 1 \end{bmatrix} = (-1)(-1)^{1+1} \det \begin{bmatrix} 0 & 4 \\ 1 & 1 \end{bmatrix} + 4(-1)^{1+2} \det \begin{bmatrix} 3 & 4 \\ -2 & 1 \end{bmatrix}$$

$$+ 1(-1)^{1+3} \det \begin{bmatrix} 3 & 0 \\ -2 & 1 \end{bmatrix}$$
$$= -1(-4) - 4(3+8) + 1(3+0) = 4 - 44 + 3 = -37.$$

Thus the value of the determinant of the given matrix is

$$1(45) + 0 + 5(-43) - 2(-37) = -96.$$

Note: The determinant of matrices of higher orders, say n, are defined in the same manner in terms of the determinant of matrices of order $n - 1$. ■

6.1 Exercises

Evaluate the determinant of each of the following matrices:

1. (Drill 6.1) $\begin{bmatrix} 1 & 0 & 0 & 0 \\ 2 & 1 & 2 & -1 \\ 0 & 0 & 4 & 5 \\ 0 & 0 & 0 & 6 \end{bmatrix}.$

2. $\begin{bmatrix} 5 & 0 & 6 & 0 \\ -1 & 7 & 0 & 1 \\ 0 & 2 & 3 & 1 \\ 0 & 1 & 2 & 3 \end{bmatrix}.$

3. $\begin{bmatrix} 5 & 3 & 7 & 9 \\ 0 & 5 & 2 & 1 \\ 0 & 0 & 5 & 1 \\ 0 & 0 & 0 & 5 \end{bmatrix}.$

4. $\begin{bmatrix} 2 & 0 & 0 & 0 \\ 3 & 6 & 0 & 0 \\ 4 & 7 & 9 & 0 \\ 5 & 8 & 0 & 10 \end{bmatrix}.$

5. $\begin{bmatrix} -1 & 0 & 0 & 0 \\ 0 & 1 & 0 & 0 \\ 0 & 0 & -1 & 0 \\ 0 & 0 & 0 & 1 \end{bmatrix}.$

6. $\begin{bmatrix} a & 0 & 0 & 0 \\ 0 & b & 0 & 0 \\ 0 & 0 & c & 0 \\ 0 & 0 & 0 & d \end{bmatrix}.$

7. $\begin{bmatrix} 1 & 2 & 1 \\ 2 & 0 & 1 \\ 1 & -1 & 1 \end{bmatrix}.$

8. $\begin{bmatrix} 1 & 1 & 2 & 1 \\ 0 & 1 & 4 & 1 \\ 2 & 1 & 3 & 0 \\ 2 & 2 & 1 & 2 \end{bmatrix}.$

9. $\begin{bmatrix} 2 & 1 & -1 & 2 \\ 3 & 0 & 0 & 1 \\ 2 & 1 & 2 & 0 \\ 3 & 1 & 1 & 2 \end{bmatrix}.$

10. $\begin{bmatrix} 1 & 2 & 1 & 3 \\ 0 & 4 & 1 & 2 \\ 0 & 0 & 3 & 1 \\ 0 & 0 & 0 & 2 \end{bmatrix}.$

11. $\begin{bmatrix} a & 0 & 0 & 2 \\ 0 & b & 0 & 0 \\ 0 & 0 & c & -1 \\ 0 & 0 & 0 & d \end{bmatrix}.$

12. Compute $\det(A + A^T)$ and $\det(AA^T)$, where A is as in Exercise 1. Conclude that the determinant function is not additive, that is, $\det(A + B)$ need not be equal to $\det A + \det B$. Note, however, that $\det(AB) = \det A \cdot \det B$ (see Property 6.2.5 and Theorem 10.6.5).

13. Use matlab commands $hilb(3)$, $zeros(3)$, $ones(3)$, and $eye(3)$ for the 3×3 Hilbert matrix, the 3×3 zero matrix, the 3×3 matrix with all entries 1, and the 3×3 identity matrix, respectively, to enter the matrix

$$A = \begin{bmatrix} hilb(3) & zeros(3) \\ ones(3) & eye(3) \end{bmatrix}.$$

Compute $\det A$, and repeat for $n = 4, 5$.

The $n \times n$ Hilbert matrix $H = (h_{ij})$ is defined by $h_{ij} = \dfrac{1}{i + j - 1}.$

6.2 Properties of the Determinant

The following are some basic properties of the determinant, which are stated without proof. The proofs of some of them will be given in a later chapter (Chapter 10).

6.2.1 Property If a row or a column of a matrix consists entirely of zeros, then the determinant is 0.

For example, if

$$A = \begin{bmatrix} 0 & 0 \\ 2 & 3 \end{bmatrix},$$

then det $A = 0$.

6.2.2 Property If two rows or two columns of a matrix are interchanged, then the determinant changes sign.

For example, if

$$A = \begin{bmatrix} 2 & 3 \\ 5 & 6 \end{bmatrix},$$

then det $A = -3$, but for the matrix

$$B = \begin{bmatrix} 5 & 6 \\ 2 & 3 \end{bmatrix},$$

obtained from the matrix A by interchanging its rows, we have det $B = 3$.

6.2.3 Property If two rows or two columns of a matrix are identical, then the determinant is 0.

For example, if

$$A = \begin{bmatrix} 5 & 6 \\ 5 & 6 \end{bmatrix},$$

then det $A = 0$.

6.2.4 Property If the matrix B is obtained from the matrix A by multiplying every element in one row or in one column by α, then det $B = \alpha$ det A.

For example, if

$$A = \begin{bmatrix} 2 & 3 \\ 5 & 6 \end{bmatrix},$$

then det $A = -3$, but for the matrix

$$B = \begin{bmatrix} 10 & 15 \\ 5 & 6 \end{bmatrix},$$

obtained from the matrix A by multiplying its first row by 5, det $B = -15 = 5(-3) = 5$ det A.

6.2.5 Property If a matrix B is obtained from a matrix A by adding to a row (or a column) of A a multiple of another row (or another column) of A, then det $B = \det A$.
For example, if

$$A = \begin{bmatrix} 2 & 3 \\ 5 & 6 \end{bmatrix},$$

then det $A = -3$, and for the matrix

$$B = \begin{bmatrix} 2 & 3 \\ 11 & 15 \end{bmatrix},$$

obtained from the matrix A by adding to its second row 3 times the first row, det $B = -3 = \det A$.

6.2.6 Property det $A = \det A^T$.
For example, if

$$A = \begin{bmatrix} 2 & 3 \\ 5 & 6 \end{bmatrix},$$

then det $A = -3$, and for the matrix

$$B = \begin{bmatrix} 2 & 5 \\ 3 & 6 \end{bmatrix},$$

obtained from the matrix A by taking its transpose, det $B = -3 = \det A$.

6.2.7 Property The determinant of a triangular matrix is the product of its diagonal entries.
For example, if

$$A = \begin{bmatrix} 3 & 5 & 8 \\ 0 & 2 & 7 \\ 0 & 0 & 5 \end{bmatrix},$$

then det $A = (3)(2)(5) = 30$.

6.2.8 Property $\det(AB) = (\det A)(\det B)$.

6.2.9 Property An $n \times n$ matrix A is invertible if and only if det $A \neq 0$, if and only if rank $A = n$.

6.2.10 Property An $n \times n$ homogeneous LS $Ax = 0$ has a nontrivial solution if and only if det $A = 0$.
In view of Properties 6.2.2, 6.2.4, 6.2.5, and 6.2.7, a practical method to find the determinant of a matrix is given below. This method is particularly useful when the entries of a matrix are numbers.

Reduce the matrix to row echelon form. For each elementary row operation performed, multiply the resulting matrix by a suitable number, namely,

i. -1 if "an interchange of row" is performed (Property 6.2.2),

ii. $\frac{1}{\alpha}$ if "a certain row is multiplied by $\alpha \neq 0$" (Property 6.2.4),

iii. 1 if "to some row a scalar multiple of another row is added" (Property 6.2.5).

Then the value of the determinant of the original matrix is the product of the diagonal entries of the row echelon matrix multiplied by the numbers obtained in (i), (ii), and (iii).

6.2.11
EXAMPLES

a. Compute

$$\det \begin{bmatrix} 0 & 1 & 2 \\ 2 & 1 & 1 \\ -1 & 3 & 1 \end{bmatrix}.$$

$$\det \begin{bmatrix} 0 & 1 & 2 \\ 2 & 1 & 1 \\ -1 & 3 & 1 \end{bmatrix} \overset{R_1 \longleftrightarrow R_2}{=\!=\!=} (-1) \det \begin{bmatrix} 2 & 1 & 1 \\ 0 & 1 & 2 \\ -1 & 3 & 1 \end{bmatrix}$$

(Note the multiplication by -1, in view of Property 6.2.2.)

$$\overset{\frac{1}{2} R_1}{=\!=\!=} (-1)(2) \det \begin{bmatrix} 1 & \frac{1}{2} & \frac{1}{2} \\ 0 & 1 & 2 \\ -1 & 3 & 1 \end{bmatrix}$$

(Note the multiplication by 2, in view of Property 6.2.4.)

$$\overset{R_3 + (1) R_1}{=\!=\!=} (-1)(2)(1) \det \begin{bmatrix} 1 & \frac{1}{2} & \frac{1}{2} \\ 0 & 1 & 2 \\ 0 & \frac{7}{2} & \frac{3}{2} \end{bmatrix}$$

$$\overset{R_3 + (-\frac{7}{2}) R_2}{=\!=\!=} (-1)(2)(1)(1) \det \begin{bmatrix} 1 & \frac{1}{2} & \frac{1}{2} \\ 0 & 1 & 2 \\ 0 & 0 & -\frac{11}{2} \end{bmatrix}$$

$$= (-2)(-11/2) = 11.$$

b. 3×3 Vandermonde determinant:

$$V_3 = \det \begin{bmatrix} 1 & x_1 & x_1^2 \\ 1 & x_2 & x_2^2 \\ 1 & x_3 & x_3^2 \end{bmatrix} = (x_2 - x_1)(x_3 - x_1)(x_3 - x_2).$$

Multiply column 2 by x_1 and subtract from column 3, and multiply column 1 by x_1 and subtract from column 2 to get

$$V_3 = \det \begin{bmatrix} 1 & 0 & 0 \\ 1 & x_2 - x_1 & x_2(x_2 - x_1) \\ 1 & x_3 - x_1 & x_3(x_3 - x_1) \end{bmatrix} \quad \text{(by Property 6.2.5 of the determinant)}$$

$$= \det \begin{bmatrix} x_2 - x_1 & x_2(x_2 - x_1) \\ x_3 - x_1 & x_3(x_3 - x_1) \end{bmatrix} \quad \text{(by the definition of } 3 \times 3 \text{ determinant)}$$

$$= (x_2 - x_1)(x_3 - x_1) \det \begin{bmatrix} 1 & x_2 \\ 1 & x_3 \end{bmatrix} \quad \text{(by Property 6.2.4 of the determinant)}$$

$$= (x_2 - x_1)(x_3 - x_1)(x_3 - x_2) \quad \text{(by the definition of } 2 \times 2 \text{ determinant).}$$

■

6.2   Exercises

Compute the determinants of the matrices in Exercises 1–6 by reducing them to row echelon form (if not already in a lower or upper triangular form).

1. $\begin{bmatrix} 3 & 0 & 2 \\ -1 & 5 & 0 \\ 1 & 9 & 6 \end{bmatrix}$

2. $\begin{bmatrix} 1 & 2 & -1 \\ 0 & 1 & 0 \\ 2 & 6 & 0 \end{bmatrix}$

 3. (Drill 6.2) $\begin{bmatrix} 2 & 0 & -1 & 3 \\ 4 & 0 & 1 & -1 \\ -3 & 1 & 0 & 1 \\ 1 & 4 & 1 & 1 \end{bmatrix}$

4. $\begin{bmatrix} 2 & 2 \\ 4 & 3 \end{bmatrix}$

5. $\begin{bmatrix} 1 & 2 & 3 \\ 4 & 5 & 6 \\ 7 & 8 & 9 \end{bmatrix}$

6. Exercises 1–5 in Section 6.1.

7. Show that

$$\det \begin{bmatrix} 1 & x_1 & x_1^2 & x_1^3 \\ 1 & x_2 & x_2^2 & x_2^3 \\ 1 & x_3 & x_3^2 & x_3^3 \\ 1 & x_4 & x_4^2 & x_4^3 \end{bmatrix} = (x_2 - x_1)(x_3 - x_1)(x_3 - x_2)(x_4 - x_1)(x_4 - x_2)(x_4 - x_3).$$

 8. (Drill 6.3) Verify the following properties of the determinant of a matrix by choosing a random matrix A.

    a. If the rows and columns of a matrix are interchanged, then the sign of the determinant changes.

    b. If the matrix B is obtained from the matrix A by multiplying every element in one row or one column by some number x, then $\det B = x \det A$.

    c. Choosing A to be triangular, verify that $\det A$ is the product of the entries on the diagonal.

    d. Verify that $\det A = \det A^T$.

 9. (Drill 6.4) Find the determinant of the following matrix by first reducing it to row echelon form:

$$A = \begin{bmatrix} 4 & 5 & 0 & 1 & 0 \\ 0 & 0 & 0 & 0 & 1 \\ 4 & 1 & 8 & 2 & 0 \\ 1 & 0 & 0 & 1 & 0 \\ 4 & 8 & 0 & 1 & 0 \end{bmatrix}.$$

 10. (Drill 6.5) Find the values of x such that the determinant of the matrix $A - xI$ is zero, where

$$A = \begin{bmatrix} 0 & -3 & 4 \\ 0 & 5 & 0 \\ 1 & -2 & 0 \end{bmatrix}$$

and I is the 3×3 identity matrix.

 11. (Drill 6.6) Verify the theorem that $\det(AB) = \det A \cdot \det B$, by taking random square matrices of the same size.

 12. (Drill 6.7) By taking a random matrix A, verify whether the following inequalities hold.

$$\det A \leqslant \prod_{j=1}^{n} \sqrt{\sum_{i=1}^{n} a_{ij}^2} \qquad \text{and} \qquad \det A \leqslant \sum_{i=1}^{n} \sqrt{\prod_{j=1}^{n} a_{ij}^2}.$$

6.3 Cofactors and Inverse of a Matrix

6.3.1
DEFINITION

Cofactor
If $A = (a_{ij})$ is an $n \times n$ matrix, then the *cofactor* of any (p, q) entry a_{pq} is defined to be $(-1)^{p+q} \det$ [the $(n-1) \times (n-1)$ matrix obtained by deleting the pth row and qth column of A and denoted by A_{pq}].

6.3.2
EXAMPLE

For

$$A = \begin{bmatrix} 1 & 2 & 1 \\ 3 & 4 & 5 \\ 6 & 0 & 1 \end{bmatrix},$$

the cofactor A_{23} of the $(2, 3)$ entry 5 is

$$(-1)^{2+3} \det \begin{bmatrix} 1 & 2 \\ 6 & 0 \end{bmatrix} = 12.$$

■

The next theorem, stated without proof, provides a method to evaluate the determinant of a matrix in terms of cofactors.

6.3.3 **THEOREM**	Let $A = (a_{ij})$ be an $n \times n$ matrix. Then $$\det A = a_{i1}A_{i1} + a_{i2}A_{i2} + \cdots + a_{in}A_{in}. \qquad (1)$$ Also, $$\det A = a_{1j}A_{1j} + a_{2j}A_{2j} + \cdots + a_{nj}A_{nj}, \qquad (2)$$ where $i, j = 1, 2, \ldots, n$. Formula 1 is called the expansion of $\det A$ by the ith row, and formula 2 is called the expansion of $\det A$ by the jth column.

The next theorem (again stated without proof) provides a method to compute the inverse of a matrix.

6.3.4 **THEOREM**	If $A = (a_{ij})$ is an $n \times n$ matrix such that $\det A \neq 0$, then $$A^{-1} = \frac{1}{\det A} \begin{bmatrix} A_{11} & A_{12} & \ldots & A_{1n} \\ A_{21} & A_{22} & \ldots & A_{2n} \\ \vdots & \vdots & & \vdots \\ A_{n1} & A_{n2} & \ldots & A_{nn} \end{bmatrix}^T.$$

6.3.5 **DEFINITION**	**Adjoint of a Matrix** The matrix on the right-hand side of the formula above for A^{-1} without the factor $\frac{1}{\det A}$ is called the *adjoint* of A and written adj(A), that is, $$\text{adj}(A) = \begin{bmatrix} A_{11} & A_{12} & \ldots & A_{1n} \\ A_{21} & A_{22} & \ldots & A_{2n} \\ \vdots & \vdots & & \vdots \\ A_{n1} & A_{n2} & \ldots & A_{nn} \end{bmatrix}^T.$$

We observe that the following is an equivalent version of Theorem 6.3.4.

6.3.6 **THEOREM**	$A(\mathrm{adj}\,A) = (\det A)I = (\mathrm{adj}\,A)A$, where I is the identity matrix of the same order as that of A.

6.3.7
EXAMPLE

Let

$$A = \begin{bmatrix} 1 & 2 \\ 3 & 4 \end{bmatrix}.$$

Then

$$\det A = (1)(4) - (2)(3) = -2 \neq 0,$$
$$A_{11} = (-1)^{1+1}4 = 4 \, , \, A_{12} = (-1)^{1+2}3 = -3,$$
$$A_{21} = (-1)^{2-1}2 = -2 \, , \, A_{22} = (-1)^{2+2}1 = 1.$$

So

$$A^{-1} = -\frac{1}{2}\begin{bmatrix} 4 & -3 \\ -2 & 1 \end{bmatrix}^{T} = -\frac{1}{2}\begin{bmatrix} 4 & -2 \\ -3 & 1 \end{bmatrix}.$$

Verification:

$$AA^{-1} = \begin{bmatrix} 1 & 2 \\ 3 & 4 \end{bmatrix}\left(-\frac{1}{2}\begin{bmatrix} 4 & -2 \\ -3 & 1 \end{bmatrix}\right)$$

$$= -\frac{1}{2}\begin{bmatrix} 1 & 2 \\ 3 & 4 \end{bmatrix}\begin{bmatrix} 4 & -2 \\ -3 & 1 \end{bmatrix}$$

$$= -\frac{1}{2}\begin{bmatrix} -2 & 0 \\ 0 & -2 \end{bmatrix} = \begin{bmatrix} 1 & 0 \\ 0 & 1 \end{bmatrix}.$$

So $AA^{-1} =$ identity matrix, as expected.   ■

6.3 Exercises

1. Find A^{-1}, if it exists, by computing cofactors and determinants:

a. $A = \begin{bmatrix} 1 & 1 \\ 2 & 2 \end{bmatrix}.$

b. $A = \begin{bmatrix} -1 & 0 \\ 0 & 1 \end{bmatrix}.$

c. $A = \begin{bmatrix} 2 & 3 \\ 1 & 1 \end{bmatrix}.$

d. $A = \begin{bmatrix} 1 & 2 & 3 \\ 4 & 5 & 6 \\ 7 & 8 & 9 \end{bmatrix}.$

e. $A = \begin{bmatrix} 1 & 0 & 1 \\ 0 & -1 & 1 \\ 2 & 3 & 4 \end{bmatrix}$.

f. $A = \begin{bmatrix} 0 & 1 & 2 \\ 0 & -1 & 1 \\ 1 & 5 & -4 \end{bmatrix}$.

g. $A = \begin{bmatrix} 2 & 1 & -1 & 2 \\ 3 & 0 & 0 & 1 \\ 2 & 1 & 2 & 0 \\ 3 & 1 & 1 & 2 \end{bmatrix}$.

 h. (Drill 6.8) $A = \begin{bmatrix} 1 & -1 & 1 & 2 \\ 1 & 0 & 1 & 3 \\ 0 & 0 & 2 & 4 \\ 1 & 1 & -1 & 1 \end{bmatrix}$.

2. Show that a matrix that has a zero row or a zero column has no inverse.

 3. (Drill 6.9) Find a and b such that the matrix

$$A = \begin{bmatrix} a & 1 & 1 & 1 \\ 1 & b & 2 & -1 \\ 2 & 2 & 3 & 1 \\ -1 & 1 & 1 & 1 \end{bmatrix}$$

is invertible. (Hint: Use Property 9, Section 6.2.)

6.4 Cramer's Rule

Theorem 6.3.3 has a useful application, particularly for symbolic matrices, in solving an $n \times n$ linear system of equations whose coefficient matrix is invertible.

6.4.1
THEOREM

Cramer's Rule

If $A = (a_{ij})$ is an invertible $n \times n$ matrix and

$$b = \begin{bmatrix} b_1 \\ \vdots \\ b_n \end{bmatrix},$$

then the solution

$$x = \begin{bmatrix} x_1 \\ x_2 \\ \vdots \\ x_n \end{bmatrix} \text{ of } Ax = b$$

is given by

$$x_1 = \frac{1}{\det A} \det \begin{bmatrix} b_1 & a_{12} & \cdots & a_{1n} \\ b_2 & a_{22} & \cdots & a_{2n} \\ \vdots & \vdots & & \vdots \\ b_n & a_{n2} & \cdots & a_{nn} \end{bmatrix},$$

where the matrix is obtained by replacing the first column of A by b;

$$x_2 = \frac{1}{\det A} \det \begin{bmatrix} a_{11} & b_1 & a_{13} & \cdots & a_{1n} \\ a_{21} & b_2 & a_{23} & \cdots & a_{2n} \\ \vdots & \vdots & \vdots & & \vdots \\ a_{n1} & b_n & a_{n3} & \cdots & a_{nn} \end{bmatrix},$$

where the matrix is obtained by replacing the second column of A by b; and so on to

$$x_n = \frac{1}{\det A} \det \begin{bmatrix} a_{11} & a_{12} & \cdots & a_{1,n-1} & b_1 \\ a_{21} & a_{22} & \cdots & a_{2,n-1} & b_2 \\ \vdots & \vdots & & \vdots & \vdots \\ a_{n1} & a_{n2} & \cdots & a_{n,n-1} & b_n \end{bmatrix},$$

where the matrix is obtained by replacing the nth column of A by b.

6.4.2 EXAMPLE Consider the linear system

$$2x_1 + 3x_2 - x_3 = 2,$$
$$x_1 + 2x_2 + x_3 = -1,$$
$$2x_1 + x_2 - 6x_3 = 4.$$

Here,

$$A = \begin{bmatrix} 2 & 3 & -1 \\ 1 & 2 & 1 \\ 2 & 1 & -6 \end{bmatrix}, \qquad b = \begin{bmatrix} 2 \\ -1 \\ 4 \end{bmatrix}, \qquad \text{and} \qquad \det A = 1.$$

So

$$x_1 = \det \begin{bmatrix} 2 & 3 & -1 \\ -1 & 2 & 1 \\ 4 & 1 & -6 \end{bmatrix} = -23,$$

$$x_2 = \det \begin{bmatrix} 2 & 2 & -1 \\ 1 & -1 & 1 \\ 2 & 4 & -6 \end{bmatrix} = 14,$$

$$x_3 = \det \begin{bmatrix} 2 & 3 & 2 \\ 1 & 2 & -1 \\ 2 & 1 & 4 \end{bmatrix} = -6.$$

6.4 Exercises

Solve each of the following linear systems using Cramer's rule:

1. $5x - y = 9$,
$3x - 3y + z = 20$,
$x + y + z = 2$.

2. $2x + y - z = 5$,
$4x - 2y - 4z = 10$,
$x - y + z = -6$.

3. $3x_1 - x_2 + x_3 = 1$,
$x_1 - 2x_2 + x_3 = 2$,
$2x_1 + x_2 + 3x_3 = 0$.

4. $x + y - z = 5$,
$2x + y - 3z = 10$,
$3x + 4y + 5z = 1$.

5. $2x - y + z = 7$,
$3x + 2y + z = 3$,
$2x - y + 3z = 9$.

  6. (Drill 6.10) $2x_1 + 3x_2 + x_3 - x_4 = 2$,
$x_1 + 2x_2 + 5x_3 + 3x_4 = 5$,
$-x_1 + 3x_3 + x_4 = 1$,
$x_1 - 2x_2 + x_3 = -2$.

6.5 Chapter Review Questions and Projects

1. What is the value of the determinant of a triangular matrix? (Section 6.2)

2. What is the value of the determinant of a square matrix with two identical rows or columns? (Section 6.2)

3. How does the value of the determinant change when two rows or columns are interchanged? (Section 6.2)

4. Perform an elementary row operation $R_i + cR_j$ $(i \neq j)$ on a square matrix A. What is the relation between the determinant of the new matrix and det A? (Section 6.2)

5. If each row of an $n \times n$ matrix A is multiplied by a scalar c, write down the formula for the determinant of the new matrix in terms of det A. (Section 6.2)

6. Suppose A is invertible matrix. Show that

$$\det(A^{-1}) = \frac{1}{\det A}.$$

(Hint: Use product theorem of determinants.) (Section 6.2)

7. Suppose A is a square matrix such that $A^2 = 0$. Show det $A = 0$. (Section 6.2)

8. Suppose A is a matrix with $A^2 = A$. Show that $\det A = 0$ or 1. (Section 6.2)

9. Evaluate the determinant of each of the following matrices by expanding in two ways, using the indicated rows and columns. Observe that you get the same answer both ways. (Sections 6.1 and 6.2)

a. $\begin{bmatrix} 0 & 2 & 5 \\ 1 & 2 & 3 \\ 4 & 1 & 2 \end{bmatrix}$ (1st row and 1st column)

b. $\begin{bmatrix} 1 & -1 & 0 \\ 0 & 2 & 1 \\ 1 & 2 & 3 \end{bmatrix}$ (2nd row and 3rd column)

10. Solve for x: (Section 6.1)

$$\det \begin{bmatrix} 1-x & -2 \\ -2 & 1-x \end{bmatrix} = 0$$

11. Evaluate the determinant of each of the following matrices by reducing it into row echelon form: (Section 6.2)

a. $\begin{bmatrix} 2 & 4 & 6 & -2 & 16 \\ 0 & 0 & 4 & 2 & -1 \\ 0 & -5 & 5 & 3 & 7 \\ 0 & 0 & 0 & 1 & 6 \\ 1 & 2 & 3 & -2 & -9 \end{bmatrix}$

b. $\begin{bmatrix} 1 & 1 & 1 & 1 \\ 1 & -i & -1 & i \\ 1 & -1 & 1 & -1 \\ 1 & i & -1 & -i \end{bmatrix}$

12. Determine whether the following matrices have inverses. If a matrix has an inverse, use the cofactor method to calculate it. (Sections 6.2 and 6.3)

a. $\begin{bmatrix} 1 & 2 \\ 3 & 4 \end{bmatrix}$

b. $\begin{bmatrix} 1 & 4 & -1 \\ 1 & 6 & -1 \\ 0 & 2 & 0 \end{bmatrix}$

c. $\begin{bmatrix} 1 & 3 & 1 \\ 2 & 5 & 1 \\ 1 & 2 & 3 \end{bmatrix}$

13. Write down the condition that must be satisfied to apply the Cramer's rule to solve the following linear system: (Section 6.4)

$$a_1 x + a_2 y = b_1,$$
$$a_1' x + a_2' y = b_2.$$

Chapter Projects

1. a. Let $A = (a_{ij})$ be an $n \times n$ matrix such that

$$a_{ij} = \begin{cases} -1, & \text{if } i > j; \\ 1, & \text{if } i \leqslant j. \end{cases}$$

Find the determinant of A for $n = 2, 3, 4$, and 5. Conjecture the value of det A for any n.

b. Suppose that the matrix $A = (a_{ij})$ is given by the definition $a_{ij} = -j$ if $j < i$ and 1 otherwise. Find the determinant of A for $n = 2, 3, 4$, and 5, and conjecture the value of det A in general.

2. A symmetric $n \times n$ matrix whose entries are real numbers is called positive definite if the determinant of the submatrix formed by the first r rows and the first r columns is positive for all values of $r = 1, 2, 3, \ldots, n$. Construct 2×2, 3×3, and 4×4 positive definite matrices, and verify the following properties of this class of matrices A.

a. $A = LDL^T$, where L is a lower triangular matrix with all entries on the diagonal equal to 1 and D is a diagonal matrix.

b. (Hadamard Inequality) (i) det $A \leqslant \prod_{i=1}^{n} a_{ii}$, (ii) det $A \leqslant \prod_{i=1}^{n} \sqrt{\left(\sum_{j=1}^{n} a_{ij}^2 \right)}$.

Can you experiment to determine when the inequality in (i) or in (ii) would become an equality?

Key Words

Determinant of a Square Matrix
Cofactor of an Entry
 of a Matrix
Adjoint of a Matrix
Cramer's Rule

Elementary Row Operations
Row Echelon Form
Triangular Matrix
Inverse of a Matrix
Transpose of a Matrix

Key Phrases

■ The determinant changes its value when elementary row operations are performed.

■ The determinant of a triangular matrix (for example, of a matrix in row echelon form) is equal to the product of diagonal entries.

■ A square matrix is invertible if and only if its determinant is not zero, if and only if its rank is equal to its size.

7 Eigenvalue Problems

Introduction

In this chapter, we consider the concepts of eigenvalues and eigenvectors of a square matrix. The importance of eigenvalues and eigenvectors can hardly be overemphasized in view of their abundant applications in all sciences. Considering an $n \times n$ matrix A as a function acting on $\mathbf{R}^n$ to $\mathbf{R}^n$, suppose a nonzero vector x is transformed to a scalar multiple λ of x. Then x is called an eigenvector corresponding to the eigenvalue λ. For instance, if we want to find the points of relative maxima or relative minima of a quadratic function, say $ax_1^2 + 2hx_1x_2 + bx_2^2$, subject to the condition that the points lie on a unit circle $x_1^2 + x_2^2 = 1$, then these points are given by $Ax = \lambda x$, where

$$A = \left[\begin{array}{cc} a & h \\ h & b \end{array} \right], \qquad x = \left[\begin{array}{c} x_1 \\ x_2 \end{array} \right].$$

7.1 Eigenvalues and Eigenvectors

As usual, let F^n denote the n-dimensional vector space over F, consisting of all n-tuples of elements of F. Recall that the elements of F^n may be treated as row vectors ($1 \times n$ matrices) or column vectors ($n \times 1$ matrices). In this chapter, unless otherwise stated, all vectors will be column vectors.

7.1.1
DEFINITION

Eigenvalue and Eigenvector
Let A be an $n \times n$ matrix. A number λ is called an *eigenvalue* of A if there exists a nonzero vector $v \in F^n$ such that $Av = \lambda v$. The vector v is then called an *eigenvector* of A corresponding to the eigenvalue λ.

For example, let $A = \begin{bmatrix} 1 & 2 \\ 4 & 3 \end{bmatrix}$. Then taking $\lambda = 5$ and $v = \begin{bmatrix} 1 \\ 2 \end{bmatrix}$, we have

$$Av = \begin{bmatrix} 1 & 2 \\ 4 & 3 \end{bmatrix} \begin{bmatrix} 1 \\ 2 \end{bmatrix} = \begin{bmatrix} 5 \\ 10 \end{bmatrix} = 5 \begin{bmatrix} 1 \\ 2 \end{bmatrix} = \lambda v.$$

So 5 is an eigenvalue of A, and the vector $\begin{bmatrix} 1 \\ 2 \end{bmatrix}$ is an eigenvector of A corresponding to the eigenvalue 5.

We emphasize that eigenvalues and eigenvectors are defined for square matrices only. The problem of finding all the eigenvalues of a given square matrix and all eigenvectors corresponding to these eigenvalues is referred to as the *eigenvalue problem*.

7.1.2
DEFINITION

Spectrum

Let A be an $n \times n$ matrix. The set of all eigenvalues of A is called the *spectrum* of A.

The following result gives a necessary and sufficient condition for a number λ to be an eigenvalue of A.

7.1.3
THEOREM

(Theorem 1, Section 7.6) Let A be an $n \times n$ matrix. A number λ is an eigenvalue of A if and only if

$$\det(A - \lambda I) = 0,$$

where I denotes the $n \times n$ identity matrix.

7.2 Characteristic Polynomial

By a polynomial of degree k in x, we mean an expression of the form

$$a_k x^k + a_{k-1} x^{k-1} + \cdots + a_1 x + a_0,$$

where $a_0, a_1, \ldots, a_k \in F$ and $a_k \neq 0$. (The symbol x is referred to as the *indeterminate*, and $a_0, a_1, \ldots, a_k$ are called *coefficients*.) For example, $5x^3 + 2x^2 - 3x + 6$ is a polynomial of degree 3 in x.

Any number c can also be considered as a polynomial in x and is referred to as a *constant polynomial*. If $c \neq 0$, then it is a polynomial of degree 0. If $c = 0$, it is called the zero polynomial, and its degree is undefined.

We will use the notation $p(x)$, $f(x)$, and so on for any polynomial in x. Two polynomials $f(x)$ and $g(x)$ can be added or multiplied, and the result is another polynomial. (The rules for adding and multiplying two polynomials are taught in high school, and the reader is no doubt familiar with them.)

Given a polynomial

$$p(x) = a_k x^k + \cdots + a_1 x + a_0,$$

we can replace x everywhere by a given number λ and so obtain

$$p(\lambda) = a_k\lambda^k + \cdots + a_1\lambda + a_0.$$

The right-hand side in this equation is clearly a number. This number $p(\lambda)$ is called the value of the polynomial $p(x)$ at $x = \lambda$.

7.2.1
DEFINITION

Root of a Polynomial

Let $p(x)$ be a polynomial in x. A number λ is called a *root* of $p(x)$ if $p(\lambda) = 0$. In other words, a root of the polynomial $p(x)$ is simply a solution of the equation $p(x) = 0$.

Let A be an $n \times n$ matrix. For any given positive integer r, we can form the product $AA \cdots A$ (r factors), which we denote by A^r. Clearly, A^r is again an $n \times n$ matrix. Given a polynomial

$$p(x) = a_k x^k + \cdots + a_1 x + a_0,$$

we define

$$p(A) = a_k A^k + \cdots + a_1 A + a_0 I.$$

Notice that we have replaced x by A throughout and have also multiplied the constant term a_0 by the $n \times n$ identity matrix I. Now each term on the right-hand side is an $n \times n$ matrix, and therefore $p(A)$ is an $n \times n$ matrix.

We have so far considered matrices in which the entries are elements of F. They are called matrices over F. Likewise, we have considered polynomials in which the coefficients are elements of F. They are called polynomials over F. We shall now generalize these concepts and consider matrices in which the entries are polynomials and also consider polynomials in which the coefficients are matrices. For example,

$$A = \begin{bmatrix} 2x^2 - 5x + 6 & 2 & -x + 3 \\ x - 2 & x & x^2 \\ 5 & x^2 & -x \end{bmatrix}$$

is a matrix in which the entries are polynomials in x over F. (Recall that any constant c is also a polynomial in x.) On the other hand,

$$p(x) = \begin{bmatrix} 2 & 1 \\ -1 & 3 \end{bmatrix} x^3 + \begin{bmatrix} 1 & -1 \\ 2 & 1 \end{bmatrix} x^2 + \begin{bmatrix} 3 & 2 \\ 1 & 4 \end{bmatrix} x + \begin{bmatrix} 1 & -1 \\ 0 & 1 \end{bmatrix}$$

is a polynomial in which the coefficients are 2×2 matrices over F.

Interestingly, these two concepts (a matrix with polynomial entries and a polynomial with matrix coefficients) are equivalent. Each can be expressed in the form of the other. For example, the matrix A written above can be expressed as

$$A = \begin{bmatrix} 2x^2 & 0 & 0 \\ 0 & 0 & x^2 \\ 0 & x^2 & 0 \end{bmatrix} + \begin{bmatrix} -5x & 0 & -x \\ x & x & 0 \\ 0 & 0 & -x \end{bmatrix} + \begin{bmatrix} 6 & 2 & 3 \\ -2 & 0 & 0 \\ 5 & 0 & 0 \end{bmatrix}$$

$$= \begin{bmatrix} 2 & 0 & 0 \\ 0 & 0 & 1 \\ 0 & 1 & 0 \end{bmatrix} x^2 + \begin{bmatrix} -5 & 0 & -1 \\ 1 & 1 & 0 \\ 0 & 0 & -1 \end{bmatrix} x + \begin{bmatrix} 6 & 2 & 3 \\ -2 & 0 & 0 \\ 5 & 0 & 0 \end{bmatrix}.$$

Likewise, for the polynomial $p(x)$ written above, we have

$$p(x) = \begin{bmatrix} 2x^3 & x^3 \\ -x^3 & 3x^3 \end{bmatrix} + \begin{bmatrix} x^2 & -x^2 \\ 2x^2 & x^2 \end{bmatrix} + \begin{bmatrix} 3x & 2x \\ x & 4x \end{bmatrix} + \begin{bmatrix} 1 & -1 \\ 0 & 1 \end{bmatrix}$$

$$= \begin{bmatrix} 2x^3 + x^2 + 3x + 1 & x^3 - x^2 + 2x - 1 \\ -x^3 + 2x^2 + x & 3x^3 + x^2 + 4x + 1 \end{bmatrix}.$$

Let A be an $n \times n$ matrix whose entries are polynomials in x. We evaluate the determinant of A in the same manner as we did earlier for an $n \times n$ matrix over F. For example, let

$$A = \begin{bmatrix} 2x^2 & -x & x+2 \\ -2 & x & x-1 \\ x^2 & -1 & 2x+3 \end{bmatrix}.$$

Then

$$\det A = 2x^2 \det \begin{bmatrix} x & x-1 \\ -1 & 2x+3 \end{bmatrix} + x \det \begin{bmatrix} -2 & x-1 \\ x^2 & 2x+3 \end{bmatrix} + (x+2) \det \begin{bmatrix} -2 & x \\ x^2 & -1 \end{bmatrix}$$

$$= 2x^2(2x^2 + 4x - 1) + x(-x^3 + x^2 - 4x - 6) + (x+2)(-x^3 + 2)$$

$$= 2x^4 + 7x^3 - 6x^2 - 4x + 4.$$

Thus we see that if A is an $n \times n$ matrix whose entries are polynomials in x, then $\det A$ is also a polynomial in x.

7.2.2
DEFINITION

Characteristic Polynomial

Let $A = (a_{ij})$ be an $n \times n$ matrix. Then the polynomial

$$\det(A - xI_n) = \det \begin{bmatrix} a_{11} - x & a_{12} & \cdots & a_{1n} \\ a_{21} & a_{22} - x & \cdots & a_{2n} \\ \vdots & \vdots & & \vdots \\ a_{n1} & a_{n2} & \cdots & a_{nn} - x \end{bmatrix}$$

is called the *characteristic polynomial* of A.

We will write the characteristic polynomial of A as $C_A(x)$. It follows from the definition of an eigenvalue that a number λ is an eigenvalue of A if and only if λ is a root of $C_A(x)$.

We note that if A is an $n \times n$ matrix, then $C_A(x)$ is a polynomial of degree n with leading coefficient (the coefficient of x^n) $(-1)^n$; also, the constant term of $C_A(x)$ is $\det A$, which is obtained by putting $x = 0$ in $C_A(x) = \det(A - xI)$.

If $C_A(x) = \det(A - xI)$ is the characteristic polynomial of A, then the equation $\det(A - xI) = 0$ (or equivalently, $\det(xI - A) = 0$) is called the *characteristic equation* of A.

> **7.2.3**
> **DEFINITION**
>
> ### Multiplicity of an Eigenvalue
> If an eigenvalue λ occurs k times as a root of the characteristic polynomial $C_A(x)$, then k is called the *multiplicity* of the eigenvalue λ.

7.2.4
EXAMPLES

In the problems worked out below, we find eigenvalues and eigenvectors by using their definitions. In the next section, we provide an alternative method that does not use determinants, and indeed one can study the concept of eigenvalues after *Chapter 3*. We redo these same examples so that the reader can compare two solutions.

a. Let

$$A = \begin{bmatrix} 3 & -1 & -1 \\ -12 & 0 & 5 \\ 4 & -2 & -1 \end{bmatrix}.$$

Then

$$\det(A - xI) = \det \begin{bmatrix} 3-x & -1 & -1 \\ -12 & -x & 5 \\ 4 & -2 & -1-x \end{bmatrix}$$

$$= (3-x)[(x^2 + x) + 10] + 1[(12x + 12) - 20] - 1(24 + 4x)$$

(expanding by the first row)

$$= -x^3 + 2x^2 + x - 2$$

$$= -(x+1)(x-1)(x-2).$$

So the eigenvalues of A are -1, 1, and 2. To find an eigenvector corresponding to eigenvalue -1, we solve the LS $(A - (-1)I)x = 0$, that is,

$$\begin{bmatrix} 4 & -1 & -1 \\ -12 & 1 & 5 \\ 4 & -2 & 0 \end{bmatrix} \begin{bmatrix} x_1 \\ x_2 \\ x_3 \end{bmatrix} = \begin{bmatrix} 0 \\ 0 \\ 0 \end{bmatrix}.$$

We now reduce the coefficient matrix to row echelon form:

$$\begin{bmatrix} 4 & -1 & -1 \\ -12 & 1 & 5 \\ 4 & -2 & 0 \end{bmatrix} \xrightarrow[R_3+(-1)R_1]{R_2+3R_1} \begin{bmatrix} 4 & -1 & -1 \\ 0 & -2 & 2 \\ 0 & -1 & 1 \end{bmatrix} \xrightarrow{R_3+(-\frac{1}{2})R_2} \begin{bmatrix} 4 & -1 & -1 \\ 0 & -2 & 2 \\ 0 & 0 & 0 \end{bmatrix}.$$

Thus the linear system becomes

$$4x_1 - x_2 - x_3 = 0,$$
$$-2x_2 + 2x_3 = 0.$$

Solving for x_1, x_2, and x_3 we get the set of solutions $x_3 = t, x_2 = t, x_1 = \frac{1}{2}t$, where t is arbitrary, which we may choose to be 1. Thus

$$x = \begin{bmatrix} \frac{1}{2} \\ 1 \\ 1 \end{bmatrix}$$

is an eigenvector corresponding to eigenvalue -1. Note that any other eigenvector corresponding to eigenvalue -1 is a multiple of $\begin{bmatrix} \frac{1}{2} \\ 1 \\ 1 \end{bmatrix}$. Thus $\begin{bmatrix} \frac{1}{2} \\ 1 \\ 1 \end{bmatrix}$ by itself forms a maximal set of linearly independent eigenvectors corresponding to eigenvalue -1.

Similarly, we obtain $\begin{bmatrix} 3 \\ -1 \\ 7 \end{bmatrix}$ as an eigenvector corresponding to eigenvalue 1 and

$\begin{bmatrix} 1 \\ -1 \\ 2 \end{bmatrix}$ as an eigenvector corresponding to eigenvalue 2.

b. Let

$$A = \begin{bmatrix} 0 & -3 & 4 \\ 0 & 5 & 0 \\ 1 & -2 & 0 \end{bmatrix}.$$

Then

$$C_A(x) = \det(A - xI)$$

$$= \det \begin{bmatrix} -x & -3 & 4 \\ 0 & 5-x & 0 \\ 1 & -2 & -x \end{bmatrix}$$

$$= -x^3 + 5x^2 + 4x - 20$$

$$= -(x+2)(x-2)(x-5).$$

Thus the eigenvalues are -2, 2, and 5. Let $x = \begin{bmatrix} a \\ b \\ c \end{bmatrix}$ be an eigenvector corresponding to eigenvalue -2. Then $(A + 2I)x = 0$ yields

$$\begin{bmatrix} 2 & -3 & 4 \\ 0 & 7 & 0 \\ 1 & -2 & 2 \end{bmatrix} \begin{bmatrix} a \\ b \\ c \end{bmatrix} = \begin{bmatrix} 0 \\ 0 \\ 0 \end{bmatrix}.$$

To solve for a, b, and c, we reduce the coefficient matrix to echelon form. Now,

$$\begin{bmatrix} 2 & -3 & 4 \\ 0 & 7 & 0 \\ 1 & -2 & 2 \end{bmatrix} \xrightarrow{R_1 \longleftrightarrow R_3} \begin{bmatrix} 1 & -2 & 2 \\ 0 & 7 & 0 \\ 2 & -3 & 4 \end{bmatrix} \xrightarrow{R_3 + (-2)R_1} \begin{bmatrix} 1 & -2 & 2 \\ 0 & 7 & 0 \\ 0 & 1 & 0 \end{bmatrix}$$

$$\xrightarrow{R_3 + (-\frac{1}{7})R_2} \begin{bmatrix} 1 & -2 & 2 \\ 0 & 7 & 0 \\ 0 & 0 & 0 \end{bmatrix}.$$

The linear system is then transformed to

$$a - 2b + 2c = 0,$$
$$7b = 0.$$

Therefore $b = 0$, and $a + 2c = 0$. Put $c = t$. Then $a = -2t$. Thus the eigenvectors corresponding to -2 are

$$\begin{bmatrix} -2t \\ 0 \\ t \end{bmatrix} = t \begin{bmatrix} -2 \\ 0 \\ 1 \end{bmatrix},$$

where t is an arbitrary nonzero number. This implies that $\begin{bmatrix} -2 \\ 0 \\ 1 \end{bmatrix}$ is an eigenvector

corresponding to eigenvalue -2 and that any other eigenvector is a scalar multiple

of $\begin{bmatrix} -2 \\ 0 \\ 1 \end{bmatrix}$.

Similar calculations yield that any eigenvector corresponding to 2 is a scalar multiple

of $\begin{bmatrix} 2 \\ 0 \\ 1 \end{bmatrix}$ and any eigenvector corresponding to 5 is a scalar multiple of $\begin{bmatrix} -23 \\ 21 \\ -13 \end{bmatrix}$.

It can be checked that

$$\begin{bmatrix} -2 \\ 0 \\ 1 \end{bmatrix}, \quad \begin{bmatrix} 2 \\ 0 \\ 1 \end{bmatrix}, \quad \text{and} \quad \begin{bmatrix} -23 \\ 21 \\ -13 \end{bmatrix}$$

are linearly independent.

c. Let v_1, v_2 be eigenvectors corresponding to distinct eigenvalues of a matrix A. Then for all $\alpha_1 \neq 0$, $\alpha_2 \neq 0$, $\alpha_1 v_1 + \alpha_2 v_2$ is not an eigenvector.

We will make use the fact that if λ_1 and λ_2 are distinct eigenvalues, then the corresponding eigenvectors v_1 and v_2 are linearly independent (Theorem 6, Section 7.6).

Suppose $A(\alpha_1 v_1 + \alpha_2 v_2) = k(\alpha_1 v_1 + \alpha_2 v_2)$, $k \in F$. This gives $\alpha_1 \lambda_1 v_1 + \alpha_2 \lambda_2 v_2 = k\alpha_1 v_1 + k\alpha_2 v_2$, that is, $\alpha_1 (\lambda_1 - k)v_1 + \alpha_2 (\lambda_2 - k)v_2 = 0$. Since v_1 and v_2 are linearly independent, this gives $\alpha_1 (\lambda_1 - k) = 0$ and $\alpha_2 (\lambda_2 - k) = 0$. Since $\alpha_1 \neq 0$ and $\alpha_2 \neq 0$, this gives $\lambda_1 = \lambda_2$, which is not true.

Later on in this chapter, we will describe another method without using determinants to find the characteristic polynomial, eigenvalues, and eigenvectors of a matrix. ∎

7.2.5 DEFINITION

Maximal Set of Linearly Independent Eigenvectors

A linearly independent set S of eigenvectors of a matrix A is said to be a *maximal set of linearly independent eigenvectors* if the number of vectors in any other set of linearly independent eigenvectors does not exceed the number of vectors in S.

7.2.6 Suppose $\lambda_1, \ldots, \lambda_k$ are distinct eigenvalues of an $n \times n$ matrix A. Corresponding to each
Remark eigenvalue λ_i, $i = 1, \ldots, k$, find a maximal set of linearly independent eigenvectors.
Then the set of all such eigenvectors obtained by considering each eigenvalue λ_i,
$i = 1, \ldots, k$, yields a maximal set of linearly independent eigenvectors of A.

7.2.7 By using the rank-nullity theorem (*Chapter 4*), the number of linearly independent
Remark eigenvectors corresponding to an eigenvalue λ is equal to $n - \mathrm{rank}(A - \lambda I)$, ($=$
the number of unknowns that can take arbitrary value), where A is the $n \times n$ matrix.

7.2 Exercises

1. Find the eigenvalues and corresponding eigenvectors for each of the following
   matrices:

   a. $\begin{bmatrix} 1 & 2 \\ -1 & 4 \end{bmatrix}$

   b. $\begin{bmatrix} 2 & 1 & 0 \\ 0 & 3 & 1 \\ 0 & 0 & 1 \end{bmatrix}$

   c. $\begin{bmatrix} 1 & 3 & 6 \\ 0 & 2 & -1 \\ 0 & 0 & 7 \end{bmatrix}$

    d. (Drill 7.1) $\begin{bmatrix} 1 & -3 & 3 \\ 0 & -1 & 2 \\ 0 & -3 & 4 \end{bmatrix}$

   e. (Drill 7.2) $\begin{bmatrix} 1 & -2 & -2 & -2 \\ -2 & 1 & -2 & -2 \\ -2 & -2 & 1 & -2 \\ -2 & -2 & -2 & 1 \end{bmatrix}$

2. Find a maximal set of linearly independent eigenvectors for each of the following
   matrices:

   a. $\begin{bmatrix} 2 & 2 \\ 3 & 1 \end{bmatrix}$

   b. $\begin{bmatrix} 1 & 2 & 1 \\ 1 & -1 & 1 \\ 2 & 0 & 1 \end{bmatrix}$

   c. (Drill 7.3) $\begin{bmatrix} 1 & 1 & 2 \\ -1 & 2 & 1 \\ 0 & 1 & 3 \end{bmatrix}$

d. (Drill 7.4) $\begin{bmatrix} 1 & 1 & 1 & 1 \\ 0 & 0 & 1 & 1 \\ 0 & 0 & 1 & 0 \\ 0 & 0 & 1 & 2 \end{bmatrix}$

e. $\begin{bmatrix} 4 & 1 & 1 & 2 \\ 0 & 4 & 1 & 1 \\ 0 & 0 & 1 & 0 \\ 0 & 0 & 0 & 1 \end{bmatrix}$

f. $\begin{bmatrix} 1 & -1 & 0 \\ -1 & 2 & -1 \\ 0 & -1 & 1 \end{bmatrix}$

3. If A is a matrix such that $A = A^2$, prove that 0 and 1 are the only possible distinct eigenvalues of A.

4. If A is a matrix with $A^2 = 0$, prove that 0 is the only distinct eigenvalue of A.

5. Show that if λ is an eigenvalue of A, then λ^2 is an eigenvalue of A^2.

6. Show that if A is a nonsingular matrix and λ is an eigenvalue of A, then $\frac{1}{\lambda}$ is an eigenvalue of A^{-1}.

7. Prove that any eigenvalue of A is also an eigenvalue of A^T.

8. Show that if A is a noninvertible matrix, then the number 0 must be one of the eigenvalues of A. (Use the fact that the constant term of $C_A(x)$ is det A.)

9. (Drill 7.5) Given the matrix

$$A = \begin{bmatrix} 1 & 1 & 2 \\ -1 & 2 & 1 \\ 0 & 1 & 3 \end{bmatrix},$$

find the eigenvalues of A, A^2, A^3, A^4, and A^{-1}, and find relations between the eigenvalues of A, A^n, and A^{-1}.

10. Prove that λ is an eigenvalue of AA^T if and only if λ is an eigenvalue of $A^T A$.

Calculating Eigenvalues and Eigenvectors (Another Approach) and the Cayley-Hamilton Theorem

7.3

We saw in the previous section that the eigenvalues of a square matrix A are the roots of the characteristic polynomial of A. So the eigenvalues are found by solving the characteristic equation $\det(A - xI) = 0$. To find the eigenvectors corresponding to an eigenvalue λ, we then solve the homogeneous LS $(A - \lambda I)x = 0$. We will now give another method for finding the eigenvalues and eigenvectors without using determinants. This method suggests that the eigenvalues can be introduced earlier in the course as has been recommended by the Linear Algebra Curriculum Committees. As a by-product of this method, we give a new and short proof of the Cayley-Hamilton Theorem.

A general description of this method for finding the eigenvalues and eigenvectors of a given $n \times n$ matrix A now follows.

We start with an arbitrary nonzero vector u, which is referred to as the seed vector. A convenient choice for the seed vector is

$$u = \begin{bmatrix} 1 \\ 0 \\ \vdots \\ 0 \end{bmatrix}.$$

We then compute $Au, A^2u, \ldots$ and check at each stage whether the vectors $u, Au, A^2u, \ldots$ are linearly independent. Since there cannot be more than n linearly independent vectors in F^n, we will arrive at some integer $r \leqslant n$ such that $u, Au, \ldots, A^{r-1}u$ are linearly independent but $u, Au, \ldots, A^ru$ are not linearly independent. So there exist scalars $\alpha_0, \alpha_1, \ldots, \alpha_r$ (not all zero) such that

$$\alpha_0 u + \alpha_1 Au + \cdots + \alpha_r A^r u = 0.$$

Now, α_r cannot be zero, for otherwise, $u, Au, \ldots, A^{r-1}u$ would be linearly dependent. Without loss of generality, we may take $\alpha_r = 1$. Thus we have

$$\alpha_0 u + \alpha_1 Au + \cdots + \alpha_{r-1} A^{r-1}u + A^r u = 0. \tag{1}$$

Equation (1) is called the first dependence equation. Since we can factor any polynomial over the field of complex numbers into linear factors, we write

$$(\alpha_0 I + \alpha_1 A + \cdots + \alpha_{r-1} A^{r-1} + A^r)u = (A - \lambda_1 I)(A - \lambda_2 I) \cdots (A - \lambda_r I)u.$$

Thus $(A - \lambda_1 I)(A - \lambda_2 I) \cdots (A - \lambda_r I)u = 0$. Setting $v_i = (A - \lambda_1 I) \cdots (A - \lambda_{i-1}I)(A - \lambda_{i+1}I) \cdots (A - \lambda_r I)u$ (note that the factor $(A - \lambda_i I)$ is omitted in defining v_i), we obtain that $(A - \lambda_i I)v_i = 0$. This gives that λ_i is an eigenvalue and v_i is an eigenvector corresponding to λ_i. This gives us r eigenvalues and corresponding eigenvectors. If $r = n$, we are done because then all the eigenvalues are the roots of the polynomial (called characteristic polynomial), given by

$$C_A(x) = \alpha_0 + \alpha_1 x + \cdots + \alpha_{n-1}x^{n-1} + x^n,$$

where $\alpha_0, \ldots, \alpha_{n-1}$ are the coefficients occurring in the dependence Equation (1) obtained above. If $r < n$, we introduce another seed vector v that is not linearly dependent on $u, Au, \ldots, A^{n-1}u$. A convenient choice for v is

$$\begin{bmatrix} 0 \\ 1 \\ \vdots \\ 0 \end{bmatrix}$$

in F^n. We next compute $Av, A^2v, \ldots$ until we arrive at some integer $s \leqslant n-r$ such that $u, Au, \ldots, A^{r-1}u, v, Av, \ldots, A^{s-1}v$ are linearly independent but $u, Au, \ldots, A^{r-1}u, v, Av, \ldots, A^s v$ are linearly dependent. Thus we get a second dependence equation:

$$\alpha_0' u + \alpha_1' Au + \cdots + \alpha_{r-1}' A^{r-1}u + \beta_0 v + \beta_1 Av + \cdots + \beta_{s-1}A^{s-1}v + A^s v = 0. \tag{2}$$

Write the second dependency relation (2) as

$$(\alpha_0' I + \alpha_1' A + \cdots + \alpha_{r-1}' A^{r-1})u + (\beta_0 I + \beta_1 A + \cdots + \beta_{s-1}A^{s-1} + A^s)v = 0,$$

and multiply both sides by the polynomial $\alpha_0 I + \alpha_1 A + \cdots + \alpha_{r-1} A^{r-1} + A^r$. Since any two polynomials in A commute, we get, by using the relation (1), that

$$(\alpha_0 + \alpha_1 A + \cdots + \alpha_{r-1} A^{r-1} + A^r)(\beta_0 + \beta_1 A + \cdots + \beta_{s-1} A^{s-1} + A^s)v = 0.$$

Factoring the polynomial $\beta_0 I + \beta_1 A + \cdots + \beta_{s-1} A^{s-1} + A^s$ into linear factors, we obtain as before additional s eigenvalues and corresponding eigenvectors. If $r + s = n$, then

$$C_A(x) = (\alpha_0 + \alpha_1 x + \cdots + \alpha_{r-1} x^{r-1} + x^r)(\beta_0 + \beta_1 x + \cdots + \beta_{s-1} x^{s-1} + x^s),$$

where $\alpha_0, \ldots, \alpha_{r-1}, \beta_0, \ldots, \beta_{s-1}$ are the coefficients in the first and second dependence equations (1) and (2). We continue likewise until we have obtained the factors whose product is a polynomial in A of degrees n. This procedure gives n eigenvalues and corresponding eigenvectors using seed vectors $u, v, \ldots$.

In the general case, in which several seed vectors are introduced,

$$C_A(x) = (\alpha_0 + \alpha_1 x + \cdots + x^r)(\beta_0 + \beta_1 x + \cdots + x^s)(\gamma_0 + \gamma_1 x + \cdots + x^t) \cdots.$$

Thus the characteristic polynomial $C_A(x)$ can be immediately written down after the dependence equations have been obtained.

As a by-product of the above discussion, we derive a quick proof of the Cayley-Hamilton Theorem that is probably not available in the literature.

7.3.1 THEOREM

Cayley-Hamilton Theorem If $C_A(x)$ is the characteristic polynomial of A, then $C_A(A) = 0$.

Proof By the above discussion, $C_A(A)x = 0$, where x runs over the seed vectors used in obtaining all eigenvalues. Note first that one can start with an arbitrary seed vector y and obtain the same net result that $C_A(A)y = 0$ and second that $C_A(A)y$ is the ith column of the matrix $C_A(A)$, where y is a member of the standard basis of F^n with ith entry 1. This implies that all the columns of the matrix $C_A(A)$ are 0. Thus $C_A(A) = 0$. ■

7.3.2 EXAMPLES We now illustrate the method described above by two examples that were already solved by the first method in Examples 7.2.4. The reader should compare the two solutions.

a. Find a maximal set of linearly independent eigenvectors of

$$A = \begin{bmatrix} 3 & -1 & -1 \\ -12 & 0 & 5 \\ 4 & -2 & -1 \end{bmatrix}.$$

Let us start with a nonzero seed vector

$$u = \begin{bmatrix} 1 \\ 0 \\ 0 \end{bmatrix}.$$

We calculate Au and test u and Au for linear dependence. Note that Au is the first column of A. Clearly, u and Au are linearly independent. So we proceed to calculate

$$A^2 u = A(Au) = \begin{bmatrix} 3 & -1 & -1 \\ -12 & 0 & 5 \\ 4 & -2 & -1 \end{bmatrix} \begin{bmatrix} 3 \\ -12 \\ 4 \end{bmatrix} = \begin{bmatrix} 17 \\ -16 \\ 32 \end{bmatrix}.$$

We consider the matrix whose columns are u, Au, $A^2 u$ and reduce it to row echelon form:

$$\begin{bmatrix} 1 & 3 & 17 \\ 0 & -12 & -16 \\ 0 & 4 & 32 \end{bmatrix} \overset{R_3 + \frac{1}{3} R_2}{\rightarrow} \begin{bmatrix} 1 & 3 & 17 \\ 0 & -12 & -16 \\ 0 & 0 & * \end{bmatrix},$$

where $*$ denotes some nonzero number, showing that u, Au, and $A^2 u$ are linearly independent because the rank is 3. Now we proceed to compute

$$A^3 u = A(Au) = \begin{bmatrix} 3 & -1 & -1 \\ -12 & 0 & 5 \\ 4 & -2 & -1 \end{bmatrix} \begin{bmatrix} 17 \\ -16 \\ 32 \end{bmatrix} = \begin{bmatrix} 35 \\ -44 \\ 68 \end{bmatrix}.$$

The four vectors u, Au, $A^2 u$ and $A^3 u$ must be linearly dependent, since any four or more vectors in F^3 must be linearly dependent. To find the dependency relation, we set $M = $ the matrix with columns u, Au, $A^2 u$ and $A^3 u$ and reduce M to row echelon form. Note that M is the coefficient matrix of the linear system

$$\alpha u + \beta Au + \gamma A^2 u + \delta A^3 u = 0.$$

So we have

$$M = \begin{bmatrix} 1 & 3 & 17 & 35 \\ 0 & -12 & -16 & -44 \\ 0 & 4 & 32 & 68 \end{bmatrix} \overset{R_2 \longleftrightarrow R_3}{\rightarrow} \begin{bmatrix} 1 & 3 & 17 & 35 \\ 0 & 4 & 32 & 68 \\ 0 & -12 & -16 & -44 \end{bmatrix}$$

$$\overset{R_3 + 3R_2}{\rightarrow} \begin{bmatrix} 1 & 3 & 17 & 35 \\ 0 & 4 & 32 & 68 \\ 0 & 0 & 80 & 160 \end{bmatrix}.$$

The equivalent linear system is

$$\alpha + 3\beta + 17\gamma + 35\delta = 0,$$
$$4\beta + 32\gamma + 68\delta = 0,$$
$$80\gamma + 160\delta = 0.$$

We solve the linear system by backward substitution. Put $\delta = 1$. Then $\gamma = -2$, $\beta = -1$, $\alpha = 2$. This gives the dependency relation

$$2u - Au - 2A^2 u + A^3 u = 0. \tag{1}$$

Equation (1) can be rewritten as

$$(A - I)(A - 2I)(A + I)u = 0.$$

Set $v = (A - 2I)(A + I)u$. Then $(A - I)v = 0$, which gives that 1 is an eigenvalue and corresponding eigenvector is v. By setting $w = (A - I)(A + I)u$, and $p = (A - I)(A - 2I)u$ we get $(A - 2I)w = 0$ and $(A + I)p = 0$. Thus w and p are eigenvectors corresponding to eigenvalues 2 and -1, respectively.

By actual computations,

$$v = (A - 2I)(A + I)u = \begin{bmatrix} 12 \\ -4 \\ 28 \end{bmatrix},$$

$$w = (A - I)(A + I)u = \begin{bmatrix} 16 \\ -16 \\ 32 \end{bmatrix},$$

$$p = (A - I)(A - 2I)u = \begin{bmatrix} 10 \\ 20 \\ 20 \end{bmatrix}.$$

b. Find the eigenvalues and a maximal set of linearly independent eigenvectors of the matrix

$$A = \begin{bmatrix} 0 & -3 & 4 \\ 0 & 5 & 0 \\ 1 & -2 & 0 \end{bmatrix}.$$

We start with a nonzero vector u and call it the first seed vector. For simplicity of calculation, we take

$$u = \begin{bmatrix} 1 \\ 0 \\ 0 \end{bmatrix}.$$

We calculate Au, which is simply the first column of A, namely, $\begin{bmatrix} 0 \\ 0 \\ 1 \end{bmatrix}$ and clearly u, and Au are linearly independent. So we calculate A^2u and test whether u, Au, A^2u is a linearly dependent list. Now

$$A^2u = A\,Au = \begin{bmatrix} 0 & -3 & 4 \\ 0 & 5 & 0 \\ 1 & -2 & 0 \end{bmatrix} \begin{bmatrix} 0 \\ 0 \\ 1 \end{bmatrix} = \begin{bmatrix} 4 \\ 0 \\ 0 \end{bmatrix}.$$

To find a dependency relation, if any, among u, Au, and A^2u, we reduce the matrix M whose columns are u, Au, and A^2u to row echelon form. Note the matrix M is the coefficient matrix of the linear system $\alpha u + \beta Au + \gamma A^2u = 0$:

$$M = \begin{bmatrix} 1 & 0 & 4 \\ 0 & 0 & 0 \\ 0 & 1 & 0 \end{bmatrix} \xrightarrow[\;\longrightarrow\;]{R_2 \longleftrightarrow R_3} \begin{bmatrix} 1 & 0 & 4 \\ 0 & 1 & 0 \\ 0 & 0 & 0 \end{bmatrix}.$$

The linear system of equations becomes

$$\alpha + 0\beta + 4\gamma = 0,$$
$$\beta + 0\gamma = 0.$$

Thus $\beta = 0$, and if we choose $\gamma = 1$, then $\alpha = -4$.

Since u, Au, and A^2u are linearly dependent, we stop. The dependence relation is

$$(-4I + 0A + 1A^2)u = 0. \tag{1}$$

We rewrite Equation (1) as $(A - 2I)(A + 2I)u = 0$. By setting $v_1 = (A + 2I)u$ and $v_2 = (A - 2I)u$, we get $(A - 2I)v_1 = 0$ and $(A + 2I)v_2 = 0$, respectively. So v_1 and v_2 are eigenvectors corresponding to eigenvalues 2 and -2, respectively.

Now $v_1 = (A + 2I)u = $ first column of $(A + 2I) = \begin{bmatrix} 2 \\ 0 \\ 1 \end{bmatrix}$ and $v_2 = (A - 2I)u = $

first column of $(A - 2I) = \begin{bmatrix} -2 \\ 0 \\ 1 \end{bmatrix}$.

We drop A^2u and choose another seed vector v such that u, Au, and v are linearly independent. Here we may choose

$$v = \begin{bmatrix} 0 \\ 1 \\ 0 \end{bmatrix}.$$

We compute Av. The vectors u, Au, v, and Av are linearly dependent, since any set containing more than three vectors in F^3 must be linearly dependent. To find dependency relation, we reduce the matrix

$$\begin{bmatrix} 1 & 0 & 0 & -3 \\ 0 & 0 & 1 & 5 \\ 0 & 1 & 0 & -2 \end{bmatrix}$$

whose columns are u, Au, v, and Av to row echelon form because this matrix is the coefficient matrix of the linear system $\alpha u + \beta Au + \gamma v + \delta Av = 0$:

$$\begin{bmatrix} 1 & 0 & 0 & -3 \\ 0 & 0 & 1 & 5 \\ 0 & 1 & 0 & -2 \end{bmatrix} \overset{R_2 \longleftrightarrow R_3}{\rightarrow} \begin{bmatrix} 1 & 0 & 0 & -3 \\ 0 & 1 & 0 & -2 \\ 0 & 0 & 1 & 5 \end{bmatrix}.$$

This gives the following linear system:

$$\alpha - 3\delta = 0,$$
$$\beta - 2\delta = 0,$$
$$\gamma + 5\delta = 0.$$

Solving the above system, we get, by putting $\delta = 1$, $\gamma = -5$, $\beta = 2$, and $\alpha = 3$.

The dependence relation is

$$3u + 2Au - 5v + Av = 0. \tag{2}$$

Rewrite Equation (2) as $(3I + 2A)u - (5I - A)v = 0$, and premultiply both sides by $(-4I + A^2)$. Using Equation (1), we get $(-4I + A^2)(5I - A)v = 0$. Thus by setting $p = (-4I + A^2)v$, we get $(5I - A)p = 0$. Therefore p is an eigenvector corresponding to eigenvalue 5. Now $p = (-4I + A^2)v$ equals the second column of the matrix

$$-4I + A^2 = \begin{bmatrix} -23 \\ 21 \\ -13 \end{bmatrix}.$$

The process of starting with a new seed vector will stop here, since we have obtained three eigenvalues of A. ∎

7.3 Exercises

1. Do as many problems in Exercise 7.2 as you need to practice by the method explained in this section. It is suggested that you at least do Exercises 1(a), 1(d), 2(a), 2(b), and 2(c) .

7.4 Applications of the Cayley-Hamilton Theorem

As was stated in Theorem 7.3.1, the characteristic polynomial

$$C_A(x) = x^n + a_{n-1}x^{n-1} + \cdots + a_1 x + a_0$$

has a remarkable property that

$$A^n + a_{n-1}A^{n-1} + \cdots + a_1 A + a_0 I = 0.$$

This result is known as *Cayley-Hamilton Theorem*. A shorter proof of this theorem was provided in Theorem 7.3.1 as a by-product of the discussion regarding computing eigenvalues without using determinants. Another proof is given in Section 7.6, Theorem 2.

The following formula gives a method of computing A^{-1} by using the Cayley-Hamilton Theorem.

7.4.1 THEOREM

(Theorem 2 Section 7.6)

Let A be a nonsingular $n \times n$ matrix, and let its characteristic polynomial be

$$C_A(x) = x^n + a_{n-1}x^{n-1} + \cdots + a_1 x + a_0.$$

Then

$$A^{-1} = -\frac{1}{a_0}(A^{n-1} + a_{n-1}A^{n-2} + \cdots + a_1 I).$$

By the Cayley-Hamilton Theorem,

$$A^n + a_{n-1}A^{n-1} + \cdots + a_1 A + a_0 I = 0.$$

As was explained above, after the definition of characteristic polynomial, $a_0 = (-1)^n \det A \neq 0$, since A is nonsingular. So the above equation can be written as

$$I = -\frac{1}{a_0}(A^n + a_{n-1}A^{n-1} + \cdots + a_1 A)$$

$$= -\frac{1}{a_0}(A^{n-1} + a_{n-1}A^{n-2} + \cdots + a_1 I)A.$$

Hence

$$A^{-1} = -\frac{1}{a_0}(A^{n-1} + a_{n-1}A^{n-2} + \cdots + a_1 I).$$

**7.4.2
EXAMPLE** For

$$A = \begin{bmatrix} 1 & 2 \\ -4 & 3 \end{bmatrix},$$

compute A^{-1}.

$$C_A(x) = \det \begin{bmatrix} 1-x & 2 \\ -4 & 3-x \end{bmatrix} = (1-x)(3-x) + 8$$

$$= x^2 - 4x + 11.$$

By the Cayley-Hamilton Theorem, $A^2 - 4A + 11I = 0$. Thus $I = -\frac{1}{11}A^2 + \frac{4}{11}A$, and so

$$A^{-1} = -\frac{1}{11}A + \frac{4}{11}I = -\frac{1}{11}\begin{bmatrix} 1 & 2 \\ -4 & 3 \end{bmatrix} + \frac{4}{11}\begin{bmatrix} 1 & 0 \\ 0 & 1 \end{bmatrix}$$

$$= \begin{bmatrix} \frac{3}{11} & -\frac{2}{11} \\ \frac{4}{11} & \frac{1}{11} \end{bmatrix} = \frac{1}{11}\begin{bmatrix} 3 & -2 \\ 4 & 1 \end{bmatrix}.$$

∎

7.4.3 Computing Powers Using the Cayley-Hamilton Theorem

Another useful application of the Cayley-Hamilton Theorem is in finding $f(A)$, in particular powers of A where A is a square matrix and $f(x)$ is a polynomial of any degree, howsoever large it may be. We explain the method by examples in which we make use of the following well-known result, known as *Euclid's Division Algorithm*.

Let $f(x)$, $g(x) \neq 0$ be polynomials. Then there exist polynomials $q(x)$ and $r(x)$ such that $f(x) = g(x)q(x) + r(x)$ and $\deg r(x) < \deg g(x)$ or $r(x) = 0$. ($q(x)$ is called the quotient, and $r(x)$ is called the remainder.)

7.4.4
EXAMPLES

a. Let

$$A = \begin{bmatrix} 2 & 5 \\ 1 & -2 \end{bmatrix} \quad \text{and} \quad f(x) = x^{735}.$$

We want to find $f(A) = A^{735}$. We have

$$C_A(x) = \det \begin{bmatrix} 2 - x & 5 \\ 1 & -2 - x \end{bmatrix} = x^2 - 9.$$

So the eigenvalues are 3 and -3. The division algorithm applied to the polynomials x^{735}, $x^2 - 9$ will give an equation of the form

$$x^{735} = (x^2 - 9)q(x) + (a_0 + a_1 x), \tag{1}$$

where $a_0 + a_1 x$ is the remainder obtained by dividing x^{735} by $x^2 - 9$. Note that the remainder is of degree less than the degree of the divisor $x^2 - 9$.

By the Cayley-Hamilton Theorem, $A^2 - 9I = 0$. Inserting A for x in Equation (1), we get

$$A^{735} = a_0 I + a_1 A. \tag{2}$$

So if we know a_0 and a_1, we easily find A^{735} by Equation (2).

The eigenvalues of A are 3 and -3. Inserting 3 and -3 for x successively in Equation (1), we get

$$3^{735} = a_0 + 3a_1$$
$$(-3)^{735} = a_0 - 3a_1$$

This gives $a_0 = 0$, $a_1 = 3^{734}$. Then

$$A^{735} = a_0 I + a_1 A$$

gives

$$A^{735} = 3^{734} A.$$

Note: For practical purposes, we could disregard the term $(x^2 - 9)q(x)$ in Equation (1) and simply work with $x^{735} = a_0 + a_1 x$.

b. Let

$$A = \begin{bmatrix} -2 & 4 & 3 \\ 0 & 0 & 0 \\ -1 & 5 & 2 \end{bmatrix} \quad \text{and} \quad f(x) = x^{593} - 2x^{15}.$$

We calculate $A^{593} - 2A^{15}$. Then

$$C_A(x) = \det \begin{bmatrix} -2 - x & 4 & 3 \\ 0 & -x & 0 \\ -1 & 5 & 2 - x \end{bmatrix}$$

$$= (-2 - x)(x^2 - 2x) - 1(3x) \quad \text{(expanding by the first column)}$$

$$= -x^3 + x.$$

So by the Cayley-Hamilton Theorem, $-A^3 + A = 0$. The eigenvalues of A are 0, 1, and -1. The division algorithm applied to the polynomials $f(x) = x^{593} - 2x^{15}$ and $g(x) = -x^3 + x$ will give an equation of the form

$$x^{593} - 2x^{15} = (-x^3 + x)q(x) + a_0 + a_1 x + a_2 x^2, \tag{1}$$

where $a_0 + a_1 x + a_2 x^2$ is the remainder. Note that the remainder must be of degree less than that of the divisor $-x^3 + x$. By inserting successively 0, 1, and -1 for x in Equation (1), we get

$$0 = a_0, \tag{2}$$
$$-1 = a_0 + a_1 + a_2, \tag{3}$$
$$1 = a_0 - a_1 + a_2. \tag{4}$$

Solving Equations (2), (3), and (4) for a_0, a_1, and a_2, we get $a_0 = 0$, $a_1 = -1$, and $a_2 = 0$. Thus

$$A^{593} - 2A^{15} = a_0 I + a_1 A + a_2 A^2$$
$$= -A = \begin{bmatrix} 2 & -4 & -3 \\ 0 & 0 & 0 \\ 1 & -5 & -2 \end{bmatrix}.$$

Note: Again, as was stated in Example (a), we could disregard the term $-(x^3 - x)q(x)$ in Equation (1) and simply work with $x^{593} - 2x^{15} = a_0 + a_1 x + a_2 x^2$.

The problems that follow illustrate some interesting applications of eigenvalues.

c. Suppose $a_0, a_1, a_2, \ldots$ is a sequence of positive integers such that $a_{k+1} = a_k + 2a_{k-1}$ for all $k \geqslant 1$. If $a_0 = 0$, $a_1 = 1$, find a_k.

$$a_{k+1} = a_k + 2a_{k-1},$$
$$a_k = a_k + 0a_{k-1}.$$

Thus

$$\begin{bmatrix} a_{k+1} \\ a_k \end{bmatrix} = \begin{bmatrix} 1 & 2 \\ 1 & 0 \end{bmatrix} \begin{bmatrix} a_k \\ a_{k-1} \end{bmatrix}. \tag{1}$$

Set

$$A = \begin{bmatrix} 1 & 2 \\ 1 & 0 \end{bmatrix}.$$

Putting $k = 1, 2, \ldots$ in Equation (1), we get

$$\begin{bmatrix} a_2 \\ a_1 \end{bmatrix} = \begin{bmatrix} 1 & 2 \\ 1 & 0 \end{bmatrix} \begin{bmatrix} a_1 \\ a_0 \end{bmatrix}, \quad \begin{bmatrix} a_3 \\ a_2 \end{bmatrix} = \begin{bmatrix} 1 & 2 \\ 1 & 0 \end{bmatrix} \begin{bmatrix} a_2 \\ a_1 \end{bmatrix} = A^2 \begin{bmatrix} a_1 \\ a_0 \end{bmatrix}, \tag{2}$$

and so on.

This yields

$$\begin{bmatrix} a_{k+1} \\ a_k \end{bmatrix} = A^k \begin{bmatrix} a_1 \\ a_0 \end{bmatrix} = A^k \begin{bmatrix} 1 \\ 0 \end{bmatrix}, \tag{3}$$

since $a_1 = 1$, $a_0 = 0$. We now proceed to compute A^k in order to determine a_k. The eigenvalues of A are 2 and -1. As was stated in the notes following Examples (a) and (b), we write $x^k = a_0 + a_1 x$ and substitute $x = 2, -1$ in succession, yielding

$$2^k = a_0 + 2a_1,$$
$$(-1)^k = a_0 - a_1.$$

Subtracting, we get

$$a_1 = \frac{2^k - (-1)^k}{3},$$

and so

$$a_0 = \frac{2^k + 2(-1)^k}{3}.$$

Thus

$$A^k = \frac{2^k + 2(-1)^k}{3} I + \frac{2^k - (-1)^k}{3} A,$$

and

$$A^k \begin{bmatrix} 1 \\ 0 \end{bmatrix} = \frac{2^k + 2(-1)^k}{3} \begin{bmatrix} 1 \\ 0 \end{bmatrix} + \frac{2^k - (-1)^k}{3} \begin{bmatrix} 1 \\ 1 \end{bmatrix}. \tag{4}$$

Since from Equation (3),

$$A^k \begin{bmatrix} 1 \\ 0 \end{bmatrix} = \begin{bmatrix} a_{k+1} \\ a_k \end{bmatrix},$$

we obtain from Equation (4), by equating the (2, 1) entries of $A^k \begin{bmatrix} 1 \\ 0 \end{bmatrix}$,

$$a_k = \frac{2^k - (-1)^k}{3}.$$

In particular, $a_0 = 0$, $a_1 = 1$, $a_2 = 1$, $a_3 = 3$, $a_4 = 5$, $a_5 = 11$, and so on.

d. Two grocery store chains, Kroger and Big Bear, have been competing with each other in a city with a population of 100,000 customers. Each month Kroger loses 10% of its customers to Big Bear and gains 10% of Big Bear's customers. If the number of customers remains constant, find the number of customers patronizing each store after 10 months, assuming that the present numbers of Kroger and Big Bear customers are 60,000 and 40,000, respectively.

If $\begin{bmatrix} x_0 \\ y_0 \end{bmatrix}$ represents Kroger and Big Bear customers at present and $\begin{bmatrix} x_1 \\ y_1 \end{bmatrix}$ represents their customers after one month, then

$$x_1 = 0.9 x_0 + 0.1 y_0,$$
$$y_1 = 0.1 x_0 + 0.9 y_0.$$

This yields

$$\begin{bmatrix} x_1 \\ y_1 \end{bmatrix} = A \begin{bmatrix} x_0 \\ y_0 \end{bmatrix},$$

where

$$A = \begin{bmatrix} 0.9 & 0.1 \\ 0.1 & 0.9 \end{bmatrix}.$$

Similarly,

$$\begin{bmatrix} x_2 \\ y_2 \end{bmatrix} = A \begin{bmatrix} x_1 \\ y_1 \end{bmatrix} = A^2 \begin{bmatrix} x_0 \\ y_0 \end{bmatrix}, \dots, \begin{bmatrix} x_{10} \\ y_{10} \end{bmatrix} = A \begin{bmatrix} x_9 \\ y_9 \end{bmatrix} = \cdots = A^{10} \begin{bmatrix} x_0 \\ y_0 \end{bmatrix}.$$

We now compute A^{10}. The eigenvalues of A are 1 and 0.8. As was stated in the notes following Examples (a) and (b), write $x^{10} = a_0 + a_1 x$ and substitute $x = 1$, 0.8 in succession to find a_0 and a_1. We have

$$1 = a_0 + a_1 \qquad \text{and} \qquad (0.8)^{10} = a_0 + (0.8)a_1.$$

These equations yield

$$a_1 = 5(1 - (0.8)^{10}), \qquad a_0 = -4 + 5(0.8)^{10}.$$

Therefore $A^{10} = a_0 I + a_1 A$ with a_0 and a_1 as above.
Since

$$A^{10} \begin{bmatrix} x_0 \\ y_0 \end{bmatrix} = \begin{bmatrix} x_{10} \\ y_{10} \end{bmatrix},$$

we get

$$(a_0 I + a_1 A) \begin{bmatrix} x_0 \\ y_0 \end{bmatrix} = \begin{bmatrix} x_{10} \\ y_{10} \end{bmatrix}.$$

Equating the $(1, 1)$ entries on both sides, we get

$$a_0 x_0 + a_1 (0.9\, x_0 + 0.1\, y_0) = x_{10},$$

that is,

$$x_{10} = [-4 + 5(0.8)^{10}][60{,}000] + [5 - 5(0.8)^{10}][54{,}000 + 4{,}000]$$
$$= 51{,}074 \quad \text{(approximately)}.$$

e. Let

$$A = \begin{bmatrix} 4 & 1 & 3 \\ 0 & 2 & 0 \\ -4 & 1 & -4 \end{bmatrix}, \qquad \text{and} \qquad f(x) = x^{14} - 3x^{13}.$$

We have

$$C_A(x) = \det \begin{bmatrix} 4 - x & 1 & 3 \\ 0 & 2 - x & 0 \\ -4 & 1 & -4 - x \end{bmatrix}$$

$$= (2 - x)\,[(4 - x)(-4 - x) + 12] \qquad \text{(expanding by the second row)}$$
$$= (2 - x)(x - 2)(x + 2).$$

The eigenvalues of A are 2, 2, and -2. Thus eigenvalue 2 occurs with multiplicity 2. We write

$$x^{14} - 3x^{13} = (x - 2)^2 (x + 2) \, q(x) + (a_0 + a_1 x + a_2 x^2). \qquad (2)$$

Since the eigenvalue 2 is repeated two times, differentiating both sides of Equation (2) twice, we get

$$14x^{13} - 39x^{12} = [2(x - 2)(x + 2) + (x - 2)^2]q(x) + (x - 2)^2(x + 2)q'(x)$$
$$+ (a_1 + 2a_2 x), \qquad (3)$$

Putting $x = 2$ in Equations (2) and (3), we get

$$2^{14} - 3 \cdot 2^{13} = a_0 + 2a_1 + 4a_2, \qquad (4)$$
$$14 \cdot 2^{13} - 39 \cdot 2^{12} = a_1 + 4a_2. \qquad (5)$$

Since $x = -2$ is a nonrepeated eigenvalue, we put $x = -2$ in Equation (2) only and obtain

$$2^{14} + 3 \cdot 2^{13} = a_0 - 2a_1 + 4a_2. \qquad (6)$$

We now solve Equations (4), (5) and (6) for a_0, a_1, and a_2. Subtracting Equation (6) from Equation (4), we get

$$-6 \cdot 2^{13} = 4a_1.$$

Thus $a_1 = -3 \cdot 2^{12}$. Substituting the value of a_1 in Equation (5), we get

$$4a_2 = 14 \cdot 2^{13} - 39 \cdot 2^{12} + 3 \cdot 2^{12}$$
$$= 14 \cdot 2^{13} - 36 \cdot 2^{12},$$

giving

$$a_2 = 7 \cdot 2^{12} - 9 \cdot 2^{12} = -2 \cdot 2^{12} = -2^{13}.$$

Finally, by substituting the values of a_1 and a_2 in Equation (4), we get

$$a_0 = 2^{14} - 3 \cdot 2^{13} - 2 \cdot (-3) \cdot 2^{12} - 4 \cdot (-2^{13})$$
$$= 2^{14} - 3 \cdot 2^{13} + 3 \cdot 2^{13} + 2^{15}$$
$$= 2^{14} + 2^{15}.$$

Inserting A for x on both sides of Equation (2) and using the Cayley-Hamilton Theorem, that is, $(A - 2I_3)^2(A + 2I_3) = 0$, we get

$$A^{14} - 3A^{13} = a_0 I_3 + a_1 A + a_2 A^2 = (2^{14} + 2^{15})I_3 - 3 \cdot 2^{12}A - 2^{13}A^2.$$

Note: We could, as was stated in Examples (a) and (b), disregard $(x - 2)^2(x + 2)q(x)$ in Equation (2) and write

$$x^{14} - 3x^{13} = a_0 + a_1 x + a_2 x^2.$$

Then follow the same steps as above. ∎

Summary of Steps for Finding Powers of Matrices

Let A be an $n \times n$ matrix. Suppose we want to compute A^m.

Step 1 Find the characteristic polynomial, say $g(x)$, and eigenvalues of A. Note that the degree of $g(x)$ is equal to n.

Step 2 Write

$$x^m = a_0 + a_1 x + \cdots + a_{n-1} x^{n-1} \quad \text{(division algorithm).} \qquad (1)$$

Insert successively for x the n distinct eigenvalues in Equation (1) and get n equations to solve for $a_0, a_1, \ldots, a_{n-1}$. However, if some eigenvalue λ repeats k times, then we will get fewer equations to solve for the n unknowns $a_0, a_1, \ldots, a_n$. To get $(k-1)$ more equations, we differentiate Equation (1) $(k-1)$ times and insert $x = \lambda$ in all these $(k-1)$ equations to get a total of n equations.

Step 3 Solve the n equations for $a_0, a_1, \ldots, a_{n-1}$.

Step 4 Then

$$A^m = a_0 I + a_1 A + \cdots + a_{n-1} A^{n-1}.$$

7.4.5 Remark If we want to find not just some power of A, but a linear combination of powers of A, say $c_0 I + c_1 A + \cdots + c_m A^m$, then we replace x^m in Step 2 by $c_0 + c_1 x_1 + \cdots + c_m x^m$, write

$$c_0 + c_1 x + \cdots + c_m x^m = a_0 + a_1 x + \cdots + a_{n-1} x^{n-1},$$

and proceed as before to solve for $a_0, a_1, \ldots, a_{n-1}$.

7.4 Exercises

1. Verify the Cayley-Hamilton theorem and compute the inverse by the Cayley-Hamilton theorem for each of the following matrices:

   a. $\begin{bmatrix} 1 & 2 \\ 3 & 4 \end{bmatrix}$

   b. $\begin{bmatrix} 0 & 1 \\ 1 & -1 \end{bmatrix}$

    c. (Drill 7.6) $\begin{bmatrix} 1 & -1 & 2 \\ 0 & 3 & 2 \\ 2 & 1 & 2 \end{bmatrix}$

   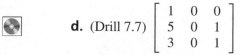 d. (Drill 7.7) $\begin{bmatrix} 1 & 0 & 0 \\ 5 & 0 & 1 \\ 3 & 0 & 1 \end{bmatrix}$

e. $\begin{bmatrix} 0 & 1 & 0 \\ 5 & 0 & 1 \\ 3 & 0 & -1 \end{bmatrix}$

2. Find A^{100} for $A = \begin{bmatrix} -3 & -4 \\ 2 & 3 \end{bmatrix}$ and also for $A = \begin{bmatrix} 2 & -1 \\ 3 & -2 \end{bmatrix}$.

3. Find $A^{520} + 3A^{70} - 7I$ for $A = \begin{bmatrix} -2 & 4 & 3 \\ 0 & 0 & 0 \\ -1 & 5 & 2 \end{bmatrix}$.

4. Find A^{212} for $A = \begin{bmatrix} 2 & -2 & 4 \\ 0 & -2 & 4 \\ 0 & 0 & 2 \end{bmatrix}$.

5. Find A^{700} for $A = \begin{bmatrix} 3 & 2 & 4 \\ 0 & 1 & 0 \\ -1 & -3 & -1 \end{bmatrix}$.

6. Find A^{15} for $A = \begin{bmatrix} 10 & 16 \\ -4 & -10 \end{bmatrix}$.

7. Find $A^{1000} - 5A^{100}$ for $A = \begin{bmatrix} 0 & 1 & 2 & 3 \\ 0 & 0 & 4 & 5 \\ 0 & 0 & 0 & 6 \\ 0 & 0 & 0 & 0 \end{bmatrix}$.

8. The Fibonacci numbers $a_0, a_1, a_2, a_3, \ldots$ are defined by $a_0 = 0$, $a_1 = 1$, and for $k > 1$, a_k is the sum of the previous two numbers. Find a_k.

7.5 Properties of Eigenvalues, Diagonalizability, and Triangularizability

We now give some properties of the eigenvalues of an $n \times n$ matrix A, whose proofs are contained in Section 7.6 (Theorems 3–7).

7.5.1 Property The sum of the eigenvalues of a matrix A equals the sum of the diagonal entries of A. (The sum of diagonal entries of a matrix is called its *trace*.)

7.5.2 Property The product of the eigenvalues of A equals the determinant of A.

7.5.3 Property The eigenvalues of a triangular matrix are precisely the entries of the diagonal.

7.5.4 Property If $\lambda_1, \ldots, \lambda_k$ are distinct eigenvalues of A with corresponding eigenvectors $v_1, \ldots, v_k$, respectively, then $v_1, \ldots, v_k$ are linearly independent.

7.5.5 Property If an $n \times n$ matrix A has n distinct eigenvalues $\lambda_1, \lambda_2, \ldots, \lambda_n$, then there exists an invertible matrix P such that

$$
P^{-1}AP = \begin{bmatrix} \lambda_1 & 0 & \cdots & 0 \\ 0 & \lambda_2 & \cdots & 0 \\ \vdots & \vdots & \ddots & \vdots \\ 0 & 0 & \cdots & \lambda_n \end{bmatrix}.
$$

7.5.6 **DEFINITION**	**Diagonalizable**
	An $n \times n$ matrix A is said to be *diagonalizable* if there exists an invertible matrix P such that $P^{-1}AP$ is a diagonal matrix.

Diagonalizable matrices play an important role in linear algebra. By Property 7.5.5, an $n \times n$ matrix with n distinct eigenvalues is diagonalizable. However, matrices with repeated eigenvalues may also be diagonalizable. A necessary and sufficient condition for an $n \times n$ matrix to be diagonalizable is that it possess n linearly independent eigenvectors. The proof of this fact is omitted. In one direction, it is a simple exercise that if there exist n linearly independent eigenvectors, then by choosing P whose columns are the linearly independent eigenvectors, $AP = PD$, where D is a diagonal matrix with diagonal entries as eigenvalues appearing in the same order on the diagonal as the corresponding eigenvectors appearing as columns in P. Since P is invertible, it follows that $P^{-1}AP = D$. (See the details of this statement in *Theorem 7, Section 7.6*.)

We give an example of a matrix that is not diagonalizable. Let

$$
A = \begin{bmatrix} 1 & 1 \\ 0 & 1 \end{bmatrix}.
$$

Its eigenvalues are 1 and 1 (Property 7.5.3). It has only one linearly independent eigenvector $\begin{bmatrix} 1 \\ 0 \end{bmatrix}$ or any nonzero multiple of $\begin{bmatrix} 1 \\ 0 \end{bmatrix}$. Thus A is not diagonalizable.

Whereas not every matrix A is diagonalizable, it is always possible to find an invertible matrix P such that $P^{-1}AP$ is a triangular matrix. The proof of this result is essentially contained in the algorithm given below for obtaining a triangular form. We speak of this fact by saying that A is triangularizable.

7.5.7 **DEFINITION**	**Triangularizable**
	A square matrix A is called *triangularizable* if there exists an invertible matrix P such that $P^{-1}AP$ is a triangular matrix.

We give below the steps that one may follow for obtaining a triangular form of a given matrix. Note that although the row echelon form of any matrix is triangular, it is not equal to $P^{-1}AP$. To be precise, the row echelon form of a matrix A is indeed equal to EA, where E is the product of suitable elementary matrices as described in *Section 4.3*.

Algorithm for Triangularizing an $n \times n$ Matrix

Step 1 Find an eigenvector v_1 corresponding to some eigenvalue λ_1, say.

Step 2 Form a linearly independent set containing n vectors that includes v_1. This can always be done by choosing suitably all but one vector of the standard basis of F^n and v_1.

Step 3 Form the $n \times n$ matrix U, say, whose columns are the n linearly independent vectors obtained in Step 2. It can then be seen that

$$U^{-1}AU = \begin{bmatrix} \lambda_1 & * \\ 0 & B \end{bmatrix}$$

where the entries $*$, 0, and B are matrix blocks of sizes $1 \times n-1, n-1 \times 1$, and $n-1 \times n-1$, respectively.

Step 4 Repeat the process on the $n-1 \times n-1$ matrix B, then on the matrix of lower order $n-2 \times n-2$, and so on. This will reduce A to a desired triangular form.

7.5.8 DEFINITION

Similar Matrices

Matrices A and B are called similar if there exists an invertible matrix P such that $B = P^{-1}AP$.

For properties of similar matrices, see Exercises 6, 10, and 11 in *Section 7.5*.

7.5.9 EXAMPLES

a. Let

$$A = \begin{bmatrix} 1 & 1 & 2 \\ -1 & 2 & 1 \\ 0 & 1 & 3 \end{bmatrix}.$$

We reduce A into a diagonal matrix similar to it.

The characteristic polynomial of A is $C_A(x) = -x^3 + 6x^2 - 11x + 6 = -(x-1)(x-2)(x-3)$. Thus 1, 2, and 3 are eigenvalues of A, and so by *Theorem 1.2*, if x_1, x_2, and x_3 are eigenvectors corresponding to 1, 2, and 3, respectively, then $P = \lfloor x_1 \ x_2 \ x_3 \rfloor$ is the required matrix. Now,

$$x_1 = \begin{bmatrix} a \\ b \\ c \end{bmatrix}$$

must satisfy

$$\begin{bmatrix} 1 & 1 & 2 \\ -1 & 2 & 1 \\ 0 & 1 & 3 \end{bmatrix} \begin{bmatrix} a \\ b \\ c \end{bmatrix} - \begin{bmatrix} a \\ b \\ c \end{bmatrix}$$

Therefore $a + b + 2c = a$, $-a + 2b + c = b$, and $b + 3c = c$. These equations give

$2a = b = -2c$. Thus $a \begin{bmatrix} 1 \\ 2 \\ -1 \end{bmatrix}$ for any nonzero a is an eigenvector corresponding to 1. Choosing $a = 1$, we have $x_1 = \begin{bmatrix} 1 \\ 2 \\ -1 \end{bmatrix}$. Similarly,

$$x_2 = \begin{bmatrix} 1 \\ -1 \\ 1 \end{bmatrix}, \qquad x_3 = \begin{bmatrix} 1 \\ 0 \\ 1 \end{bmatrix}.$$

Therefore

$$P = \begin{bmatrix} 1 & 1 & 1 \\ 2 & -1 & 0 \\ -1 & 1 & 1 \end{bmatrix}.$$

Actual computations give

$$P^{-1} = \frac{1}{2} \begin{bmatrix} 1 & 0 & -1 \\ 2 & -2 & -2 \\ -1 & 2 & 3 \end{bmatrix} \quad \text{and} \quad P^{-1}AP = \begin{bmatrix} 1 & 0 & 0 \\ 0 & 2 & 0 \\ 0 & 0 & 3 \end{bmatrix}.$$

b. We reduce

$$A = \begin{bmatrix} 1 & -2 & 0 \\ 1 & -1 & 2 \\ 0 & 1 & 1 \end{bmatrix}$$

to a triangular matrix similar to it.

The eigenvalues of A are 1, 1, and -1, and

$$v = \begin{bmatrix} 2 \\ 0 \\ -1 \end{bmatrix}$$

is an eigenvector corresponding to eigenvalue 1. We choose vectors $\begin{bmatrix} 0 \\ 1 \\ 0 \end{bmatrix}$ and $\begin{bmatrix} 0 \\ 0 \\ 1 \end{bmatrix}$

so as to obtain three linearly independent vectors that include $v = \begin{bmatrix} 2 \\ 0 \\ -1 \end{bmatrix}$. Set

$$P = \begin{bmatrix} 2 & 0 & 0 \\ 0 & 1 & 0 \\ -1 & 0 & 1 \end{bmatrix}$$

whose columns are the vectors chosen above. It can be verified that

$$P^{-1}AP = \begin{bmatrix} 1 & -1 & 0 \\ 0 & -1 & 2 \\ 0 & 0 & 1 \end{bmatrix},$$

which is already in the desired form. If the 2×2 block on the right-hand corner were not triangular, we would have repeated the steps for that block. ∎

7.5 Exercises

1. (Drill 7.8) Find the sum and product of the eigenvalues of $\begin{bmatrix} 2 & 0 & 1 \\ 1 & 0 & -1 \\ 1 & 5 & 3 \end{bmatrix}$ without actually finding the eigenvalues.

2. If

$$A = \begin{bmatrix} 2 & 3 & 1 \\ 0 & a & 0 \\ 0 & 0 & 1 \end{bmatrix}$$

and the sum of the eigenvalues is 3, find the value of a.

3. In Exercise 2, if the product of the eigenvalues is 3, find a.

4. Let

$$A = \begin{bmatrix} 5 & 0 & 0 \\ 6 & a & 0 \\ 7 & 8 & b \end{bmatrix}.$$

Find a and b if the trace of A is 5 and the product of the eigenvalues is -5.

5. Let

$$A = \begin{bmatrix} a & 5 \\ 1 & b \end{bmatrix}.$$

Find a and b, if the trace of A is 5 and the product of the eigenvalues is 1.

6. Recall that an $n \times n$ matrix B is said to be similar to an $n \times n$ matrix A if $B = P^{-1}AP$ for some invertible matrix P. Prove that (a) if B is similar to A, then A is similar to B and (b) similar matrices have the same eigenvalues.

7. For each of the following matrices A, find an invertible matrix P such that $P^{-1}AP$ is diagonal. Use Matlab to verify your answer.

a. $\begin{bmatrix} -2 & 3 \\ -1 & 2 \end{bmatrix}$

b. (Drill 7.9) $\begin{bmatrix} 1 & 1 & 2 \\ -1 & 2 & 1 \\ 0 & 1 & 3 \end{bmatrix}$

c. (Drill 7.10) $\begin{bmatrix} 4 & -1 & -2 \\ 2 & 1 & -2 \\ 1 & -1 & 1 \end{bmatrix}$

d. $\begin{bmatrix} 1 & 1 & -1 \\ 0 & 1 & 0 \\ 1 & 0 & 1 \end{bmatrix}$

8. Show that

$$A = \begin{bmatrix} 5 & 1 & 2 \\ 0 & 3 & 0 \\ 2 & 1 & 5 \end{bmatrix}$$

has three linearly independent eigenvectors x, y, and z. Also show that if

$$P = [x \ \ y \ \ z],$$

then $P^{-1}AP$ is diagonal.

9. Show that

$$A = \begin{bmatrix} 1 & -2 & -2 & -2 \\ -2 & 1 & -2 & -2 \\ -2 & -2 & 1 & -2 \\ -2 & -2 & -2 & 1 \end{bmatrix}$$

has four linearly independent eigenvectors, say x, y, z, and t. Also, show that if

$$P = [x \ \ y \ \ z \ \ t],$$

then $P^{-1}AP$ is diagonal.

10. Let A be an $n \times n$ matrix. Show that for any $n \times n$ nonsingular matrix B, (a) AB is similar to BA, (b) AB and BA have the same characteristic polynomial.

11. Let A and B be similar matrices. Show that if A satisfies the equation $A^3 - 3A + I = 0$, then B also satisfies a similar equation, $B^3 - 3B + I = 0$.

12. For each of the following matrices, use the method described in Section 7.3 to find eigenvectors. Find a basis of $\mathbf{R}^3$ consisting of the eigenvectors and a matrix P such that $P^{-1}AP$ is a diagonal matrix.

a. $A = \begin{bmatrix} 3 & -1 & -1 \\ -12 & 0 & 5 \\ 4 & -2 & -1 \end{bmatrix}$.

b. $A = \begin{bmatrix} 1 & 1 & 1 \\ 0 & 3 & 3 \\ -2 & 1 & 1 \end{bmatrix}$.

13. (Drill 7.11) If the product of the eigenvalues of the matrix

$$A = \begin{bmatrix} 4 & 3 & 5 & 7 \\ 2 & x & 0 & 1 \\ 2 & 2 & 3 & 8 \\ 1 & 3 & 6 & 2 \end{bmatrix}$$

is 355, find the value of x.

14. (Drill 7.12) (Gerschgorin Theorem) Verify that for any symmetric matrix $A = (a_{ij})$, all the eigenvalues lie in the union of the intervals given by

$$|z - a_{ii}| \leqslant \sum_{\substack{j=1 \\ j \neq i}}^{n} |a_{ij}|$$

15. Reduce each of the following matrices to a triangular matrix similar to it:

a. $\begin{bmatrix} 1 & -3 & 3 \\ 0 & -1 & 2 \\ 0 & -3 & 4 \end{bmatrix}$

b. $\begin{bmatrix} 4 & 0 & 1 \\ 2 & 2 & 3 \\ -1 & 0 & 2 \end{bmatrix}$

16. Reduce the matrix $\begin{bmatrix} 0 & 1 & 0 & 0 \\ 0 & 0 & 1 & 0 \\ 0 & 0 & 0 & 1 \\ -1 & 4 & -6 & 4 \end{bmatrix}$ to a triangular matrix similar to it, given that its eigenvalues are 1, 1, 1, and 1.

7.6   Proofs of Facts

THEOREM 1

Let A be an $n \times n$ matrix. A number λ is an eigenvalue of A if and only if

$$\det(A - \lambda I) = 0,$$

where I denotes the $n \times n$ identity matrix.

Proof Let λ be an eigenvalue of A. Then there exists a nonzero vector $v \in F^n$ such that $Av = \lambda v$, that is, $(A - \lambda I)v = 0$. This means that the homogeneous LS $(A - \lambda I)x = 0$ has a nontrivial solution $x = v$. Hence by Property 6.2.10, $\det(A - \lambda I) = 0$.

Conversely, suppose that $\det(A - \lambda I) = 0$. Then again by the property just cited, the homogeneous LS $(A - \lambda I)x = 0$ has a nontrivial solution $x = v$ (say). So v is a nonzero vector such that $(A - \lambda I)v = 0$, that is, $Av = \lambda v$. Hence λ is an eigenvalue of A. ■

We give two proofs of the following Cayley-Hamilton Theorem. The first proof (observed in our discussions with S. R. Nagpaul) seems quite straightforward once we realize that every matrix can be triangularized. Since it is a simple matter to triangularize any matrix, this proof seems to have value in its being transparent and simple. We are not aware of this proof being in any book, although it may be known to some. The proof depends upon multiplication of matrices with block entries. The reader should consider block multiplication to be just like the usual multiplication, bearing in mind that all multiplications and additions of blocks involved in computations are meaningful. The second proof is the traditional proof given in all books. Another very short proof of the Cayley-Hamilton Theorem was given in Section 7.3 (see Theorem 7.3.1).

THEOREM 2

Cayley-Hamilton
Every square matrix A satisfies its characteristic equation. That is to say, if

$$C_A(x) = a_n x^n + a_{n-1} x^{n-1} + \cdots + a_1 x + a_0,$$

then

$$a_n A^n + a_{n-1} A^{n-1} + \cdots + a_1 A + a_0 I = 0.$$

Proof (First Proof) We know, as was remarked just preceding the definition of triangularizability, that for any matrix A, there exists an invertible matrix P such that $T = P^{-1}AP$ is a triangular matrix. Furthermore, if the matrix T satisfies a polynomial $a_n x^n + a_{n-1} x^{n-1} + \cdots + a_1 x + a_0$, then it is trivial to see that A also satisfies the same polynomial by noting that $(P^{-1}AP)^k = P^{-1}A^k P$ for any positive integer k. Thus it is enough to prove the theorem for triangular matrices. We also know that the eigenvalues of a triangular matrix are the entries on its diagonal. Let $\lambda_1, \lambda_2, \ldots, \lambda_n$ be eigenvalues of T. Now we can write

$$T = \begin{bmatrix} \lambda_1 & X \\ 0 & B \end{bmatrix},$$

where $X, 0$, and B are matrix blocks of sizes $1 \times n-1$, $n-1 \times 1$, and $n-1 \times n-1$, respectively.

Now

$$(T - \lambda_1 I)(T - \lambda_2 I) = \begin{bmatrix} 0 & X \\ 0 & B - \lambda_1 I \end{bmatrix} \begin{bmatrix} \lambda_1 - \lambda_2 & X \\ 0 & B - \lambda_2 I \end{bmatrix}$$

$$= \begin{bmatrix} 0 & X(B - \lambda_2 I) \\ 0 & (B - \lambda_1 I)(B - \lambda_2 I) \end{bmatrix},$$

by block multiplication of matrices. Next,

$$(T - \lambda_1 I)(T - \lambda_2 I)(T - \lambda_3 I) = \begin{bmatrix} 0 & X(B - \lambda_2 I) \\ 0 & (B - \lambda_1 I)(B - \lambda_2 I) \end{bmatrix} \begin{bmatrix} \lambda_1 - \lambda_3 & X \\ 0 & B - \lambda_3 I \end{bmatrix}$$

$$= \begin{bmatrix} 0 & X(B - \lambda_2 I)(B - \lambda_3 I) \\ 0 & (B - \lambda_1 I)(B - \lambda_2 I)(B - \lambda_3 I) \end{bmatrix}.$$

In general,

$$\prod_{i=1}^{n} (T - \lambda_i I) = \begin{bmatrix} 0 & X \prod_{i=2}^{n}(B - \lambda_i I) \\ 0 & \prod_{i=1}^{n}(B - \lambda_i I) \end{bmatrix}. \tag{1}$$

We may now complete the proof by using induction on n. Assume that the result is true for all square matrices of order less than n. Then $\prod_{i=2}^{n}(B - \lambda_i I) = 0$. Thus by Equation (1), $\prod_{i=1}^{n}(T - \lambda_i I) = 0$, completing the proof. ∎

Second Proof (Traditional) Consider the adjoint matrix $\mathrm{adj}(A - xI)$. Each entry of $\mathrm{adj}(A - xI)$ is the determinant of an $(n-1) \times (n-1)$ submatrix of $(A - xI)$ and is therefore a polynomial of degree $\leqslant (n-1)$ in x. Thus $\mathrm{adj}(A - xI)$ can be expressed as

$$\mathrm{adj}(A - xI) = x^{n-1} B_{n-1} + x^{n-2} B_{n-2} + \cdots + B_0, \tag{1}$$

where $B_{n-1}, \ldots, B_0$ are $n \times n$ matrices (see Section 7.2). Let

$$\det(A - xI) = C_A(x) = a_n x^n + a_{n-1} x^{n-1} + \cdots + a_1 x + a_0. \qquad (2)$$

From Theorem 3 of Section 6.3,

$$(A - xI)\operatorname{adj}(A - xI) = \det(A - xI)I. \qquad (3)$$

From Equations (1), (2), and (3), we have

$$(A - xI)(x^{n-1} B_{n-1} + \cdots + x B_1 + B_0) = (a_n x^n + a_{n-1} x^{n-1} + \cdots + a_1 x + a_0)I. \qquad (4)$$

Equating the coefficients of the various powers of x on both sides of Equation (4), we get

$$-B_{n-1} = a_n I,$$
$$-B_{n-2} + A B_{n-1} = a_{n-1} I,$$
$$-B_{n-2} + A B_{n-2} = a_{n-2} I,$$
$$\vdots$$
$$A B_0 = a_0 I.$$

On multiplying the above $n + 1$ equations by $A^n, A^{n-1}, \ldots, I$, respectively, and adding, we get

$$0 = a_n A^n + a_{n-1} A^{n-1} + \cdots + a_1 A + a_0 I,$$

as desired.   ■

THEOREM 3   The sum of the eigenvalues of a matrix A equals the sum of the diagonal entries of A.

> (The sum of diagonal entries of a matrix is called its trace.)

Proof   The sum of the eigenvalues is the sum of the roots of the equation $\det(A - xI) = 0$. Let $\det(A - xI) = a_n x^n + a_{n-1} x^{n-1} + \cdots + a_1 x + a_0$. Thus the sum of the eigenvalues is equal to

$$-\frac{a_{n-1}}{a_n} = -\frac{\text{coefficient of } x^{n-1}}{\text{coefficient of } x^n}.$$

If we expand $\det(A - xI)$ by the first column, then the terms x^{n-1} and x^n appear only in the expansion of the product $(a_{11} - x) \cdots (a_{nn} - x)$, where $A = (a_{ij})$ is the given $n \times n$ matrix. Thus the sum of the eigenvalues is

$$-\frac{(-1)^{n-1}}{(-1)^n}(a_{11} + a_{22} + \cdots + a_{nn}) = a_{11} + a_{22} + \cdots + a_{nn}.$$

■

THEOREM 4

The product of the eigenvalues of A equals the determinant of A.

Proof $C_A(x) = \det(A - xI)$. Thus $C_A(0) = \det A$. Since the product of the roots of a polynomial is the constant term, the result follows. ∎

THEOREM 5

The eigenvalues of a triangular matrix are precisely the entries of the diagonal.

Proof $C_A(x) = \det(A - xI)$. Since $A - xI$ is also triangular, by Property 6.2.7, $C_A(x) = (\alpha_1 - x)(\alpha_2 - x) \cdots (\alpha_n - x)$, where $\alpha_1, \alpha_2, \ldots, \alpha_n$ are diagonal entries of A. ∎

THEOREM 6

If $\lambda_1, \ldots, \lambda_k$ are distinct eigenvalues of A with corresponding eigenvectors $v_1, \ldots, v_k$, respectively, then $v_1, \ldots, v_k$ are linearly independent.

Proof We prove the theorem for the special case $k = 2$ only. If λ and μ are distinct eigenvalues of A with corresponding eigenvectors x and y, respectively, then x and y are linearly independent. We have $Ax = \lambda x$ and $Ay = \mu y$. Suppose

$$\alpha x + \beta y = 0. \tag{1}$$

Then $\alpha Ax + \beta Ay = 0$, yielding

$$\alpha \lambda x + \beta \mu y = 0. \tag{2}$$

Multiply Equation (1) by λ and subtract from Equation (2) to obtain $(\beta\mu - \beta\lambda)y = 0$, that is, $\beta = 0$, since $(\mu - \lambda)y \neq 0$. Similarly, $\alpha = 0$. This proves the theorem. ∎

THEOREM 7

If an $n \times n$ matrix A has n distinct eigenvalues $\lambda_1, \lambda_2, \ldots, \lambda_n$, then there exists an invertible matrix P such that

$$P^{-1}AP = \begin{bmatrix} \lambda_1 & 0 & 0 & 0 \\ 0 & \lambda_2 & 0 & 0 \\ \vdots & \vdots & \ddots & \vdots \\ 0 & 0 & 0 & \lambda_n \end{bmatrix}.$$

Proof Let $x_1, \ldots, x_n$ be eigenvectors corresponding to eigenvalues $\lambda_1, \ldots, \lambda_n$, respectively. Let $P = [x_1 \ x_2 \cdots x_n]$ be the matrix whose columns are $x_1, \ldots, x_n$. Since $x_1, x_2, \ldots, x_n$ are linearly independent, rank $P = n$, and hence P^{-1} exists. It is a simple verification that

$$AP = P \begin{bmatrix} \lambda_1 & 0 & 0 & 0 \\ 0 & \lambda_2 & 0 & 0 \\ \vdots & \vdots & \ddots & \vdots \\ 0 & 0 & 0 & \lambda_n \end{bmatrix},$$

and so

$$P^{-1}AP = \begin{bmatrix} \lambda_1 & 0 & 0 & 0 \\ 0 & \lambda_2 & 0 & 0 \\ \vdots & \vdots & \ddots & \vdots \\ 0 & 0 & 0 & \lambda_n \end{bmatrix}.$$

∎

7.7 Chapter Review Questions and Projects

1. What is meant by the statement that λ is an eigenvalue of a square matrix A? Consider a triangular matrix A, and find its eigenvalues. (Section 7.1)

2. What is meant by the statement that x is an eigenvector of a square matrix A corresponding to an eigenvalue λ? Can an eigenvector be equal to zero? Find which of the following vectors are eigenvectors of the matrix $\begin{bmatrix} 2 & 4 \\ 2 & 4 \end{bmatrix}$:

$$\begin{bmatrix} 2 \\ 2 \end{bmatrix}, \quad \begin{bmatrix} 0 \\ 0 \end{bmatrix}, \quad \begin{bmatrix} 1 \\ -1 \end{bmatrix}, \quad \begin{bmatrix} 1 \\ 1 \end{bmatrix}. \text{ (Section 7.1)}$$

3. Find an eigenvector of the matrix $\begin{bmatrix} \frac{1}{3} & \frac{1}{2} \\ \frac{2}{3} & \frac{1}{2} \end{bmatrix}$ corresponding to eigenvalue 1. Write down the other eigenvalue without actually computing the characteristic polynomial. Find also an eigenvector corresponding to the second eigenvalue. (Sections 7.5 and 7.2)

4. Comment on the statement "Eigenvalues of a matrix may be found by reducing the matrix into row echelon form."

5. Using the definition of an eigenvalue of a matrix, show that (a) if $A^2 = 0$, then 0 is the only eigenvalue of A, (b) if $A^2 = A$, then 0 and 1 are the only eigenvalues of A. What conclusions can you draw about the eigenvalues of a matrix A if it satisfies a polynomial $a_0 + a_1x + \cdots + a_rx^r$, where $a_i \subset \mathbf{R}$? (Section 7.2)

6. Find a maximal set of linearly independent eigenvectors for each of the following matrices:

a. $\begin{bmatrix} 0 & 1 & 0 \\ 4 & 0 & 0 \\ 0 & 0 & -2 \end{bmatrix}$

b. $\begin{bmatrix} -1 & 0 & -3 \\ -1 & 1 & -2 \\ 1 & -1 & 2 \end{bmatrix}$

c. $\begin{bmatrix} 1 & 2 & 2 \\ -1 & -2 & -1 \\ 1 & 1 & 0 \end{bmatrix}$. (Section 7.2)

7. Find the eigenvalues of the matrix A^{-100}, where

$$A = \begin{bmatrix} 1 & 2 & 3 & 4 \\ 0 & 2 & 3 & 5 \\ 0 & 0 & 3 & 0 \\ 0 & 0 & 0 & -1 \end{bmatrix}. \text{ (Section 7.2 and Section 7.5)}$$

8. What is meant by the statement that a square matrix is diagonalizable? Find which of the matrices in Question 6 are diagonalizable. (Section 7.5)

9. For each of the following matrices A, find a matrix P such that $P^{-1}AP$ is a diagonal matrix: (Section 7.5)

a. $\begin{bmatrix} 0 & 0 & 4 \\ -2 & -4 & -2 \\ -2 & 0 & -6 \end{bmatrix}$
b. $\begin{bmatrix} -1 & 1 & 0 \\ 0 & 4 & -2 \\ 0 & 0 & 2 \end{bmatrix}$.

10. State the Cayley-Hamilton Theorem, and verify it for each of the following matrices:

a. $\begin{bmatrix} -1 & 1 & 0 \\ 0 & 4 & -2 \\ 0 & 0 & 2 \end{bmatrix}$
b. $\begin{bmatrix} 1 & 2 \\ 3 & 4 \end{bmatrix}$.

Also compute the inverse of each of the above matrices by using the Cayley-Hamilton Theorem. (Section 7.4)

11. For the matrix

$$A = \begin{bmatrix} 1 & 2 \\ 2 & 4 \end{bmatrix},$$

compute $A^{50} + A^{10} + A$. (Section 7.4)

Chapter Projects

1. Consider the Pacific Ocean near Vancouver, British Columbia, where the populations of both types A and B of a certain species of fish change each year. It is found that if the type A fish kill the type B fish at a certain rate, both types will continue to exist. Experiments indicate that

$$S_{i+1} = 0.7 \, S_i + 0.00004 \, F_i,$$
$$F_{i+1} = k \, S_i + 1.2 \, F_i,$$

where S_i and F_i represent the populations of type A and type B, respectively, at the end of i years and k is the rate at which the type B fish are killed by the type A fish. Write the above system of equations as a difference equation $P_{i+1} = AP_i$, where

$$A = \begin{bmatrix} 0.7 & 0.00004 \\ k & 1.2 \end{bmatrix}, \qquad P_i = \begin{bmatrix} S_i \\ F_i \end{bmatrix}.$$

Suppose $S_0 = 10,000$ and $F_0 = 50,000$ are the present populations of the two types and $k = -0.02$.

a. Find the eigenvalues and eigenvectors of A.

b. Write the present population vector as a linear combination of the eigenvectors of A.

c. Use the linear combination obtained in part (b) to determine the populations after 5, 10, 13, 14, and 15 years.

d. Can you conclude whether either one of the types would become extinct after a while?

2. Three gas stations serve in a locality. Customers change from one gas station to the other according to the gas prices. The probability matrix A of shift of the customers at the end of each month is given by

$$A = \begin{bmatrix} 0.44 & 0.35 & 0.35 \\ 0.14 & 0.35 & 0.10 \\ 0.42 & 0.30 & 0.55 \end{bmatrix},$$

where the (i, j) entry represents the probability of a customer moving from the gas station j to the gas station i.

a. Find powers of A and their ranks. Observe that the sequence of matrices A, A^2, $A^3, \ldots$ becomes stationary, say, at the mth stage and the rank$(A^m) = 1$.

b. The equilibrium vector is a vector representing the fraction of customers that remains the same from one month to the next month. Note that the equilibrium vector is an eigenvector of A. What is the corresponding eigenvalue? Find the equilibrium vector.

c. Compute $\lim_{n \to \infty} A^n$. What is the relation that you observe between the matrix $\lim_{n \to \infty} A^n$ and the equilibrium vector?

d. If on April 1, the market share of the customers going to gas stations I, II, and III is given by $\begin{bmatrix} \frac{1}{3} & \frac{1}{2} & \frac{1}{6} \end{bmatrix}$, find the market shares on May 1, and December 1 of the same year.

Key Words

Eigenvalues of a Square Matrix
Eigenvectors of a Square Matrix
Characteristic Polynomial of a Matrix
Spectrum of a Matrix
Trace of a Matrix
Multiplicity of an Eigenvalue

Similar Matrices
Maximal Set of Linearly Independent
  Vectors
Diagonalizable Matrix
Triangularizable Matrix
Cayley-Hamilton Theorem

Key Phrases

■ An $n \times n$ matrix is diagonalizable if and only if it has n linearly independent eigenvectors.

■ Every $n \times n$ matrix is triangularizable.

■ Every $n \times n$ matrix satisfies the characteristic polynomial.

8 Inner Product Spaces

Introduction

In this chapter, we define the inner product (also called dot or scalar product) of two n-dimensional vectors. As in the case of vectors in two-dimensional and three-dimensional spaces, we say that two n-dimensional vectors are perpendicular if their inner (dot) product is zero, and we prove the usual Pythagorean Theorem. Among several applications, methods to obtain approximate solutions of inconsistent linear systems are also given.

8.1 Gram-Schmidt Orthogonalization Process

Consider the two perpendicular line segments OP and OQ:

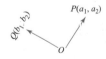

We know that $OP \perp OQ$ if (slope of OP)(slope of OQ) = -1; that is,

$$\frac{a_2}{a_1}\frac{b_2}{b_1} = -1,$$

which gives $a_2 b_2 = -a_1 b_1$, or $a_1 b_1 + a_2 b_2 = 0$. Using matrix notation, this means that

$$[a_1 \ a_2] \begin{bmatrix} b_1 \\ b_2 \end{bmatrix} = 0.$$

Equivalently, $x^T y = 0$, where

$$x = \begin{bmatrix} a_1 \\ a_2 \end{bmatrix}, \qquad y = \begin{bmatrix} b_1 \\ b_2 \end{bmatrix}.$$

However, if we use row vector notation, that is, if $x = [a_1 \ a_2]$ and $y = [b_1 \ b_2]$, then $x \perp y$ means that $xy^T = 0$.

The above concrete idea of perpendicularity of two line segments in a plane is generalized to any two vectors x and y in F^n for any positive integer n.

Throughout, unless otherwise stated, F denotes the field of real numbers, and all vectors are column vectors.

8.1.1
DEFINITION

Orthogonal (or Perpendicular) Vectors

A vector $x \in F^n$ is called *orthogonal* (or *perpendicular*) to a vector $y \in F^n$ if $x^T y = 0$. If x is orthogonal to y, we write $x \perp y$.

Note that $x \perp y$ implies that $y \perp x$, since $x^T y = 0$ if and only if $y^T x = 0$. Thus saying that x is orthogonal to y is equivalent to saying that x and y are orthogonal to each other.

For example,

$$\text{if } e_1 = \begin{bmatrix} 1 \\ 0 \end{bmatrix} \quad \text{and} \quad e_2 = \begin{bmatrix} 0 \\ 1 \end{bmatrix}$$

in $\mathbf{R}^2$, then $e_1 \perp e_2$. Also, if

$$v_1 = \begin{bmatrix} 2 \\ 3 \end{bmatrix} \quad \text{and} \quad v_2 = \begin{bmatrix} -3 \\ 2 \end{bmatrix},$$

then $v_1 \perp v_2$.

More generally, vectors $v_1, \ldots, v_m \in F^n$ are said to be *orthogonal* if $v_i \perp v_j$, for all $i, j, i \neq j$.

8.1.2
DEFINITION

Inner, Dot, or Scalar Product of Vectors

If

$$x = \begin{bmatrix} x_1 \\ x_2 \\ \vdots \\ x_n \end{bmatrix} \quad \text{and} \quad y = \begin{bmatrix} y_1 \\ y_2 \\ \vdots \\ y_n \end{bmatrix},$$

then the product $x^T y = x_1 y_1 + \cdots + x_n y_n$, denoted by $\langle x, y \rangle$, is called the *inner (dot, or scalar) product* of x and y. (Note: We identify the 1×1 matrix $[a]$ with the number a itself.)

If F is the field of complex numbers, the inner product of x and y is defined to be $x^T \overline{y} = x_1 \overline{y_1} + \cdots + x_n \overline{y_n}$.

Note that $x \perp y$ if and only if $\langle x, y \rangle = 0$.

Recall that the length of the line segment OP in the figure is $\sqrt{a_1^2 + a_2^2} = \sqrt{x^T x}$, where

$$x = \begin{bmatrix} a_1 \\ a_2 \end{bmatrix}.$$

We give below the definition of the length of a vector in F^n.

8.1.3 Remark If x and y are row vectors, then the inner product $\langle x, y \rangle$ is defined to be xy^T.

8.1.4 DEFINITION

Length of a Vector
If $x \in F^n$, then the nonnegative real number $\sqrt{\langle x, x \rangle} = \sqrt{x^T x}$ is called the *length* of x.

The length of a vector x is denoted by $\|x\|$ and is 0 if and only if $x = 0$.

8.1.5 DEFINITION

Orthogonal and Orthonormal Bases
A basis $\{v_1, v_2, \ldots, v_m\}$ of a subspace W of F^n is called an *orthogonal basis* if

$$\langle v_i, v_j \rangle = v_i^T v_j = 0, \qquad i, j = 1, \ldots, m, i \neq j. \tag{1}$$

If in addition

$$\|v_i\| = 1, \qquad i = 1, \ldots, m, \tag{2}$$

then it is called an *orthonormal basis*.

8.1.6 EXAMPLES

a. $e_1 = \begin{bmatrix} 1 \\ 0 \end{bmatrix}$ and $e_2 = \begin{bmatrix} 0 \\ 1 \end{bmatrix}$ form an orthonormal basis of $\mathbf{R}^2$.

b. $v_1 = \begin{bmatrix} 3 \\ 0 \\ 4 \end{bmatrix}$ and $v_2 = \begin{bmatrix} -4 \\ 0 \\ 3 \end{bmatrix}$ form an orthogonal basis of the subspace of $\mathbf{R}^3$ spanned by v_1 and v_2.

Let us first prove the familiar Pythagorean Theorem in the context of the inner product. ■

8.1.7 THEOREM

Pythagorean Theorem
Let u and v be two orthogonal vectors in F^n. Then $\|u + v\|^2 = \|u\|^2 + \|v\|^2$.

Proof $\|u+v\|^2 = (u+v)^T(u+v) = (u^T + v^T)(u+v) = u^T u + u^T v + v^T u + v^T v = \|u\|^2 + \|v\|^2$, since $u \perp v$. ■

We now give a method for finding an orthonormal basis of a subspace of F^n, known as Gram-Schmidt method.

8.1.8 Gram-Schmidt Method

For $x, y \in F^n$, $\langle x, y \rangle$ will denote the real number $x^T y$, the inner product as defined above.

Let W be a subspace of F^n with a basis $\{v_1, v_2, \ldots, v_m\}$. We construct an orthogonal basis $w_1, w_2, \ldots, w_m$ and an orthonormal basis $\{u_1, u_2, \ldots, u_m\}$ as follows:

Set

$$w_1 = v_1, u_1 = \frac{w_1}{\|w_1\|},$$

$$w_2 = v_2 - \langle v_2, u_1 \rangle u_1, u_2 = \frac{w_2}{\|w_2\|},$$

$$w_3 = v_3 - \langle v_3, u_1 \rangle u_1 - \langle v_3, u_2 \rangle u_2, u_3 = \frac{w_3}{\|w_3\|},$$

$$\vdots$$

$$w_m = v_m - \langle v_m, u_1 \rangle u_1 - \cdots - \langle v_m, u_{m-1} \rangle u_{m-1}, u_m = \frac{w_m}{\|w_m\|}.$$

Then it can be shown that $\{u_1, u_2, \ldots, u_m\}$ is an orthonormal basis of W.

8.1.9 Warning

If a list $v_1, \ldots, v_m$ of vectors generates W but is not linearly independent, then the above process would lead to some $w_i = 0$, and so the corresponding u_i would not be obtainable. We then disregard the vector v_i because it is a linear combination of v_1, $v_2, \ldots, v_{i-1}$ and proceed as if v_i were absent.

8.1.10 EXAMPLES

a. Let W be the subspace of $\mathbf{R}^4$ generated by

$$v_1 = \begin{bmatrix} 1 \\ 0 \\ 0 \\ 1 \end{bmatrix}, \qquad v_2 = \begin{bmatrix} 2 \\ 0 \\ 1 \\ 2 \end{bmatrix}, \qquad \text{and} \qquad v_3 = \begin{bmatrix} 0 \\ 0 \\ 1 \\ 0 \end{bmatrix}.$$

Show that

$$u_1 = \frac{1}{\sqrt{2}} \begin{bmatrix} 1 \\ 0 \\ 0 \\ 1 \end{bmatrix} \qquad \text{and} \qquad u_2 = \begin{bmatrix} 0 \\ 0 \\ 1 \\ 0 \end{bmatrix}$$

form an orthonormal basis of W.

It can easily be seen that $2v_1 + (-1)2v_2 + v_3 = 0$. Thus the set $\{v_1, v_2, v_3\}$ is a linearly dependent set. But $\{v_1, v_2\}$ is clearly a linearly independent set, and so

$\{v_1, v_2\}$ is a basis of W. We now apply the Gram-Schmidt method to find an orthonormal basis:

$$u_1 = \frac{v_1}{\|v_1\|} = \frac{1}{\sqrt{2}}v_1 = \frac{1}{\sqrt{2}}\begin{bmatrix} 1 \\ 0 \\ 0 \\ 1 \end{bmatrix},$$

$$w_2 = v_2 - \langle v_2, u_1 \rangle u_1 = \begin{bmatrix} 0 \\ 0 \\ 1 \\ 0 \end{bmatrix},$$

$$u_2 = \frac{w_2}{\|w_2\|} = \begin{bmatrix} 0 \\ 0 \\ 1 \\ 0 \end{bmatrix}.$$

Note: We could have applied the Gram-Schmidt process without first checking for linear independence and omitted the vector v_3 as explained in the warning preceding this example.

b. Let W be the subspace of $\mathbf{R}^4$ generated by

$$v_1 = \begin{bmatrix} 1 \\ 1 \\ 1 \\ 1 \end{bmatrix}, \qquad v_2 = \begin{bmatrix} 1 \\ 0 \\ 1 \\ 0 \end{bmatrix}, \qquad v_3 = \begin{bmatrix} 0 \\ 0 \\ 1 \\ 0 \end{bmatrix}.$$

Find an orthonormal basis of W. Observing that v_1, v_2, and v_3 are linearly independent, we can apply the Gram-Schmidt method to the set $\{v_1, v_2, v_3\}$ to find an orthonormal basis of W. The vectors u_1, u_2, and u_3 forming an orthonormal basis are obtained as follows:

$$u_1 = \frac{v_1}{\|v_1\|} = \frac{1}{2}\begin{bmatrix} 1 \\ 1 \\ 1 \\ 1 \end{bmatrix},$$

$$w_2 = v_2 - \langle v_2, u_1 \rangle u_1 = \frac{1}{2}\begin{bmatrix} 1 \\ -1 \\ 1 \\ -1 \end{bmatrix},$$

$$u_2 = \frac{w_2}{\|w_2\|} = \frac{1}{2}\begin{bmatrix} 1 \\ -1 \\ 1 \\ 1 \end{bmatrix},$$

$$w_3 = v_3 - \langle v_3, u_1 \rangle u_1 - \langle v_3, u_2 \rangle u_2 = \frac{1}{2} \begin{bmatrix} -1 \\ 0 \\ 1 \\ 0 \end{bmatrix},$$

$$u_3 = \frac{1}{\sqrt{2}} \begin{bmatrix} -1 \\ 0 \\ 1 \\ 0 \end{bmatrix}.$$

8.1 Exercises

1. Determine which of the following sets of vectors are orthogonal:

    a. $(1, 1, 0), (1, 0, 1), (0, 1, 1)$
    b. $(4, 1, 3), (2, 0, -3), (1, 4, 1)$
    c. $(2, 0, 2), (0, 1, 0), (-2, 0, 2)$
    d. $(1, 0, -1, 2), (-3, 1, -3, 0)$

2. Find x and y such that the vector

$$\begin{bmatrix} x \\ y \\ 1 \end{bmatrix} \in \mathbf{R}^3$$

    is orthogonal to the vectors $\begin{bmatrix} 1 \\ 2 \\ 3 \end{bmatrix}$ and $\begin{bmatrix} 1 \\ 0 \\ 1 \end{bmatrix}$.

3. Find x and y such that the vector $\begin{bmatrix} 2x \\ 1 \\ 3y \end{bmatrix} \in \mathbf{R}^3$ is orthogonal to the vectors $\begin{bmatrix} 2 \\ 0 \\ 1 \end{bmatrix}$

    and $\begin{bmatrix} -1 \\ 1 \\ 2 \end{bmatrix}$.

4. (Drill 8.1) Find a, b, c, d such that the vectors $[1\ 0\ 1\ 2]$, $[1\ 1\ -1\ 0]$, $[1\ -2\ -1\ 0]$, and $[a\ b\ c\ d]$ are mutually orthogonal.

5. Find a, b, c such that $\left\{ \begin{bmatrix} 2 \\ 1 \\ -1 \end{bmatrix}, \begin{bmatrix} a \\ 1 \\ -1 \end{bmatrix}, \begin{bmatrix} b \\ 3 \\ c \end{bmatrix} \right\}$ forms an orthogonal basis

    of $\mathbf{R}^3$.

6. (Drill 8.2) Find a, b, c such that $\{[2\ b\ 1], [a\ 1\ -1], [1\ 3\ c]\}$ forms an orthogonal basis of $\mathbf{R}^3$.

7. Use the Gram-Schmidt orthonormalization method to find an orthonormal basis for each of the subspaces spanned by the following vectors. Use the Matlab command $[Q, R] = qr(A)$ to find a matrix Q whose columns form an orthonormal basis of the column space of the matrix A formed by the given column vectors. The reader may note that R is upper triangular and $A = QR$.

    a. (Drill 8.3) $\begin{bmatrix} 1 \\ 2 \\ 1 \end{bmatrix}, \begin{bmatrix} 1 \\ 0 \\ 1 \end{bmatrix}, \begin{bmatrix} 3 \\ 1 \\ 0 \end{bmatrix}$

    b. $\begin{bmatrix} 1 \\ 1 \\ 0 \end{bmatrix}, \begin{bmatrix} -1 \\ 0 \\ 1 \end{bmatrix}$

    c. (Drill 8.4) $\begin{bmatrix} 0 \\ 0 \\ 1 \\ 0 \end{bmatrix}, \begin{bmatrix} 1 \\ 0 \\ 1 \\ 1 \end{bmatrix}, \begin{bmatrix} 1 \\ 1 \\ 2 \\ 1 \end{bmatrix}$

    d. $[1 \ 0 \ 1 \ 2]^T, [2 \ 1 \ 0 \ 2]^T, [1 \ -1 \ 0 \ 1]^T$

8. Show that any set of nonzero orthogonal vectors is linearly independent.

9. Let A be an $n \times n$ matrix with real entries such that A^T is the inverse of A. Show that the columns of the matrix A as elements of $\mathbf{R}^n$ form an orthonormal basis of $\mathbf{R}^n$.

10. A matrix U with real entries is called orthogonal if it is invertible and $U^T = U^{-1}$. Show that if U is an $n \times n$ orthogonal matrix and x is an $n \times 1$ column vector, then $\|Ux\| = \|x\|$.

11. Show that for any $m \times n$ matrix A, $A^T Ax = 0$ implies that $\|Ax\| = 0$, and hence $Ax = 0$. Use this to conclude that rank $A = $ rank $A^T A$. Prove a similar result for AA^T.

8.2 Diagonalization of Symmetric Matrices

8.2.1
DEFINITION

Diagonalizable
Recall that a square matrix A is said to be *diagonalizable* if there exists an invertible matrix P such that $P^{-1}AP$ is a diagonal matrix.

We observe that not every square matrix is diagonalizable. For example, it can be easily shown that $\begin{bmatrix} 1 & 1 \\ 0 & 1 \end{bmatrix}$ cannot be diagonalized. Our objective in this section is to show that every symmetric matrix is diagonalizable.

8.2.2 An $n \times n$ matrix is diagonalizable if and only if it has n linearly independent eigen-
Remark vectors.

Let us recall the definition of a symmetric matrix.

8.2.3
DEFINITION

Symmetric Matrix
An $n \times n$ matrix $A = (a_{ij})$, where a_{ij} are real, is called *real symmetric* or *simply symmetric* if $A = A^T$, that is, $a_{ij} = a_{ji}$ for all $i, j = 1, 2, \ldots, n$.

We assume that the reader is familiar with the complex numbers. As a reminder, let us recapitulate the concept of complex numbers through some examples. Consider the equation $x^2 + 1 = 0$, which has no real roots. Its roots are written as $\pm\sqrt{-1}$ or as $\pm i$. Also, the roots of $x^2 + x + 1 = 0$ are $(-1 \pm \sqrt{-3})/2$, which again are not real numbers. These numbers $\sqrt{-1}$ and $-1 + \sqrt{-3}$ are examples of complex numbers. In fact, every complex number is of the form $a + ib$, where a and b are real. Furthermore, the fundamental theorem of algebra states that any polynomial of degree n has n (not necessarily distinct) roots, which may be real or complex numbers. Thus the characteristic polynomial of degree n has n roots, which may be real or complex numbers. Theorem 1 below asserts that if A is a real symmetric matrix, then every eigenvalue is real.

If $z = a + ib$ is a complex number, where a and b are real, then $\bar{z}$ denotes the complex number $a - ib$ and is called the complex conjugate of z.

If $A = (a_{ij})$ is an $m \times n$ matrix, then $\overline{A}$ denotes the matrix $(\overline{a_{ij}})$; that is, $\overline{A}$ is the matrix whose (i, j) entry is the complex conjugate of the (i, j) entry of A.

It is easy to verify that for any two matrices A and B, where AB is defined, $\overline{AB} = \overline{A}\,\overline{B}$.

8.2.4
THEOREM

(Theorems 1 and 2, Section 8.6)
The eigenvalues of a real symmetric matrix are real and eigenvectors corresponding to distinct eigenvalues are orthogonal.

The following additional facts will be used throughout this section and are stated without proof.

8.2.5 Fact

If k is the multiplicity of an eigenvalue λ of a symmetric matrix, then there are k linearly independent eigenvectors corresponding to eigenvalue λ.

8.2.6 Fact

If A is an $n \times n$ symmetric matrix, then there are n linearly independent eigenvectors of A.

8.2.7
DEFINITION

Orthogonal Matrix
A square matrix P with real entries is called *orthogonal* if $P^T = P^{-1}$.

It is an interesting fact that for any vector u, $\|Pu\| = \|u\|$ when P is an orthogonal matrix. We next state the most important property of symmetric matrices contained in the spectral theorem and outline its proof.

**8.2.8
THEOREM**

(Spectral Theorem)
If A is an $n \times n$ symmetric matrix, then there exists an orthogonal matrix P such that

$$\mathbf{P}^T \mathbf{A} \mathbf{P} = \begin{bmatrix} \lambda_1 & 0 & \cdots & 0 \\ 0 & \lambda_2 & \cdots & 0 \\ \vdots & \vdots & & \vdots \\ 0 & 0 & \cdots & \lambda_n \end{bmatrix},$$

where $\lambda_1, \ldots, \lambda_n$ are eigenvalues of A.

The matrix P in the statement of the theorem is $[x_1 \, x_2 \, \ldots \, x_n]$ whose columns $x_1, \ldots, x_n$ form an orthonormal set of eigenvectors of A.

Sketch of the Proof   Let $\lambda_1, \ldots, \lambda_m$ be distinct eigenvalues of A. If k_1 is the multiplicity of λ_1, then by Fact 8.2.5, we can select a set of k_1 linearly independent eigenvectors that can be orthogonalized by the Gram-Schmidt method. Repeat the process for every eigenvalue, so as to obtain an orthonormal set of n eigenvectors. Call these vectors $x_1, x_2, \ldots, x_n$, and set $P = [x_1 \, x_2 \cdots x_n]$. Then

$$AP = [Ax_1 \quad Ax_2 \quad \cdots \quad Ax_n]$$
$$= [\lambda_1 x_1 \quad \lambda_2 x_2 \quad \cdots \quad \lambda_n x_n]$$
$$= [x_1 \quad x_2 \quad \cdots \quad x_n]D,$$

where D is the diagonal matrix

$$\begin{bmatrix} \lambda_1 & 0 & \cdots & 0 \\ 0 & \lambda_2 & \cdots & 0 \\ \vdots & \vdots & & \vdots \\ 0 & 0 & \cdots & \lambda_n \end{bmatrix},$$

and so $P^{-1}AP = D$. An immediate computation yields $PP^T = I$, and so $P^t = P^{-1}$.
This completes the proof.

In summary, following is a procedure for diagonalizing a symmetric matrix.

Step 1 Find the eigenvalues of the matrix.

Step 2 Find a maximal set of linearly independent eigenvectors for each eigen-value. (Note that the multiplicity of the eigenvalue is the number of linearly independent eigenvectors corresponding to that eigenvalue.)

Step 3 Apply the Gram-Schmidt method to find an orthonormal set of eigenvec-tors found in Step 2 for each eigenvalue.

Step 4 The union of orthonormal sets in Step 3 yields the desired matrix P. If $x_1, x_2, \ldots, x_n$ is a complete set of orthonormal eigenvectors, then put $P = [x_1 \ x_2 \cdots x_n]$.

8.2.9 EXAMPLES

a. Let

$$A = \begin{bmatrix} 1 & 2 \\ 2 & 1 \end{bmatrix}.$$

We find an orthogonal matrix P such that $P^T A P$ is diagonal. The characteristic polynomial of A is

$$C_A(x) = \det \begin{bmatrix} 1-x & 2 \\ 2 & 1-x \end{bmatrix}$$
$$= (x^2 - 2x - 3) = (x+1)(x-3).$$

Thus the eigenvalues are -1 and 3.

Consider first the eigenvalue, -1. We proceed to find an orthonormal set of eigenvectors corresponding to eigenvalue -1. Solving $(A + I)x = 0$, we get

$$x = \begin{bmatrix} 1 \\ -1 \end{bmatrix}$$

as a particular solution, and any solution is a multiple of $\begin{bmatrix} 1 \\ -1 \end{bmatrix}$. Thus $\left\{ \begin{bmatrix} 1 \\ -1 \end{bmatrix} \right\}$ is a maximal linearly independent set of eigenvectors corresponding to eigenvalue -1. We apply the Gram-Schmidt method to transform a basis $\left\{ \begin{bmatrix} 1 \\ -1 \end{bmatrix} \right\}$ into an orthonormal basis $\left\{ \frac{1}{\sqrt{2}} \begin{bmatrix} 1 \\ -1 \end{bmatrix} \right\}$. Similarly, $\left\{ \frac{1}{\sqrt{2}} \begin{bmatrix} 1 \\ 1 \end{bmatrix} \right\}$ is an orthonormal basis of the null space of $A - 3I$. Put

$$P = \begin{bmatrix} \frac{1}{\sqrt{2}} & \frac{1}{\sqrt{2}} \\ -\frac{1}{\sqrt{2}} & \frac{1}{\sqrt{2}} \end{bmatrix}.$$

Then

$$P^T P = I \quad \text{and} \quad P^T A P = \begin{bmatrix} -1 & 0 \\ 0 & 3 \end{bmatrix}$$

as desired.

b. $A = \begin{bmatrix} 5 & 2 & 2 \\ 2 & 5 & 2 \\ 2 & 2 & 5 \end{bmatrix}.$

We find an orthogonal matrix P such that $P^T A P$ is diagonal. The characteristic polynomial $C_A(x)$ is $-(x^3 - 15x^2 + 63x - 81) = (x-3)^2(9-x)$. So the eigenvalues are 3, 3, and 9.

Consider first the eigenvalue 3, and solve $(A - 3I)x = 0$. Actual computations yield that

$$x = \begin{bmatrix} a \\ b \\ c \end{bmatrix}$$

is a solution if and only if $a + b + c = 0$. Thus

$$x = \begin{bmatrix} -b - c \\ b \\ c \end{bmatrix}$$

is an eigenvector corresponding to eigenvalue 3, where b and c are arbitrary numbers.

Rewrite

$$x = \begin{bmatrix} -b - c \\ b \\ c \end{bmatrix}$$

as

$$x = \begin{bmatrix} -b \\ b \\ 0 \end{bmatrix} + \begin{bmatrix} -c \\ 0 \\ c \end{bmatrix} = b \begin{bmatrix} -1 \\ 1 \\ 0 \end{bmatrix} + c \begin{bmatrix} -1 \\ 0 \\ 1 \end{bmatrix}.$$

This shows that $\begin{bmatrix} -1 \\ 1 \\ 0 \end{bmatrix}$ and $\begin{bmatrix} -1 \\ 0 \\ 1 \end{bmatrix}$ are eigenvectors and that any other eigenvector corresponding to eigenvalue 3 is a linear combination of these two vectors. Now, given these two linearly independent vectors

$$v_1 = \begin{bmatrix} -1 \\ 1 \\ 0 \end{bmatrix} \qquad \text{and} \qquad v_2 = \begin{bmatrix} -1 \\ 0 \\ 1 \end{bmatrix},$$

we apply the Gram-Schmidt method to find an orthonormal basis $\{u_1, u_2\}$ of the null space of $A - 3I$. It turns out that

$$u_1 = \frac{1}{\sqrt{2}} \begin{bmatrix} -1 \\ 1 \\ 0 \end{bmatrix}, \qquad u_2 = \frac{1}{\sqrt{6}} \begin{bmatrix} -1 \\ -1 \\ 2 \end{bmatrix}.$$

Next, we find that $\left\{ \begin{bmatrix} 1 \\ 1 \\ 1 \end{bmatrix} \right\}$ is a basis of the null space of $A - 9I$; that is,

$$\left\{ \begin{bmatrix} 1 \\ 1 \\ 1 \end{bmatrix} \right\}$$ is a maximal set of linearly independent eigenvectors corresponding

to eigenvalue 9. By the Gram-Schmidt method, we obtain $\left\{ \dfrac{1}{\sqrt{3}} \begin{bmatrix} 1 \\ 1 \\ 1 \end{bmatrix} \right\}$ as an orthonormal basis of the null space of $A - 9I$. Put

$$P = \begin{bmatrix} -\dfrac{1}{\sqrt{2}} & -\dfrac{1}{\sqrt{6}} & \dfrac{1}{\sqrt{3}} \\ \dfrac{1}{\sqrt{2}} & -\dfrac{1}{\sqrt{6}} & \dfrac{1}{\sqrt{3}} \\ 0 & \dfrac{2}{\sqrt{6}} & \dfrac{1}{\sqrt{3}} \end{bmatrix}.$$

Then

$$PP^T = I \quad \text{and} \quad P^T A P = \begin{bmatrix} 3 & 0 & 0 \\ 0 & 3 & 0 \\ 0 & 0 & 9 \end{bmatrix},$$

as desired. ∎

8.2 Exercises

1. Find a maximal set of orthonormal eigenvectors for the following matrices. Use the Matlab command $[X, D] = eig(A)$ to obtain a matrix X whose columns are eigenvectors and then $[Q, R] = qr(X)$ to obtain Q whose columns form an orthonormal basis of the column space of X.

   a. $\begin{bmatrix} -4 & 4 & -2 \\ 4 & 2 & -4 \\ -2 & -4 & -4 \end{bmatrix}$, given that its eigenvalues are -6, -6, and 6.

   b. (Drill 8.5) $\begin{bmatrix} 2 & 0 & 0 & 0 \\ 0 & 1 & -2 & 2 \\ 0 & -2 & 1 & 2 \\ 0 & 2 & 2 & 1 \end{bmatrix}$, given that its eigenvalues are $-3, 2, 3$, and 3.

   c. $\begin{bmatrix} 6 & 0 & 0 & 0 \\ 0 & 6 & 1 & 1 \\ 0 & 1 & 6 & 1 \\ 0 & 1 & 1 & 6 \end{bmatrix}$, given that its eigenvalues are $5, 5, 6$, and 8.

   d. $\begin{bmatrix} 1 & 1 & 0 \\ 1 & 1 & 0 \\ 0 & 0 & 2 \end{bmatrix}$

   e. (Drill 8.5) $\begin{bmatrix} 5 & 2 & 2 \\ 2 & 5 & 2 \\ 2 & 2 & 5 \end{bmatrix}$

2. a. Find an orthogonal matrix P such that $P^T A P$ is diagonal, where

$$A = \begin{bmatrix} 1 & 1 & 1 \\ 1 & 1 & 1 \\ 1 & 1 & 1 \end{bmatrix}.$$

b. Use part (a) to compute A^{100} and $\sqrt{A}$ (i.e., a matrix B such that $B^2 = A$).

3. a. For each of the following matrices A, find an orthogonal matrix P such that $P^T A P$ is diagonal:

i. (Drill 8.6) $\begin{bmatrix} 0 & 2 & 2 \\ 2 & 0 & -2 \\ 2 & -2 & 0 \end{bmatrix}$

ii. $\begin{bmatrix} 1 & 2 \\ 2 & 1 \end{bmatrix}$

iii. $\begin{bmatrix} -1 & -3 & 0 \\ -3 & -1 & 0 \\ 0 & 0 & 2 \end{bmatrix}$

iv. $\begin{bmatrix} 0 & 2 & 2 \\ 2 & 4 & 2 \\ 2 & 2 & 0 \end{bmatrix}$

b. (Drill 8.6) Use part (a) to compute A^{10}.

4. (Singular Value Decomposition) For each of the following matrices A:

i. $\begin{bmatrix} 1 & 2 \\ 1 & 2 \end{bmatrix}$

ii. $\begin{bmatrix} 2 & 2 & 0 \\ -1 & 1 & 0 \\ 0 & 0 & 1 \end{bmatrix}$

a. find the eigenvalues of AA^T and $A^T A$;

b. find an orthonormal set of eigenvectors of the matrix $A^T A$;

c. prove that $A = s_1 u_1 v_1^T + s_2 u_2 v_2^T$ and $A = s_1 u_1 v_1^T + s_2 u_2 v_2^T + s_3 u_3 v_3^T$ for matrices in (i) and (ii) respectively, where v_i's are eigenvectors, s_i's are eigenvalues of $A^T A$, and $s_i u_i = A v_i$, $i = 1, 2, 3$.
(If the eigenvalues s_1^2, s_2^2, and s_3^2 are arranged in decreasing order, then the above decomposition of A is known as singular value decomposition of A, and s_1, s_2, and s_3 are called singular values of A.)

5. For each of the following matrices A, find orthogonal matrices U and V such that $A = USV^T$, where S is a diagonal matrix whose diagonal entries are the singular values of A (see Exercise 4):

a. $A = \begin{bmatrix} 2 & -2 \\ 3 & 3 \end{bmatrix}$,

$$\textbf{b.} \ \ A = \begin{bmatrix} 6 & 3 & 0 \\ -6 & 0 & 3 \\ 0 & -6 & -6 \end{bmatrix},$$

$$\textbf{c.} \ \ A = \begin{bmatrix} 2 & 0 & 0 \\ 0 & 2 & 1 \\ 0 & 1 & 2 \end{bmatrix}.$$

8.3 Application of the Spectral Theorem

A homogeneous polynomial of degree 2 is called a *quadratic form*. For example, $f(x_1, x_2) = ax_1^2 + 2hx_1x_2 + bx_2^2$ represents a quadratic form in x_1, x_2. The graph of $f(x_1, x_2) = c$ is known to be a conic (circle, parabola, hyperbola, ellipse, or pair of straight lines). We may rewrite $f(x_1, x_2)$ as

$$ax_1^2 + 2hx_1x_2 + bx_2^2 = x^T Ax,$$

where

$$x = \begin{bmatrix} x_1 \\ x_2 \end{bmatrix} \qquad \text{and} \qquad A = \begin{bmatrix} a & h \\ h & b \end{bmatrix}$$

is a symmetric matrix.

A quadratic form of degree 3 in x_1, x_2, x_3 is

$$a_{11}x_1^2 + a_{22}x_2^2 + a_{33}x_3^2 + 2a_{12}x_1x_2 + 2a_{23}x_2x_3 + 2a_{31}x_3x_1.$$

This can also be written as $x^T Ax$, where

$$x = \begin{bmatrix} x_1 \\ x_2 \\ x_3 \end{bmatrix}$$

and $A = (a_{ij})$ is a 3×3 symmetric matrix.

In general, a quadratic form in n variables can similarly be expressed as $Q(x_1, \ldots, x_n) = x^T Ax$, where A is an $n \times n$ symmetric matrix and

$$x = \begin{bmatrix} x_1 \\ x_2 \\ \vdots \\ x_n \end{bmatrix}.$$

The spectral theorem can be used to study the nature of quadratic forms in any number of variables. For example, we can check whether the quadratic form is always positive or always negative, a question that must be answered to determine the points of maxima and minima of functions of several variables. As a consequence of the spectral theorem, we state the principal axes theorem.

8.3.1
THEOREM

(Principal Axes Theorem)
Let $Q(x_1, \ldots, x_n) = x^T A x$ be a quadratic form. Then $Q(x_1, \ldots, x_n)$ can be transformed into $\lambda_1 x_1'^2 + \cdots + \lambda_n x_n'^2$ by an orthogonal transformation, that is, by $x = Px'$, where P is an orthogonal matrix and $\lambda_1, \ldots, \lambda_n$ are eigenvalues of A.

The proof of the above theorem is straightforward and is provided here for the interested reader. Since A is symmetric, there exists an orthogonal matrix P such that $P^T A P$ is diagonal.

Consider the transformation of coordinates $x = Px'$. Then

$$
\begin{aligned}
x^T A x &= (Px')^T A P x' \\
&= x'^T P^T A P x' \\
&= \lambda_1 x_1'^2 + \lambda_2 x_2'^2 + \cdots + \lambda_n x_n'^2,
\end{aligned}
$$

where $\lambda_1, \lambda_2, \ldots, \lambda_n$ are the eigenvalues of A.

8.3.2
EXAMPLES

a. $x^2 + 4xy + y^2 = 6$ represents a hyperbola.

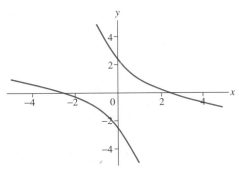

Write $x^2 + 4xy + y^2 = X^T A X$, where

$$
X = \begin{bmatrix} x \\ y \end{bmatrix} \quad \text{and} \quad A = \begin{bmatrix} 1 & 2 \\ 2 & 1 \end{bmatrix}.
$$

The eigenvalues of A are 3 and -1. So by the principal axes theorem, $x^2 + 4xy + y^2$ can be transformed into $3x'^2 - y'^2 = 6$ by the transformation $X = PX'$, where P is an orthogonal matrix.

b. The quadratic function $Q(x_1, x_2, x_3) = 5x_1^2 + 5x_2^2 + 5x_3^2 + 4x_1x_2 + 4x_2x_3 + 4x_3x_1$ is nonnegative for all x_1, x_2, and x_3.
We can rewrite $Q(x_1, x_2, x_3) = x^T A x$, where

$$
x = \begin{bmatrix} x_1 \\ x_2 \\ x_3 \end{bmatrix} \quad \text{and} \quad A = \begin{bmatrix} 5 & 2 & 2 \\ 2 & 5 & 2 \\ 2 & 2 & 5 \end{bmatrix}.
$$

The eigenvalues of A are 3, 3, and 9. By the principal axes theorem, $Q(x_1, x_2, x_3)$ can be transformed into $3x_1'^2 + 3x_2'^2 + 9x_3'^2$ by the transformation $x = Px'$, where

P is an orthogonal matrix. Since

$$3x_1'^2 + 3x_2'^2 + 9x_3'^2 \geqslant 0$$

for all values of x_1', x_2', and x_3', it follows that $Q(x_1, x_2, x_3) \geqslant 0$ for all values of x_1, x_2, and x_3. ■

8.3 Exercises

1. Find the nature of the graph represented by each of the following equations:

    a. (Drill 8.7) $x_1^2 + 4x_1x_2 + 3x_2^2 = 4$.
    b. $x^2 - 4xy + y^2 = 1$.

2. An $n \times n$ symmetric matrix A is called positive definite if $x^T Ax > 0$ for all $x \neq 0$. Show that A is positive definite if and only if all its eigenvalues are positive.

3. Check whether the following symmetric matrices are positive definite:

    a. $\begin{bmatrix} 1 & 2 \\ 2 & 1 \end{bmatrix}$
                b. $\begin{bmatrix} 2 & 0 & 1 \\ 0 & 1 & 0 \\ 1 & 0 & 1 \end{bmatrix}$

    c. $\begin{bmatrix} -1 & 0 & 2 & 2 \\ 0 & -1 & 2 & -2 \\ 2 & 2 & 1 & 0 \\ 2 & -2 & 0 & 1 \end{bmatrix}$

4. (Drill 8.8) Show that the expression

$$2x^2 + 2y^2 + 3z^2 + 2t^2 + 4xz - 2xt - 2yt$$

is positive for all x, y, z, and t (except when $x = y = z = t = 0$).

8.4 Least-Squares Solution

In *Chapters 1* and *2*, we showed how a consistent linear system can be solved by the Gauss elimination method. However, when a linear system $Ax = b$ does not have a solution, one is interested in finding a best approximate solution. By an "approximate solution," we mean a vector x^* such that $\|b - Ax^*\|$ is the smallest, that is, $\|b - Ax^*\| \leqslant \|b - Ax\|$ for all x. Suppose A is an $m \times n$ matrix and $V = \{Ax \mid x \in \mathbf{R}^n\}$, the range of the matrix A. Then the question of finding x^* such that $\|b - Ax^*\|$ is the smallest is equivalent to finding v^* in V such that $\|b - v^*\| \leqslant \|b - v\|$ for all $v \in V$. Indeed, one may visualize V as the floor of a room and b as a point in the space above the floor. Then v^* is the foot of the perpendicular dropped from the point b to the floor. v^* is called the *projection of the point (vector) b* on the space V. Since v^* is a member of the space V, it follows that $v^* = Ax^*$ for some x^* in V. Indeed, $\|b - v^*\|$ is the length of the perpendicular from the point b to the subspace V.

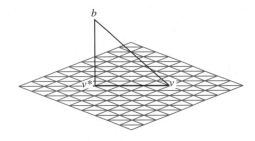

Orthonormal bases constructed by Gram-Schmidt method have a number of applications to both theoretical and numerical problems. The inequality $\|b - v^*\| \leqslant \|b - v\|$ for all $v \in V$ is an algebraic statement of the question of finding a vector in V that is nearest to the given vector b. This question can be easily answered by finding an orthonormal basis of V. From the geometric point of view, the vector $v^* - b$ is perpendicular to the space V because v^* is the foot of the perpendicular dropped from the point b to V.

So, let $e_1, e_2, \ldots, e_k$ be an orthonormal basis of V. Write $v^* = \alpha_1 e_1 + \alpha_2 e_2 + \cdots + \alpha_k e_k$. Since $v^* - b$ is perpendicular to the space V, $v^* - b$ is orthogonal to each basis vector e_i of V. Thus $\langle v^* - b, e_i \rangle = 0$. This implies that $\langle \alpha_1 e_1 + \alpha_2 e_2 + \cdots + \alpha_k e_k - b, e_i \rangle = 0$, and so $(\alpha_1 e_1 + \alpha_2 e_2 + \cdots + \alpha_k e_k - b)^T e_i = 0$. Using the fact that $e_i^T e_j = 0$ or 1 according as $i \neq j$ or $i = j$, we obtain $\alpha_i = b^T e_i$. Thus

$$v^* = (b^T e_1)e_1 + (b^T e_2)e_2 + \cdots + (b^T e_k)e_k.$$

Thus the foot v^* of the perpendicular from b is determined, and all we need to prove is the inequality $\|b - v^*\| \leqslant \|b - v\|$ for all $v \in V$. This is the content of the proof of Theorem 3 in Section 8.6.

The above discussion suggests to us steps to find the length of the perpendicular from a point to any subspace. In particular, it gives a method to find all least-squares solutions x^* of the linear system $Ax = b$. The term "least-squares solution" is suggested by the fact that minimizing $\|b - Ax\|$ is equivalent to minimizing its square $\|b - Ax\|^2 = (b - Ax)^T (b - Ax) = r_1^2 + r_2^2 + \cdots + r_k^2$, where $r_1, r_2, \ldots, r_k$ are the components of the residue vector $b - Ax$.

Recall that if A is an $m \times n$ matrix, then the set $V = \{Au \mid u \in \mathbf{R}^n\}$ is the range of A. The following steps give us a procedure to compute the vectors v^* and x^*:

Step 1 Find a basis of the subspace V, and then invoke the Gram-Schmidt formula to obtain an orthonormal basis. Let $(e_1, e_2, \ldots, e_k)$ be an orthonormal basis.

Step 2 Let b be any given point (vector). Since $b^T e_i = e_i^T b$, compute

$$v^* = (e_1^T b)e_1 + (e_2^T b)e_1 + \cdots + (e_k^T b)e_k$$

or

$$v^* = (b^T e_1)e_1 + (b^T e_2)e_2 + \cdots + (b^T e_k)e_k.$$

Step 3 Compute $\|b - v^*\|$, which gives the shortest distance of the point b from the subspace V. The linear system $Ax = v^*$ is consistent because v^* belongs to the set whose members are of the form Au. Then any solution x^* of the linear system $Ax = v^*$ is a least-squares solution of the linear system $Ax = b$.

a. Find the length of the perpendicular from the point $(1, 2)$ to the line $y = x$ as drawn below:

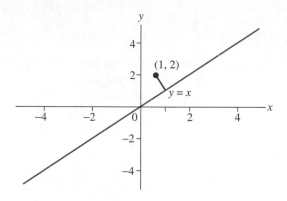

Step 1 Find a basis of the subspace $y = x$. This subspace consists of vectors $\begin{bmatrix} a \\ a \end{bmatrix}$, and so it is generated by $\begin{bmatrix} 1 \\ 1 \end{bmatrix}$. By the Gram-Schmidt method, we obtain

$$e_1 = \frac{1}{\sqrt{2}} \begin{bmatrix} 1 \\ 1 \end{bmatrix}$$

as orthonormal basis of the space.

Step 2 Compute $v^* = (e_1^T b)e_1$, where b is the vector $\begin{bmatrix} 1 \\ 2 \end{bmatrix}$ corresponding to the point $(1, 2)$. By substituting the values, we obtain

$$v^* = \frac{3}{2} \begin{bmatrix} 1 \\ 1 \end{bmatrix}.$$

Step 3 Compute the distance between the point $(1, 2)$, that is, the vector

$$b = \begin{bmatrix} 1 \\ 2 \end{bmatrix}$$

and the vector

$$v^* = \frac{3}{2} \begin{bmatrix} 1 \\ 1 \end{bmatrix}.$$

$$\|b - v^*\| = \left\| \begin{bmatrix} -0.5 \\ 0.5 \end{bmatrix} \right\| = \sqrt[2]{0.5}.$$

b. Find the length of the perpendicular from the point $(1, 2, 0)$ to the plane $z = x - y$ as drawn below:

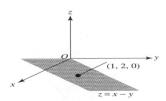

Step 1 Find a basis of the subspace $z = x - y$ that consists of the vectors

$$\begin{bmatrix} a \\ b \\ a - b \end{bmatrix} = a \begin{bmatrix} 1 \\ 0 \\ 1 \end{bmatrix} + b \begin{bmatrix} 0 \\ 1 \\ -1 \end{bmatrix}$$

and is thus generated by $\begin{bmatrix} 1 \\ 0 \\ 1 \end{bmatrix}$ and $\begin{bmatrix} 0 \\ 1 \\ -1 \end{bmatrix}$. By the Gram-Schmidt method, we obtain

$$e_1 = \frac{1}{\sqrt{2}} \begin{bmatrix} 1 \\ 0 \\ 1 \end{bmatrix} \qquad \text{and} \qquad e_2 = \frac{1}{\sqrt{6}} \begin{bmatrix} 1 \\ 2 \\ -1 \end{bmatrix}$$

as an orthonormal basis of the space given by the plane $z = x - y$.

Step 2 Compute $v^* = (e_1^T b)e_1 + (e_2^T b)e_2$, where b is the vector $\begin{bmatrix} 1 \\ 2 \\ 0 \end{bmatrix}$ corresponding to the point $(1, 2, 0)$. We obtain

$$v^* = \frac{1}{3} \begin{bmatrix} 4 \\ 5 \\ -1 \end{bmatrix}.$$

Step 3 Compute the distance between the point $(1, 2, 0)$, that is, the vector

$$b = \begin{bmatrix} 1 \\ 2 \\ 0 \end{bmatrix},$$

and the vector

$$v^* = \frac{1}{3} \begin{bmatrix} 4 \\ 5 \\ -1 \end{bmatrix}.$$

$$\|b - v^*\| = \sqrt{\frac{1}{3}}.$$

c. Find least-squares solutions of $Ax = b$, where

$$A = \begin{bmatrix} 1 & 2 & -1 \\ 2 & 3 & 1 \\ -1 & -1 & -2 \\ 3 & 5 & 0 \end{bmatrix} \quad \text{and} \quad b = \begin{bmatrix} 1 \\ 0 \\ 1 \\ 0 \end{bmatrix}.$$

Step 1 We find a basis of the range of A, that is, the column space of A. To do so, we find a basis of the row space of A^T by performing elementary row operations on it as follows:

$$A^T = \begin{bmatrix} 1 & 2 & -1 & 3 \\ 2 & 3 & -1 & 5 \\ -1 & 1 & -2 & 0 \end{bmatrix} \begin{array}{c} R_2+(-2)R_1 \\ \longrightarrow \\ R_3+1R_1 \end{array} \begin{bmatrix} 1 & 2 & -1 & 3 \\ 0 & -1 & 1 & -1 \\ 0 & 3 & -3 & 3 \end{bmatrix}$$

$$\begin{array}{c} R_3+3R_2 \\ \longrightarrow \end{array} \begin{bmatrix} 1 & 2 & -1 & 3 \\ 0 & -1 & 1 & -1 \\ 0 & 0 & 0 & 0 \end{bmatrix}.$$

Therefore a basis of the column space of A is

$$\begin{bmatrix} 1 \\ 2 \\ -1 \\ 3 \end{bmatrix}, \quad \begin{bmatrix} 0 \\ -1 \\ 1 \\ -1 \end{bmatrix}.$$

We then orthonormalize it by the Gram-Schmidt method and obtain

$$\left\{ e_1 = \frac{1}{\sqrt{15}} \begin{bmatrix} 1 \\ 2 \\ -1 \\ 3 \end{bmatrix}, e_2 = \frac{1}{\sqrt{15}} \begin{bmatrix} 2 \\ -1 \\ 3 \\ 1 \end{bmatrix} \right\}$$

as our desired basis.

Step 2 We compute

$$e_1^T b = [1 \ 2 \ -1 \ 3] \begin{bmatrix} 1 \\ 0 \\ 1 \\ 0 \end{bmatrix} = 0$$

and

$$e_2^T b = \frac{1}{\sqrt{15}} [2 \ -1 \ 3 \ 1] \begin{bmatrix} 1 \\ 0 \\ 1 \\ 0 \end{bmatrix} = \frac{5}{\sqrt{15}},$$

so

$$v^* = \frac{1}{3} \begin{bmatrix} 2 \\ -1 \\ 3 \\ 1 \end{bmatrix}.$$

Step 3 We now solve $Ax = v^*$. The augmented matrix of the LS is

$$\begin{bmatrix} 1 & 2 & -1 & \frac{2}{3} \\ 2 & 3 & 1 & -\frac{1}{3} \\ -1 & -1 & -2 & 1 \\ 3 & 5 & 0 & \frac{1}{3} \end{bmatrix},$$

whose row echelon form is

$$\begin{bmatrix} 1 & 2 & -1 & \frac{2}{3} \\ 0 & -1 & 3 & -\frac{5}{3} \\ 0 & 0 & 0 & 0 \\ 0 & 0 & 0 & 0 \end{bmatrix}.$$

The transformed LS is

$$x_1 + 2x_2 - x_3 = \frac{2}{3},$$

$$-x_2 + 3x_3 = -\frac{5}{3}.$$

We put $x_3 = t$. Then $x_2 = 3t + \frac{5}{3}$, $x_1 = -5t - \frac{8}{3}$. So the solution is

$$\begin{bmatrix} -5t - \frac{8}{3} \\ 3t + \frac{5}{3} \\ t \end{bmatrix} = t \begin{bmatrix} -5 \\ 3 \\ 1 \end{bmatrix} + \begin{bmatrix} -\frac{8}{3} \\ \frac{5}{3} \\ 0 \end{bmatrix},$$

where t is an arbitrary number.

d. Find all vectors x that minimize $\|Ax - b\|$, where

$$A = \begin{bmatrix} 1 & 2 \\ -1 & 1 \\ 1 & 3 \end{bmatrix} \quad \text{and} \quad b = \begin{bmatrix} 1 \\ 1 \\ 1 \end{bmatrix}.$$

Step 1 Here, $A^T = \begin{bmatrix} 1 & -1 & 1 \\ 2 & 1 & 3 \end{bmatrix}$, and its row echelon form is $\begin{bmatrix} 1 & -1 & 1 \\ 0 & 3 & 1 \end{bmatrix}$, and so the column space of A has a basis consisting of

$$\begin{bmatrix} 1 \\ -1 \\ 1 \end{bmatrix} \quad \text{and} \quad \begin{bmatrix} 0 \\ 3 \\ 1 \end{bmatrix}.$$

By the Gram-Schmidt method, we get an orthonormal basis:

$$\left\{ e_1 = \frac{1}{\sqrt{3}} \begin{bmatrix} 1 \\ -1 \\ 1 \end{bmatrix}, e_2 = \frac{1}{\sqrt{78}} \begin{bmatrix} 2 \\ 7 \\ 5 \end{bmatrix} \right\}.$$

Step 2

$$e_1^T b = \frac{1}{\sqrt{3}},$$

$$e_2^T b = \frac{14}{\sqrt{78}},$$

so

$$v^* = \frac{1}{\sqrt{3}} \cdot \frac{1}{\sqrt{3}} \begin{bmatrix} 1 \\ -1 \\ 1 \end{bmatrix} + \frac{14}{\sqrt{78}} \cdot \frac{1}{\sqrt{78}} \begin{bmatrix} 2 \\ 7 \\ 5 \end{bmatrix}$$

$$= \frac{1}{3} \begin{bmatrix} 1 \\ -1 \\ 1 \end{bmatrix} + \frac{7}{39} \begin{bmatrix} 2 \\ 7 \\ 5 \end{bmatrix} = \frac{1}{39} \begin{bmatrix} 27 \\ 36 \\ 48 \end{bmatrix}.$$

Step 3 Solve $Ax = v^*$, whose solution gives the desired result. It is given by

$$x^* = \frac{1}{13} \begin{bmatrix} -5 \\ 7 \end{bmatrix}.$$

∎

8.4 Exercises

1. Find least-squares solutions for each of the following linear systems:

a. (Drill 8.9) $\begin{bmatrix} 1 & 1 \\ 0 & 2 \\ 1 & 1 \end{bmatrix} x = \begin{bmatrix} 2 \\ 1 \\ 0 \end{bmatrix}.$

b. $\begin{bmatrix} 2 & 3 \\ -1 & 3 \\ 2 & 4 \end{bmatrix} x = \begin{bmatrix} 3 \\ 2 \\ 1 \end{bmatrix}.$

c. $\begin{bmatrix} 1 & 2 & 4 \\ -2 & -3 & -7 \\ 1 & 3 & 5 \end{bmatrix} x = \begin{bmatrix} 1 \\ 2 \\ 3 \end{bmatrix}.$

d. (Drill 8.10) $\begin{bmatrix} 1 & 2 & -1 \\ 3 & 5 & 5 \\ -1 & -1 & -2 \\ 3 & 5 & 0 \end{bmatrix} x = \begin{bmatrix} 1 \\ 1 \\ 1 \\ 0 \end{bmatrix}.$

2. Find the conic $ax^2 + by^2 = 1$ that best fits the points $(1, 1)$, $(0, 2)$, $(-1, 1)$, and $(-1, 2)$.

3. Find a and b such that $ax + by = 1$ best fits the points $(1, 2)$, $(2, 3)$, and $(3, 3.5)$.

4. For

$$A = \begin{bmatrix} 2 & 1 & 5 \\ 0 & 2 & 2 \\ 0 & 0 & 0 \end{bmatrix} \quad \text{and} \quad b = \begin{bmatrix} 2 \\ 0 \\ 2 \end{bmatrix},$$

find x such that $\|b - Ax\|$ is minimal. Among all such values of x, find x with minimum length.

5. Find the shortest distance of the point $(1, 1, 1)$ from each of the following planes:

a. $x + y + z = 0$.

b. $z = 2x + y$.

6. The monthly revenue figures (in thousands of dollars) of a store are as given in the following table:

Months	1	2	3	4
Revenue	10	8	12	10

a. Find the straight line that best fits the data.

b. Use this equation to predict the revenue in the fifth and sixth months.

7. The number of absentees (in hundreds) in a factory during its first four months is given by the following table:

Months	1	2	3	4
Absentees	3	4	1	2

a. Find the best fit curve of degree 1.

b. What would you predict the number of absentees to be in the eighteenth week ($4\frac{1}{2}$ months) and in the fifth month?

8.5 Generalized Inverse and Least-Squares Solution

Another method of finding least-squares solutions is to appeal to the theory of generalized inverses of matrices. Matrices that are not invertible nonetheless possess generalized inverses of various types. We do not intend to undertake an extensive study of generalized inverses. In the following, we will need the useful fact that for any matrix A, $\text{rank}(A) = \text{rank}(AA^T)$.

8.5.1
LEMMA

Let A be an $m \times n$ matrix of rank r. Let $A = FG$ be a full-rank factorization of A. Define

$$A^\dagger = G^T (F^T A G^T)^{-1} F^T.$$

Then

i. $AA^\dagger$ is symmetric.

ii. $AA^\dagger A = A$.

Proof We rewrite

$$A^\dagger = G^T (F^T F G G^T)^{-1} F^T \qquad \text{(since } A = FG\text{)}$$
$$= G^T (GG^T)^{-1} (F^T F)^{-1} F^T,$$

because each of the matrices GG^T and $F^T F$ is an $r \times r$ matrix of rank r and thus invertible. Then

i.

$$AA^\dagger = (FG)(G^T (GG^T))^{-1} (F^T F)^{-1} F^T$$
$$= F(F^T F)^{-1} F^T.$$

Thus $AA^\dagger$ is symmetric.

ii.

$$AA^\dagger A = (F(F^T F)^{-1} F^T)(FG) = FG = A.$$

∎

8.5.2
Remark

$A^\dagger$ is known as a generalized inverse (more precisely, the Moore-Penrose inverse) of the matrix A.

8.5.3
THEOREM

A least-squares solution of a linear system $Ax = b$ is given by $Ax = AA^\dagger b$. Furthermore, $x = A^\dagger b$ is a least-squares solution of minimum norm.

Sketch of Proof

$$\|Ax - b\|^2 = \|Ax - AA^\dagger b + AA^\dagger b - b\|^2$$
$$= \|Ax - AA^\dagger b\|^2 + \|AA^\dagger b - b\|^2,$$

since $(Ax - AA^\dagger b)$ can be shown to be orthogonal to $(AA^\dagger b - b)$.

Thus $\|Ax - x\|^2$ is minimal if $Ax - AA^\dagger b = 0$, in which case the minimum value is $\|AA^\dagger b - b\|^2$. This proves that $\|Ax - b\|$ is minimal for $x = x^*$, where x^* is a solution of $Ax = AA^\dagger b$. Clearly, $x^* = A^\dagger b$ is one such solution.

8.5.4 EXAMPLE Do Example 8.4.1(c) by the method of this section. First we find a full-rank factorization of A (see Section 5.3):

$$\begin{bmatrix} 1 & 2 & -1 \\ 2 & 3 & 1 \\ -1 & -1 & -2 \\ 3 & 5 & 0 \end{bmatrix} \xrightarrow[\substack{R_3+1R_1 \\ R_4+(-3)R_1}]{R_2+(-2)R_1} \begin{bmatrix} 1 & 2 & -1 \\ 0 & -1 & 3 \\ 0 & 1 & -3 \\ 0 & -1 & 3 \end{bmatrix} \xrightarrow[\substack{R_4+(-1)R_2}]{R_3+1R_2} \begin{bmatrix} 1 & 2 & -1 \\ 0 & -1 & 3 \\ 0 & 0 & 0 \\ 0 & 0 & 0 \end{bmatrix} = \begin{bmatrix} G \\ 0 \end{bmatrix},$$

where

$$G = \begin{bmatrix} 1 & 2 & -1 \\ 0 & -1 & 3 \end{bmatrix}.$$

To find F, we begin with I:

$$\begin{bmatrix} 1 & 0 & 0 & 0 \\ 0 & 1 & 0 & 0 \\ 0 & 0 & 1 & 0 \\ 0 & 0 & 0 & 1 \end{bmatrix} \xrightarrow[\substack{R_4+1R_2}]{R_3-1R_2} \begin{bmatrix} 1 & 0 & 0 & 0 \\ 0 & 1 & 0 & 0 \\ 0 & -1 & 1 & 0 \\ 0 & 1 & 0 & 1 \end{bmatrix} \xrightarrow[\substack{R_3-1R_1 \\ R_4+3R_1}]{R_2+2R_1} \begin{bmatrix} 1 & 0 & 0 & 0 \\ 2 & 1 & 0 & 0 \\ -1 & -1 & 1 & 0 \\ 3 & 1 & 0 & 1 \end{bmatrix},$$

and so

$$F = \begin{bmatrix} 1 & 0 \\ 2 & 1 \\ -1 & -1 \\ 3 & 1 \end{bmatrix}.$$

Now $A^\dagger = G^T (F^T A G^T)^{-1} F^T$.

First let us compute

$$F^T A G^T = \begin{bmatrix} 1 & 2 & -1 & 3 \\ 0 & 1 & -1 & 1 \end{bmatrix} \begin{bmatrix} 1 & 2 & -1 \\ 2 & 3 & 1 \\ -1 & -1 & -2 \\ 3 & 5 & 0 \end{bmatrix} \begin{bmatrix} 1 & 0 \\ 2 & -1 \\ -1 & 3 \end{bmatrix}$$

$$= \begin{bmatrix} 15 & 24 & 3 \\ 6 & 9 & 3 \end{bmatrix} \begin{bmatrix} 1 & 0 \\ 2 & -1 \\ -1 & 3 \end{bmatrix} = \begin{bmatrix} 60 & -15 \\ 21 & 0 \end{bmatrix}.$$

Thus

$$(F^T A G^T)^{-1} = \frac{1}{315} \begin{bmatrix} 0 & 15 \\ -21 & 60 \end{bmatrix}$$

$$= \frac{1}{105} \begin{bmatrix} 0 & 5 \\ -7 & 20 \end{bmatrix},$$

$$A^\dagger = \frac{1}{105} \begin{bmatrix} 1 & 0 \\ 2 & -1 \\ -1 & 3 \end{bmatrix} \begin{bmatrix} 0 & 5 \\ -7 & 20 \end{bmatrix} \begin{bmatrix} 1 & 2 & -1 & 3 \\ 0 & 1 & -1 & 1 \end{bmatrix}$$

$$= \frac{1}{105} \begin{bmatrix} 0 & 5 \\ 7 & -10 \\ -21 & 55 \end{bmatrix} \begin{bmatrix} 1 & 2 & -1 & 3 \\ 0 & 1 & -1 & 1 \end{bmatrix},$$

$$A^\dagger b = \frac{1}{105} \begin{bmatrix} 0 & 5 \\ 7 & -10 \\ -21 & 55 \end{bmatrix} \begin{bmatrix} 1 & 2 & -1 & 3 \\ 0 & 1 & -1 & 1 \end{bmatrix} \begin{bmatrix} 1 \\ 0 \\ 1 \\ 0 \end{bmatrix}$$

$$= \frac{1}{105} \begin{bmatrix} 0 & 5 \\ 7 & -10 \\ -21 & 55 \end{bmatrix} \begin{bmatrix} 0 \\ -1 \end{bmatrix} = \frac{1}{105} \begin{bmatrix} -5 \\ 10 \\ -55 \end{bmatrix}$$

$$= \frac{1}{21} \begin{bmatrix} -1 \\ 2 \\ -11 \end{bmatrix}$$

is one least-squares solution. Other solutions are obtained by solving

$$Ax = AA^\dagger b = \frac{1}{3} \begin{bmatrix} 2 \\ -1 \\ 3 \\ 1 \end{bmatrix}.$$

■

8.5 Exercises

Compute the Moore-Penrose inverses of matrices in Exercises 1(a), 1(b), and 4 of Section 8.4, and invoke Theorem 8.5.3 to obtain the least-squares solution of minimum norm.

8.6 Proofs of Facts

THEOREM 1

The eigenvalues of a real symmetric matrix are real.

Proof Let λ be an eigenvalue of a symmetric matrix A. So

$$Ax = \lambda x \Rightarrow A\bar{x} = \bar{\lambda}\bar{x} \qquad \text{(since } A \text{ is real)}$$

$$\Rightarrow x^T A\bar{x} = \bar{\lambda}x^T\bar{x} \Rightarrow x^T A^T\bar{x} = \bar{\lambda}x^T\bar{x} \qquad \text{(since } A = A^T\text{)}$$
$$\Rightarrow (Ax)^T\bar{x} = \bar{\lambda}x^T\bar{x} \Rightarrow (\lambda x)^T\bar{x} = \bar{\lambda}x^T\bar{x}$$
$$\Rightarrow (\lambda - \bar{\lambda})x^T\bar{x} = 0 \Rightarrow \lambda = \bar{\lambda} \qquad \text{(since } x^T\bar{x} \text{ is zero only if } x = 0\text{)}.$$

■

THEOREM 2 | Let A be a symmetric matrix. Then the eigenvectors corresponding to distinct eigenvalues of A are orthogonal.

Proof Let x and y be eigenvectors corresponding to eigenvalues λ and μ, where $\lambda \neq \mu$. Then $Ax = \lambda x$, and $Ay = \mu y$.

Now, $Ax = \lambda x \implies (Ax)^T = (\lambda x)^T \implies x^T A^T = \lambda x^T \implies x^T A = \lambda x^T$, since A is symmetric. Next, $x^T A = \lambda x^T \implies x^T Ay = \lambda x^T y \implies x^T (\mu y) = \lambda x^T y \implies (\lambda - u)x^T y = 0 \implies x^T y = 0$, since $\lambda \neq \mu$. Therefore x is orthogonal to y. ∎

THEOREM 3 | Let V be a subspace, and let b be any vector. Then there exists $v^* \in V$ such that

$$\|b - v^*\| \leqslant \|b - v\| \qquad \text{for all } v \in V.$$

Proof Let $(e_1, \ldots, e_k)$ be an orthonormal basis of V. Then any vector $v^* \in V$ can be expressed as $v^* = \alpha_1 e_1 + \cdots + \alpha_k e_k$. Choose v^* such that $(b - v^*) \perp e_i$ for all e_i. Then $(b - \alpha_1 e_1 - \cdots - \alpha_k e_k)^T e_i = 0, i = 1, 2, \ldots, k$. This gives $\alpha_1 = b^T e_1, \alpha_2 = b^T e_2, \ldots, \alpha_k = b^T e_k$.

Further, if $v \in V$, then

$$
\begin{aligned}
\|b - v\|^2 &= \|(b - v^*) - (v - v^*)\|^2 \\
&= (b - v^*)^T (b - v^*) - 2(b - v^*)^T (v - v^*) + (v - v^*)^T (v - v^*) \\
&\qquad (\text{since } \|x\|^2 = x^T x) \\
&= \|b - v^*\|^2 + \|v - v^*\|^2,
\end{aligned}
$$

since

i. $(b - v^*)$ is orthogonal to each e_i, and so $(b - v^*)$ is orthogonal to each vector in V, and

ii. $v - v^* \in V$, since V is a subspace.

Hence $\|b - v^*\| < \|b - v\|$ for all $v \in V$. ∎

THEOREM 4 | Let A be an $m \times n$ matrix. The eigenvalues of AA^T or $(A^T A)$ are nonnegative.

Proof Since AA^T is a symmetric matrix, there exists an orthonormal set $u_1, u_2, \ldots, u_m$ of eigenvectors corresponding to its eigenvalues $\lambda_1, \lambda_2, \ldots, \lambda_m$, respectively.

Now, $AA^T u_i = \lambda_i u_i$ gives that $u_i^T AA^T u_i = \lambda_i u_i^T u_i$, which in turn yields $\|A^T u_i\|^2 = \lambda_i \|u_i\|^2$. Thus $\lambda_i \geqslant 0$. Similar arguments give that the eigenvalues of $A^T A$ are nonnegative. This completes the proof. ∎

THEOREM 5

(Singular Value Decomposition) Let A be an $m \times n$ matrix, and let $s_1, s_2, \ldots, s_n$ be the singular values of A. Then $A = USV^T$, where U and V are orthogonal matrices and S is a diagonal matrix with diagonal entries $s_1, s_2, \ldots, s_n$. (Equivalently, $A = s_1 u_1 v_1^T + s_2 u_2 v_2^T + \cdots + s_n u_n v_n^T$, where u_i and v_i are the ith and jth columns of the matrices U and V, respectively, and $s_1 \geqslant s_2 \geqslant \cdots \geqslant s_n$.)

Proof Let $v_1, v_2, \ldots, v_n$ be an orthonormal set of eigenvectors of $A^T A$ corresponding to the eigenvalues $s_1^2, s_2^2, \ldots, s_n^2$ written in decreasing order, where, possibly, $s_{r+1} = \cdots = s_n = 0$. Let $u_i = \frac{1}{s_i} A v_i$, for $i = 1, 2, \ldots, r$. Now

$$u_i^T u_j = \frac{1}{s_i s_j} v_i^T A^T A v_j = \frac{1}{s_i s_j} v_i^T s_j^2 v_j = \frac{1}{s_i s_j} s_j^2 v_i^T v_j = 0, \, u_i^T u_i = \frac{1}{s_i^2} v_i^T A^T A v_i$$

$$= \frac{1}{s_i^2} v_i^T s_i^2 v_i = v_i^T v_i = 1.$$

Thus, $u_1, u_2, \ldots, u_r$ form an orthonormal set of vectors. If $r \leqslant m$, we extend the set $\{u_1, u_2, \ldots, u_r\}$ to an orthonormal set $\{u_1, u_2, \ldots, u_r, \ldots, u_m\}$ of vectors so as to form a basis of $\mathbf{R}^m$. We will prove that $AV = US$, where S is the $n \times n$ diagonal matrix whose diagonal entries are $s_1, s_2, \ldots s_r, 0, \ldots, 0$ and U and V are $m \times m$ and $n \times n$ orthogonal matrices, respectively, given by $V = [v_1, v_2, \ldots, v_n]$ and $U = [u_1, u_2, \ldots, u_r, \ldots, u_m]$.

$$
\begin{aligned}
AV &= A[v_1, v_2, \ldots, v_n] \\
&= [Av_1, Av_2, \ldots, Av_n] \\
&= [s_1 u_1, \ldots, s_r u_r, 0, \ldots, 0]
\end{aligned}
$$

$$
= U \begin{bmatrix}
s_1 & & & & & & & \\
& s_2 & & & & & & \\
& & \ddots & & & & & \\
& & & s_r & & & & \\
& & & & 0 & & & \\
& & & & & \ddots & & \\
& & & & & & 0 & \\
& & & & & & & 0
\end{bmatrix} = US,
$$

because $V^T = V^{-1}$. Thus, we obtain $A = USV^T$, as desired. ∎

Remark 6 If A is approximated by the sum of the first k terms, then this sum is called a rank-k approximation of A. For an application of singular value decomposition, see Chapter Project 2.

8.7 Chapter Review Questions and Projects

1. Show that the following vectors are mutually orthogonal and find their lengths:

   a. $\begin{bmatrix} -1 \\ 1 \\ 1 \end{bmatrix}$, $\begin{bmatrix} 1 \\ 0 \\ 1 \end{bmatrix}$, $\begin{bmatrix} 1 \\ 2 \\ -1 \end{bmatrix}$

   b. $\begin{bmatrix} 1 \\ 1 \\ 0 \end{bmatrix}$, $\begin{bmatrix} -1 \\ 1 \\ 0 \end{bmatrix}$, $\begin{bmatrix} 0 \\ 0 \\ 1 \end{bmatrix}$ (Section 8.1)

2. Find an orthonormal basis of the subspace spanned by each of the following sets of vectors:

   a. $\left\{ \begin{bmatrix} 1 \\ 2 \\ 3 \\ 4 \end{bmatrix} \right\}$

   b. $\left\{ \begin{bmatrix} 0 \\ 1 \\ 0 \\ 1 \end{bmatrix}, \begin{bmatrix} -1 \\ 0 \\ 1 \\ 1 \end{bmatrix} \right\}$

   c. $\left\{ \begin{bmatrix} 1 \\ 2 \\ -1 \end{bmatrix}, \begin{bmatrix} 1 \\ -1 \\ 1 \end{bmatrix}, \begin{bmatrix} 1 \\ 0 \\ 1 \end{bmatrix} \right\}$ (Section 8.1)

3. What is meant by an orthogonal matrix? Give examples of a 2×2 and a 3×3 orthogonal matrix other than the identity matrix. (Exercise 10, Section 8.1)

4. For the following symmetric matrices, find an orthogonal matrix P such that $P^T A P$ is diagonal:

   a. $\begin{bmatrix} 3 & 4 \\ 4 & -3 \end{bmatrix}$

   b. $\begin{bmatrix} 0 & 2 & 2 \\ 2 & 0 & 2 \\ 2 & 2 & 0 \end{bmatrix}$ (Section 8.2)

5. Find a and b such that the following matrices are positive definite.

   a. $\begin{bmatrix} a & 3 & 0 \\ 3 & a & 4 \\ 0 & 4 & a \end{bmatrix}$

   b. $\begin{bmatrix} 4 & b \\ b & 1 \end{bmatrix}$ (Exercise 2, Section 8.3)

6. Find the symmetric matrix A for each of the following quadratic forms, and determine whether it is always positive, always negative, or neither.

a. $x^2 - xy + y^2$.
b. $2x^2 - y^2 + 2z^2 - 4xy + 4yz + 2xz$.
c. $-2x^2 - y^2 - 2z^2 + 2xz$. (Exercises 2 and 4, Section 8.3)

7. Find the least-squares solution of the following LS:

a. $\begin{bmatrix} 1 & 1 \\ 2 & -1 \\ 1 & -1 \end{bmatrix} \begin{bmatrix} x_1 \\ x_2 \end{bmatrix} = \begin{bmatrix} 2 \\ 2 \\ 0 \end{bmatrix}$.

b. $\begin{bmatrix} 0 & 1 \\ 1 & 1 \\ 2 & 1 \\ 3 & 2 \\ 4 & 3 \end{bmatrix} \begin{bmatrix} a \\ b \end{bmatrix} = \begin{bmatrix} 1 \\ 1 \\ 1 \\ 1 \\ 1 \end{bmatrix}$. (Section 8.4)

8. Compute the Moore-Penrose inverse $A^\dagger$ of the coefficient matrices of the linear systems given in Problem 7, and use it to find the least-squares solution. (Section 8.5)

9. If A is an $n \times n$ orthogonal matrix and v is an n-dimensional vector show that $\|Av\| = \|v\|$. (Sections 8.1 and 8.2)

10. If A is an $n \times n$ orthogonal matrix and λ is an eigenvalue of A, then $|\lambda| = 1$. (Hint: Use Question 9.)

Chapter Projects

1. The closing stock prices of Interactive Systems are given for the last 10 days by the following table:

1	2	3	4	5	6	7	8	9	10
45.3	50.2	30.1	56.0	55.2	50.0	45.1	30.0	44.2	50.0

The company would like to predict the stock prices for the next 3 days.

a. Assuming that the data fit a linear model ($y = ax + b$), predict the stock prices for the next 3 days.
b. Assuming that the data fit a quadratic model ($y = ax^2 + bx + c$), predict the stock prices for the next 3 days.
c. Assuming that the data fit a cubic model ($y = ax^3 + bx^2 + cx + d$), predict the stock prices for the next 3 days.

2. Singular Value Decomposition: Given an $m \times n$ matrix A, there exists a decomposition such that $A = USV^T$, where U and V are orthogonal matrices and S is a diagonal matrix with nonnegative diagonal entries in decreasing order. The diagonal entries of S are the positive square roots of the eigenvalues of AA^T and are called the singular values of A (it can be proved that the eigenvalues of AA^T, where A is any matrix, are

always nonnegative, and the proof of this fact is contained in Section 8.6, Theorem 4). The decomposition $A = USV^T$ is called a singular value decomposition of A, and A can be written as $A = s_1u_1v_1^T + s_2u_2v_2^T + \cdots + s_nu_nv_n^T$, where u_i and v_i are the ith and jth columns of the matrices U and V, respectively, and $s_1 = s_2 = \cdots = s_n$. Each of the nonzero terms in this representation of A is a rank-1 matrix, and if A is approximated by the sum of the first k terms, then this sum is called a rank-k approximation of A. The reader who is interested in the proof of singular value decomposition of a matrix may go to Section 8.6, Theorem 5.

Application: The purpose of this project is to study an important application of singular value decomposition (SVD) to image processing. The idea is that by using a smaller number of vectors, one can reconstruct an image that is "closer" to the original. The clarity of the image depends on how many singular values are used to reconstruct it. Of course you do not want to use too many singular values otherwise compression may not be practical. A rank-k approximation requires k singular values and k-columns of matrices U and V, that is, $k(2n + 1)$ values. Therefore, in order to compress an image of $n \times n$ pixels, no more than $k < \frac{n^2}{2n+1}$ singular values must be selected. In this project, we will apply SVD to the following image. Figure 8.1 is a bitmap image of size 128×128 pixels. Since this is a color image, the image is stored as matrices $[\ R\ \ G\ \ B\ ]$ each of size 128×128, representing Red, Green, and Blue components of the image. In order to work with Matlab student version, we took part of the image in Figure 8.1 of size 64×64. First we need three 64×64 matrices representing Red, Green, and Blue components of the image in Figure 8.2. Both figures are available on CD or can be downloaded from the book's web site.

Figure 8.1   128 × 128

Figure 8.2 – 64 × 64

Task 1: Read in the file 'castle64.bmp' to Matlab using image command of Matlab. If you need to understand how to use the image command, type help image.
Assign the result to a 3-D array A (image) and plot the image in Matlab using imagesc(A). You should see the image shown in Figure 8.3. Note that this image is of size 64×64 pixels.

Figure 8.3 – (64 × 64) Matlab Display

Task 2: The 3-D array A obtained in Task 1 needs to be decomposed into three matrices R, G, and B, of size 64×64 each. Use Matlab command, $R = A(:, :, 1)$ to extract the Red matrix from 3-D array A. Similarly, extract matrices G and B. Apply Singular Value Decomposition to each of the three matrices R, G and B. The Matlab command for finding singular value decomposition of matrix M is $[U, S, V] = svd(M)$, where U, V are orthogonal matrices and S is the matrix of singular values.

Task 3: Reconstruct Rank-1, Rank-8, Rank-16, and Rank-32 approximations of the image A. The rank-k approximation of the matrix A is given by

$$sum(S(i, i) * U(:, i) * V(:, i)^T, i = 1 \cdots k)$$

where $S(i, i)$ is the i-th singular value and $U(:, i)$ and $V(:, i)$ are the i-th columns of matrices U and V. To construct rank-1 approximation of the matrix, first find rank-1 approximations to R, G, and B (say $R1$, $G1$ and $B1$) and reconstruct rank-1 approximation $A1$ to the original image A. Note that $A1$ is a 3-D array of size $64 \times 64 \times 3$. Therefore you need to form $A1$ using $A1(:, :, 1) = R1$, $A1(:, :, 2) = G1$ etc. Figures 8.4 and 8.5 are rank-1 and rank-8 approximations of the image-A.

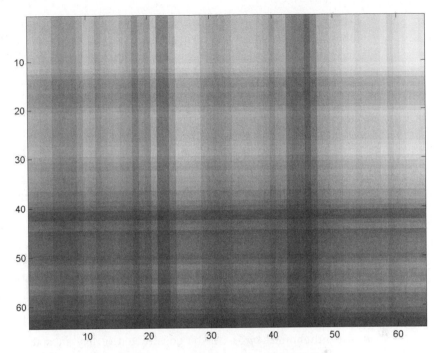

Figure 8.4 – Rank-1 Approximation

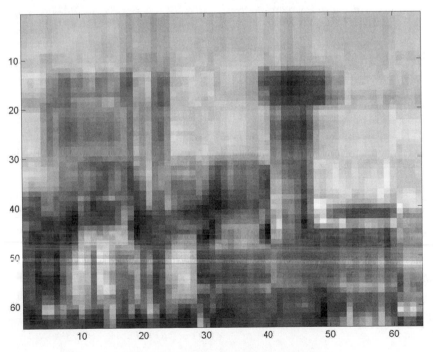

Figure 8.5 – Rank-8 Approximation

Note: Matlab student version cannot handle the original image of size 128×128. However, if you have a professional copy of Matlab or other software such as Maple or Mathematica, you may want to apply singular value decomposition to 128×128 matrix. You can also apply these rank approximations to any other bitmap file of your choice.

Key Words

Inner, Dot, or Scalar Product of Vectors
Length of Vector
Orthogonal or Perpendicular Vectors
Pythagorean Theorem
Orthogonal Basis of a Subspace
Orthonormal Basis of a Subspace
Gram-Schmidt Method
Orthogonal Matrix
Diagonalization of Symmetric Matrix

Spectral Theorem
Principle Axis Theorem
Singular Value Decomposition
Rank-k Approximation of A
Projection
Least-Squares Solutions of an Inconsistent Linear System
Generalized Inverse of a Matrix
Moore-Penrose Inverse

Key Phrases

- An orthonormal basis can be constructed of any subspace by the Gram-Schmidt method.

- Every symmetric matrix is diagonalizable.

- Singular value decomposition of any matrix provides a rank-k approximation of the matrix.

- Best approximate solutions of an inconsistent linear system $Ax = b$ can be obtained by solving $Ax = c$, where c is the projection of b on the range of A, or by using the Moore-Penrose inverse $A^{\dagger}$ of A, in which case the solution is $x = A^{\dagger}b$.

9 Vector Spaces and Linear Mappings

Introduction

In this chapter, we generalize the notion of subspace discussed in *Chapter 3* to abstract vector spaces. The concepts of linear dependence and linear independence in any vector space are also introduced. We discuss how an $m \times n$ matrix may be looked upon as a function from $\mathbf{R}^n$ to $\mathbf{R}^m$ that preserves addition of vectors and scalar multiplication with a vector. Such functions, called *linear mappings* or *linear transformation*, are of paramount importance in applications to problems in computer graphics, physics, engineering, and several other areas.

9.1 Vector Spaces

In the previous chapters, we considered matrices with entries in the field of real or complex numbers. Before we define a vector space, we define the concept of an abstract field of which the rational numbers $\mathbf{Q}$, the real numbers $\mathbf{R}$, and the complex numbers $\mathbf{C}$ are particular examples.

9.1.1 DEFINITION

Field

A *field* is an algebraic structure F consisting of a nonempty set F together with two operations, addition and multiplication, that assign to each pair of elements $\alpha, \beta \in F$ uniquely determined elements $\alpha + \beta$ and $\alpha\beta$ of F such that the following conditions are satisfied for all $\alpha, \beta, \gamma \in F$:

1. $\alpha + \beta = \beta + \alpha, \alpha\beta = \beta\alpha$.
2. $\alpha + (\beta + \gamma) = (\alpha + \beta) + \gamma, \alpha(\beta\gamma) = (\alpha\beta)\gamma$.
3. $\alpha(\beta + \gamma) = \alpha\beta + \alpha\gamma$.
4. There exists an element 0 in F such that $\alpha + 0 = \alpha$ for all $\alpha \in F$.
5. For each $\alpha \in F$, there exists an element $-\alpha \in F$ such that $\alpha + (-\alpha) = 0$.

6. There exists a nonzero element $1 \in F$ such that $\alpha 1 = \alpha$ for all $\alpha \in F$.

7. For each nonzero $\alpha \in F$, there exists an element $\alpha^{-1} \in F$ such that $\alpha \alpha^{-1} = 1$.

As was stated above, **Q**, **R**, and **C** are fields. Clearly, the set of integers **Z** is not a field. Another example of a field is the field of rational functions, whose elements are of the form $p(x)/q(x)$, where $p(x)$ and $q(x)$ are polynomials in x with coefficients in a given field, say **R**, and $q(x) \neq 0$. An example of a finite field is the field $F = \{0, 1\}$, where addition and multiplication are as usual excepting that $1 + 1 = 0$.

We considered in *Chapter 3* the vector space F^n whose elements are n-tuples, that is, $1 \times n$ or $n \times 1$ matrices with entries that are real or complex numbers. It would be unfortunate if a student completed a first course in linear algebra and had the impression that a vector is simply an $n \times 1$ or $1 \times n$ matrix. Indeed, vectors that are functions of a more general type occur frequently in mathematics. In what follows, we will briefly discuss abstract vector spaces whose members are vectors of this sort. Vector spaces whose elements are functions of a more general kind occur quite often in applications.

Recall the vector space F^n, in which we add two elements as

$$\begin{bmatrix} a_1 \\ a_2 \\ \vdots \\ a_n \end{bmatrix} + \begin{bmatrix} b_1 \\ b_2 \\ \vdots \\ b_n \end{bmatrix} = \begin{bmatrix} a_1 + b_1 \\ a_2 + b_2 \\ \vdots \\ a_n + b_n \end{bmatrix}$$

and multiply a vector by a scalar as

$$c \begin{bmatrix} a_1 \\ a_2 \\ \vdots \\ a_n \end{bmatrix} = \begin{bmatrix} ca_1 \\ ca_2 \\ \vdots \\ ca_n \end{bmatrix}.$$

We note that the following properties of addition hold. Let $x, y, z \in F^n$. Then

A1. $x + y = y + x$.

A2. $x + (y + z) = (x + y) + z$.

A3. $x + 0 = 0 + x = x$.

A4. If $x = \begin{bmatrix} x_1 \\ x_2 \\ \vdots \\ x_n \end{bmatrix}$, then $x + (-x) = 0$, where $-x = \begin{bmatrix} -x_1 \\ -x_2 \\ \vdots \\ -x_n \end{bmatrix}$.

Abstractly speaking, F^n is the set of functions on the set $\{1, 2, \ldots, n\}$ to the set F. We encounter many such functions in mathematics. We give two examples.

9.1.2 EXAMPLES

a. Let $P_2(x) = \{a_0 + a_1x + a_2x^2 \mid a_0, a_1, a_2 \in F\}$, the set of polynomials of degree less than or equal to 2. We can add two elements of $P_2(x)$ as follows:

$$(a_0 + a_1x + a_2x^2) + (b_0 + b_1 \times b_2x^2) = (a_0 + b_0) + (a_1 + b_2)x + (a_2 + b_2)x^2.$$

It can be easily verified that addition in the set $P_2(x)$ satisfies properties A1, A2, A3, and A4. Furthermore, if $c \in F$ and $a_0 + a_1 \times a_2x^2 \in P_2(x)$, then $c(a_0 + a_1x + a_2x^2) = (ca_0) + (ca_1)x + (ca_2)x^2 \in P_2(x)$.

b. Let M denote the set of $m \times n$ matrices. We defined, in *Chapter 2*, the sum of two $m \times n$ matrices and have shown that addition in the set M satisfies properties A1, A2, A3, and A4. Furthermore, if $A = (a_{ij})$ is an $m \times n$ matrix and c is a scalar, then $cA = (ca_{ij})$ is an $m \times n$ matrix.

The sets F^n, $P_2(x)$, and M are examples of a set S of functions in which we can add two elements and we can multiply an element by a scalar. Addition satisfies properties A1, A2, A3, and A4, and scalar multiplication satisfies the following properties for all $x, y \in S$ and all scalars $c, c_1, c_2 \in F$:

M1. $c(x + y) = cx + cy$.

M2. $(c_1 + c_2)x = c_1x + c_2x$.

M3. $(c_1c_2)x = c_1(c_2x)$.

M4. $1x = x$. ∎

We now give a definition of a vector space. In what follows, F will denote an arbitrary field, unless otherwise stated.

9.1.3 DEFINITION

Vector Space

Let S be a set in which any two elements of S can be added to obtain an element of S and any element of S can be multiplied by an element of a field F to get again an element of S. Then S is called a *vector space* over F if properties A1, A2, A3, and A4 (for addition) and M1, M2, M3, and M4 (for scalar multiplication) are satisfied.

The set of polynomials of degree at most n with coefficients in F and the set of $m \times n$ matrices with entries in F are examples of vector spaces. The elements of a vector space are called *vectors*.

If V is a vector space, one is often required to find whether a given subset is also a vector space.

9.1.4 DEFINITION

Subspace

If W is a nonempty subset of a vector space V such that

i. $x - y \in W$ for all $x, y \in W$, and

ii. $\alpha x \in W$ for all $\alpha \in F$ and $x \in W$,

then W is called a *subspace* of V.

9.1.5 By (ii), $-y \in W$ for all $y \in W$. So by (i), $x + y \in W$ for all $x, y \in W$. Since $W \subset V$,
Remark the elements in W satisfy properties A1–A4 and M1–M4, and so W is a vector space
in its own right.

9.1.6
EXAMPLES

a. Let V be the set of real-valued continuous functions defined on $[0, 1]$. If $f, g \in V$
then $f - g$ is also continuous, and if $\alpha \in \mathbf{R}$, then αf is also continuous. Thus V is
a vector space over $\mathbf{R}$.

b. Let V be the set of real-valued differentiable functions on $(-\infty, \infty)$. By the properties of differentiable functions, $f - g \in V$ and $\alpha f \in V$ when $f, g \in V$ and $\alpha \in \mathbf{R}$.
Thus V is a vector space over $\mathbf{R}$.

c. Let V be the set of $m \times n$ matrices whose entries are in F. Thus V is a vector space
over F and is denoted by $F^{m \times n}$.

d. Let $V = \{1, x, x^2\}$. Then V is not a vector space, since, for example, $1 - 1 = 0$,
$1 - x, \ldots$ do not belong to V, violating condition (i) for V to be a vector space.

e. Let $V = \{f \mid \frac{d^2 f}{dx^2} = 0\}$. If $f_1, f_2 \in V$ and $\alpha \in \mathbf{R}$, then

$$\frac{d^2}{dx^2}(f_1 - f_2) = \frac{d^2 f_1}{dx^2} - \frac{d^2 f_2}{dx^2} = 0,$$

and

$$\frac{d^2}{dx^2}(\alpha f) = \alpha \frac{d^2 f}{dx^2} = 0.$$

Thus V is a vector space.

f. Let $P_n(x) = \{a_0 + a_1 x + \cdots + a_m x^m \mid a_i \in R, m \leqslant n\}$ be the set of polynomials
of degree $\leqslant n$. Then $P_n(x)$ is a vector space. ∎

9.1 Exercises

1. Which of the following subsets of $C[0, 1]$ (the set of continuous functions on the
unit interval $[0, 1]$) are subspaces?

  a. The set of all $f \in C[0, 1]$ such that $f(1) = 1$.

  b. The set of all $f \in C[0, 1]$ such that $f(1) = 2$.

  c. The set of all $f \in C[0, 1]$ such that $\int_0^1 f(t)\,dt = 1$.

  d. The set of all $f \in C[0, 1]$ such that $df/dx = 0$.

  e. The set of all $f \in C[0, 1]$ such that $d^2 f/dx^2 + 5(df/dx) + 6f = 0$.

2. Let V be the vector space of all 2×2 matrices over F. Let

$$W_1 = \left\{ \begin{bmatrix} a & b \\ 0 & c \end{bmatrix} \mid a, b, c \in F \right\}$$

and

$$W_2 = \left\{ \begin{bmatrix} a & 0 \\ b & c \end{bmatrix} \mid a, b, c \in F \right\}.$$

Show that W_1 and W_2 are subspaces. Also compute $W_1 \cap W_2$ and $W_1 + W_2 = \{u + v \mid u \in W_1, v \in W_2\}$, and show that these are subspaces.

3. Let u be a fixed vector in F^n, and let $W = \{x \in F^n \mid x \perp u\}$. Prove that W is a subspace. (Recall that $x \perp u$ means $x^T u = 0$.)

4. Let W be a subspace of F^n. Prove that

$$W^\perp = \{x \in F^n \mid x \perp u \text{ for all } u \in W\}$$

is a subspace (called the orthogonal complement of W). Find $W \cap W^\perp$.

5. Show in Exercise 4 that $F^n = W + W^\perp$. (Hint: Take an orthonormal basis of W, and extend it to an orthonormal basis of F^n.)

9.2 Linear Dependence and Linear Independence

The concepts of linear combination of a finite set of vectors, linear independence, and linear dependence of a finite set of vectors are defined in exactly the same manner as in *Chapter 3*. In the examples that follow, we will need to remind ourselves of the notion of a function (also called mapping) that we learned in calculus. In particular, a function f is a zero function if its image $f(x)$ is zero for all values of x in its domain.

a. The functions $f_1(x) = x$, $f_2(x) = \sin x$ are linearly independent over **R**. For example, let, $\alpha_1 x + \alpha_2 \sin x$ be a zero function. That is, suppose

$$\alpha_1 x + \alpha_2 \sin x = 0 \qquad \text{for all } x. \tag{1}$$

Putting $x = \pi, \pi/2$ successively in Equation (1), we obtain $\alpha_1 = 0 = \alpha_2$.

b. The functions

$$f_1(x) = 1, \qquad f_2(x) = x, \qquad f_2(x) = x^2$$

are linearly independent over **R**. For example, suppose

$$\alpha_1 + \alpha_2 x + \alpha_3 x^2 = 0 \quad \text{for all } x. \tag{1}$$

Differentiating Equation (1) twice, we get

$$\alpha_2 + 2\alpha_3 x = 0 \quad \text{for all } x, \tag{2}$$
$$2\alpha_3 = 0. \tag{3}$$

From Equations (1), (2), and (3), we get

$$\alpha_1 = 0 = \alpha_2 = \alpha_3.$$

Alternatively, we could put $x = 0, 1, 2$, say, in Equation (1) successively to obtain

$$\alpha_1 = 0 = \alpha_2 = \alpha_3.$$

c. $f_1(x) = 1$, $f_2(x) = x$, $f_3(x) = 1 + x$ is a linearly dependent list because

$$\alpha_1 f_1 + \alpha_2 f_2 + \alpha_3 f_3 = 0$$

for

$$\alpha_1 = 1, \qquad \alpha_2 = 1, \qquad \alpha_3 = -1.$$

d. $f_1(x) = x$, $f_2(x) = \sin x$, $f_3(x) = \frac{1}{2}x$ are linearly dependent functions. A dependence relation is given by

$$\alpha_1 x + \alpha_2 \sin x + \alpha_3 \left(\frac{1}{2}x\right) = 0,$$

where

$$\alpha_1 = -\frac{1}{2}, \qquad \alpha_2 = 0, \qquad \text{and} \qquad \alpha_3 = 1.$$

e. Let $P_n(x)$ be the set of polynomials over $\mathbf{R}$ in x of degree $\leqslant n$. Then $\{1, x, \ldots, x^n\}$ is a basis of $P_n(x)$. First, clearly, every polynomial in $P_n(x)$ is a linear combination of $1, x, \ldots, x^n$. The linear independence of $1, x, \ldots, x^n$ is proved on the same lines as in Example (b).

f. Let $S = \{y \mid d^2y/dx^2 + 4y = 0\}$. Here, $\sin 2x$, $\cos 2x \in S$, and any solution of $d^2y/dx^2 + 4y = 0$ is a linear combination of $\sin 2x$, $\cos 2x$. Since $\sin 2x$ and $\cos 2x$ are linearly independent, a basis of S is $\{\sin 2x, \cos 2x\}$.   ∎

9.2 Exercises

1. Let $D(\mathbf{R})$ be the vector space of all differentiable functions, and let $W = \{f \in D(\mathbf{R}) \mid d^2 f/dx^2 = 0\}$. Show that W is a two-dimensional subspace.

2. Show that the following sets of functions are linearly independent:

   a. $f_1(x) = x$, $f_2(x) = e^x$, $f_3(x) = e^{2x}$.

   b. $f_1(x) = \sin x$, $f_2(x) = 1 + \cos x$.

   c. $f_1(x) = x^2 + 1$, $f_2(x) = x^3$, $f_3(x) = 1 - x$.

9.3 Linear Mappings

To study relationships between two vector spaces over a field F, we need to consider functions from one vector space to another that preserve the vector space operations. Such functions are called *linear mappings* or *linear transformations*. Linear mappings occur in almost all areas of science and in everyday life. Whenever we want to transform a picture to another picture by shifting its position and by rotating it, we construct the image by applying this kind of function. For example, the function $T(x, y) = (kx, y)$ scales the x coordinate by a factor of k, and the function $T(x, y) = (x, ky)$ scales the y coordinate by a factor of k. These functions expand or contract the figure according to whether k is bigger or smaller than 1.

Let us consider a function T from a vector space V to a vector space W:

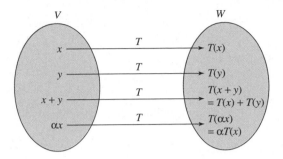

As exhibited in the diagram above, the output $T(x + y)$ of the sum of the vectors x and y is the sum $T(x) + T(y)$ of the outputs, and the output of αx, where α is some number, is α times the output of x. (Note that the output $T(x)$ when the input is x is commonly known in mathematics as the *image* of x under the mapping T.)

9.3.1
DEFINITION

Linear Mapping or Linear Transformation
A function T from a vector space V over a field F to a vector space W over F satisfying the two properties exhibited in the boxes in the diagram above is called a *linear mapping* or a *linear transformation*. In symbols, we write this mapping as

$$T : V \longrightarrow W,$$

satisfying the axioms

  i. $T(x + y) = T(x) + T(y)$,
  ii. $T(\alpha x) = \alpha T(x)$,

for all $x, y \in V$ and for all $\alpha \in F$.

9.3.2
EXAMPLES

a. (Expansion-Contraction Mappings). Let $T : \mathbf{R}^2 \longrightarrow \mathbf{R}^2$ be given by $T(x, y) = (kx, y)$ (or by (x, ky)). Consider first the former case. We have

$$(x_1, y_1) \longrightarrow (kx_1, y_1)$$
$$(x_2, y_2) \longrightarrow (kx_2, y_2)$$
$$(x_1, y_1) + (x_2, y_2) = (x_1 + x_2, y_1 + y_2) \longrightarrow (k(x_1 + x_2), y_1 + y_2)$$
$$= (kx_1, y_1) + (kx_2, y_2).$$

Thus

$$(x_1, y_1) + (x_2, y_2) \longrightarrow T(x_1, y_1) + T(x_2, y_2). \tag{1}$$

Next, we find $\alpha(x_1, y_1) = (\alpha x_1, \alpha y_1) \longrightarrow (k\alpha x_1, \alpha y_1) = \alpha(kx_1, y_1)$, and so

$$\alpha(x_1, y_1) \longrightarrow \alpha T(x_1, y_1). \tag{2}$$

It follows from Equations (1) and (2) that T is a linear mapping.

b. (Expansion-Reflection Mapping). Let $T : F^2 \longrightarrow F^2$, where

$$T\begin{bmatrix} x_1 \\ x_2 \end{bmatrix} = \begin{bmatrix} -x_1 \\ 2x_2 \end{bmatrix},$$

that is,

$$x = \begin{bmatrix} x_1 \\ x_2 \end{bmatrix} \longrightarrow \begin{bmatrix} -x_1 \\ 2x_2 \end{bmatrix} = T(x). \tag{1}$$

This mapping dilates the y coordinate to twice its original value and reflects the point in y axis.

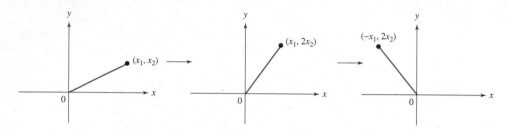

We check whether T is a linear mapping. For this, take another vector

$$y = \begin{bmatrix} y_1 \\ y_2 \end{bmatrix} \in F^2$$

and exhibit its output under T, that is,

$$y = \begin{bmatrix} y_1 \\ y_2 \end{bmatrix} \longrightarrow \begin{bmatrix} -y_1 \\ 2y_2 \end{bmatrix} = T(y). \tag{2}$$

Consider next the output of $x + y$ and of αx, where α is any scalar:

$$x + y = \begin{bmatrix} x_1 + y_1 \\ x_2 + y_2 \end{bmatrix} \longrightarrow \begin{bmatrix} -(x_1 + y_1) \\ 2(x_2 + y_2) \end{bmatrix} = T(x + y), \tag{3}$$

$$\alpha x = \begin{bmatrix} \alpha x_1 \\ \alpha x_2 \end{bmatrix} \longrightarrow \begin{bmatrix} -\alpha x_1 \\ 2\alpha x_2 \end{bmatrix} = T(\alpha x). \tag{4}$$

Is $T(x + y) = T(x) + T(y)$? Yes, by Equations (1)–(3).
Is $T(\alpha x) = \alpha T(x)$? Yes, by Equations (1) and (4).
Hence T is a linear mapping.
c. Check whether $T : F^3 \longrightarrow F^2$, where

$$T\begin{bmatrix} x_1 \\ x_2 \\ x_3 \end{bmatrix} = \begin{bmatrix} 2x_1 \\ 0 \end{bmatrix}$$

is a linear mapping. As in part (a), we consider the output of the vectors $x, y, x + y$ and αx. Note that all these vectors are in F^3 and are therefore 3×1 matrices.

However, the output vector is in F^2:

$$x = \begin{bmatrix} x_1 \\ x_2 \\ x_3 \end{bmatrix} \longrightarrow \begin{bmatrix} 2x_1 \\ 0 \end{bmatrix} = T(x), \tag{1}$$

$$y = \begin{bmatrix} y_1 \\ y_2 \\ y_3 \end{bmatrix} \longrightarrow \begin{bmatrix} 2y_1 \\ 0 \end{bmatrix} = T(y), \tag{2}$$

$$x + y = \begin{bmatrix} x_1 + y_1 \\ x_2 + y_2 \\ x_3 + y_3 \end{bmatrix} \longrightarrow \begin{bmatrix} 2x_1 + 2y_1 \\ 0 \end{bmatrix} = T(x + y), \tag{3}$$

$$\alpha x = \begin{bmatrix} \alpha x_1 \\ \alpha x_2 \\ \alpha x_3 \end{bmatrix} \longrightarrow \begin{bmatrix} 2\alpha x_1 \\ 0 \end{bmatrix} = T(\alpha x). \tag{4}$$

Is $T(x + y) = T(x) + T(y)$? Yes, by Equations (1)–(3).
Is $T(\alpha x) = \alpha T(x)$? Yes, by Equations (1) and (4).
Hence T is a linear mapping.

d. Check whether $T : F^2 \longrightarrow F^2$, where

$$T \begin{bmatrix} a \\ b \end{bmatrix} = \begin{bmatrix} a^2 \\ b + 1 \end{bmatrix}$$

is a linear mapping. As in the above examples, we consider the outputs of the vectors x, y, $x + y$, and αx, which belong to F^2:

$$x = \begin{bmatrix} x_1 \\ x_2 \end{bmatrix} \longrightarrow \begin{bmatrix} x_1^2 \\ x_2 + 1 \end{bmatrix} = T(x), \tag{1}$$

$$y = \begin{bmatrix} y_1 \\ y_2 \end{bmatrix} \longrightarrow \begin{bmatrix} y_1^2 \\ y_2 + 1 \end{bmatrix} = T(y), \tag{2}$$

$$x + y = \begin{bmatrix} x_1 + y_1 \\ x_2 + y_2 \end{bmatrix} \longrightarrow \begin{bmatrix} (x_1 + y_1)^2 \\ x_2 + y_2 + 1 \end{bmatrix} = T(x + y), \tag{3}$$

$$\alpha x = \begin{bmatrix} \alpha x_1 \\ \alpha x_2 \end{bmatrix} \longrightarrow \begin{bmatrix} \alpha^2 x_1^2 \\ \alpha x_2 + 1 \end{bmatrix} = T(\alpha x). \tag{4}$$

Is $T(x + y) = T(x) + T(y)$ for all x, y? No, since from Equations (1) and (2),

$$T(x) + T(y) = \begin{bmatrix} x_1^2 + y_1^2 \\ x_2 + y_2 + 2 \end{bmatrix} \quad \text{and} \quad T(x + y) = \begin{bmatrix} (x_1 + y_1)^2 \\ x_2 + y_2 + 1 \end{bmatrix}.$$

Choosing $x_1 = 1 = x_2 = y_1 = y_2$, we note that $T(x + y) \neq T(x) + T(y)$.

So T is not a linear mapping. Once we violate one of the two axioms for T to be a linear mapping, we can conclude that T is not a linear mapping.

e. (Canonical mapping induced by a matrix) Check whether $T : F^3 \longrightarrow F^3$, where $T(x) = Ax$, A being any 3×3 matrix with entries from F, is a linear mapping.

Here $x \in F^3$ stands for $\begin{bmatrix} x_1 \\ x_2 \\ x_3 \end{bmatrix}$.

$$x \longrightarrow Ax = T(x), \tag{1}$$

$$y \longrightarrow Ay = T(y), \tag{2}$$

$$x + y \longrightarrow A(x + y) = T(x + y), \tag{3}$$

$$\alpha x \longrightarrow A(\alpha x) = T(\alpha x). \tag{4}$$

From Equations (1)–(3), $T(x+y) = A(x+y) = Ax+Ay$ (by the distributive law for matrices) $= T(x)+T(y)$. Also, from Equations (1) and (4), $T(\alpha x) = A(\alpha x) = \alpha Ax$ (since α is a number). So $T(\alpha x) = \alpha T(x)$. Hence T is a linear mapping.

Note that in the above example, we made no special use of the fact that T is a function from F^3 to F^3. The same argument would hold if T were a function from F^n to F^m and A were an $m \times n$ matrix. This example thus shows that to each matrix A, there corresponds a linear transformation. Indeed, the converse also holds, that is, to each linear mapping there corresponds a matrix. We will discuss this correspondence briefly in Section 9.5.

f. (Projection mapping) Let $\pi : F^3 \longrightarrow F^2$ be the mapping given by $\pi[a\ b\ c] = [a\ b]$. Then π is a linear mapping. This function gives the projection of any object in a space to a plane by intersecting with the plane $z = 0$. (For example, the sphere $x^2 + y^2 + z^2 = 1$ will project onto the circle $x^2 + y^2 = 1$ in the xy plane.)

Let $x = [a\ b\ c]$, $y = [a'\ b'\ c'] \in F^3$, and $\alpha \in F$. Then

$$\begin{aligned} \pi(x + y) &= \pi \begin{bmatrix} a + a' & b + b' & c + c' \end{bmatrix} \\ &= \begin{bmatrix} a + a' & b + b' \end{bmatrix} = \begin{bmatrix} a & b \end{bmatrix} + \begin{bmatrix} a' & b' \end{bmatrix} \\ &= \pi(x) + \pi(y). \end{aligned}$$

Also,

$$\begin{aligned} \pi(\alpha x) &= \pi[\alpha a\ \alpha b\ \alpha c] = [\alpha a\ \alpha b] \\ &= \alpha[a\ b] = \alpha \pi(x). \end{aligned}$$

g. Let $f : F^3 \longrightarrow F^3$ be a linear mapping for which

$$f \begin{bmatrix} 1 \\ 0 \\ 0 \end{bmatrix} = \begin{bmatrix} 1 \\ 1 \\ 1 \end{bmatrix}, \qquad f \begin{bmatrix} 1 \\ 1 \\ 1 \end{bmatrix} = \begin{bmatrix} 0 \\ -1 \\ 1 \end{bmatrix}, \qquad \text{and} \qquad f \begin{bmatrix} 1 \\ 0 \\ 1 \end{bmatrix} = \begin{bmatrix} 0 \\ 0 \\ 0 \end{bmatrix}.$$

Find $f \begin{bmatrix} 2 \\ 3 \\ 4 \end{bmatrix}$.

Let us compute α, β, and γ such that

$$\begin{bmatrix} 2 \\ 3 \\ 4 \end{bmatrix} = \alpha \begin{bmatrix} 1 \\ 0 \\ 0 \end{bmatrix} + \beta \begin{bmatrix} 1 \\ 1 \\ 1 \end{bmatrix} + \gamma \begin{bmatrix} 1 \\ 0 \\ 1 \end{bmatrix}.$$

This gives

$$2 = \alpha + \beta + \gamma,$$
$$3 = \beta,$$
$$4 = \beta + \gamma.$$

Thus $\beta = 3$, $\gamma = 1$, and $\alpha = -2$. So

$$f \begin{bmatrix} 2 \\ 3 \\ 4 \end{bmatrix} = (-2)f \begin{bmatrix} 1 \\ 0 \\ 0 \end{bmatrix} + 3f \begin{bmatrix} 1 \\ 1 \\ 1 \end{bmatrix} + f \begin{bmatrix} 1 \\ 0 \\ 1 \end{bmatrix} = \begin{bmatrix} -2 \\ -5 \\ 1 \end{bmatrix}.$$

h. Let $V = C[0, 1]$ be the vector space of all continuous real-valued functions on $[0, 1]$ over the real field $\mathbf{R}$. The mapping

$$\phi \longrightarrow \int_0^1 \phi(t)\, dt$$

is a linear mapping from V to $\mathbf{R}$. ∎

9.3.3 Remark Throughout this chapter, an n-tuple $(x_1, x_2, \ldots, x_n)$ will denote a member of the vector space F^n, besides our usual notation of denoting the members of F^n by column or row vectors.

9.3 Exercises

1. Which of the following mappings of $\mathbf{R}^2 \longrightarrow \mathbf{R}^2$ are linear?

a. $(x_1, x_2) \longrightarrow (y_1, y_2)$, where $y_1 = 2x_1 + 3x_2 + 1$ and $y_2 = -x_1 + x_2$.

b. $(x_1, x_2) \longrightarrow (y_1, y_2)$, where $y_1 = x_1^2 + x_1$ and $y_2 = -x_1 + 3x_2$.

c. $(x_1, x_2) \longrightarrow (y_1, y_2)$, where $y_1 = 1 + x_1$ and $y_2 = x_2$.

d. $(x_1, x_2) \longrightarrow (y_1, y_2)$, where $y_1 = x_1 - x_2$ and $y_2 = 0$.

2. Let $f : \mathbf{R}^3 \longrightarrow \mathbf{R}^3$ be defined by

$$f(x_1, x_2, x_3) = (x_1, x_1 + x_2, x_1 + x_2 + x_3).$$

Show that f is a linear mapping.

3. Let $f : \mathbf{R}^3 \longrightarrow \mathbf{R}^3$ be a mapping defined by

$$f(x_1, x_2, x_3) = (x_1 - x_2, x_2 - x_3, x_3 - x_1).$$

Show that f is a linear mapping.

4. Let $f : \mathbf{R}^2 \longrightarrow \mathbf{R}^2$ be a linear mapping. Find (i) $f(-1, 1)$; (ii) $f(2, 5)$; (iii) $f(3, 0)$; and (iv) $f(a, b)$ in each of the following cases:

a. $f(1, 0) = (2, 3)$; $f(0, 1) = (-1, 1)$.

b. $f(1, 2) = (1, -1)$; $f(2, -3) = (4, 1)$.

5. (Drill 9.1) Let $f : R^3 \longrightarrow R^3$ be a linear mapping such that $f(1, 2, 1) = (1, -1, 0)$, $f(2, -3, 0) = (4, 1, 0)$, $f(0, 0, 1) = (0, 0, 1)$. Find (a) $f(-1, 1, 0)$; (b) $f(2, 5, 1)$; (c) $f(3, 0, 1)$.

6. (Drill 9.2) Let $f : F^3 \to F^3$ be a linear mapping such that $f(1, 0, 0) = (1, 1, 1)$, $f(1, 1, 1) = (0, 1, 1)$, and $f(1, 0, 1) = (0, 0, 0)$. Find $f(2, 3, 4)$.

7. Let $f : F^{n \times n} \longrightarrow F^{n \times n}$ be a mapping such that $f(A) = AB$ for all $A \in F^{n \times n}$, where B is a fixed $n \times n$ matrix. Prove that f is a linear mapping.

8. Let $T : \mathbf{R}^2 \longrightarrow \mathbf{R}^2$ be a function. Determine in each of the following cases whether T is a linear mapping:

a. $T \begin{bmatrix} x \\ y \end{bmatrix} = \begin{bmatrix} 2x \\ 5y \end{bmatrix}$. **b.** $T(x, y) = (x^2, y)$.

c. $T(x, y) = (-y, x)$. **d.** $T(x_1, x_2) = (x_1, 0)$.

9. Determine whether $T : \mathbf{R}^3 \longrightarrow \mathbf{R}^2$ is a linear mapping.

a. $T(x, y, z) = (0, 0)$.

b. $T(x, y, z) = (2, 3)$.

c. $T(x, y, z) = (x, x + y + z)$.

10. Determine whether $T : V \longrightarrow \mathbf{R}$ is a linear mapping, where V is the vector space of 2×2 matrices whose entries are real numbers.

a. $T \begin{bmatrix} a & b \\ c & d \end{bmatrix} = a + d$.

b. $T \begin{bmatrix} a & b \\ c & d \end{bmatrix} = b + c$.

c. $T \begin{bmatrix} a & b \\ c & d \end{bmatrix} = 2a - b$.

d. $T \begin{bmatrix} a & b \\ c & d \end{bmatrix} = \det \begin{bmatrix} a & b \\ c & d \end{bmatrix}$.

e. $T \begin{bmatrix} a & b \\ c & d \end{bmatrix} = a^2 + b^2 + c^2 + d^2$.

In the section that follows, we prove some properties of linear mappings. The reader who is not interested in their proofs may skim over it or just skip this section and go to *Section 9.5* without losing continuity.

9.4 Some Properties of Linear Mappings: Image and Kernel

9.4.1 Property

If $T : V \longrightarrow W$ is a linear mapping, then

i. $T(0) = 0$.
ii. $T(-x) = -T(x)$, $x \in V$.

Proof

i. $T(0) = T(0 + 0) = T(0) + T(0)$, since T is a linear mapping. Thus $T(0) = 0$.
ii. $0 = T(0) = T(x + (-x)) = T(x) + T(-x)$. Thus $T(-x) = -T(x)$.

Next we give the concept of the image of a linear mapping.

9.4.2 DEFINITION

Image
Let $T : V \longrightarrow W$ be a linear mapping. Then

$$\operatorname{Im} T = \{T(x) \mid x \in V\},$$

that is, the set of outputs of all elements of T, is called the *image* of T, denoted by $\operatorname{Im} T$.

9.4.3 EXAMPLE Let $T : F^2 \longrightarrow F^2$, where

$$T \begin{bmatrix} a \\ b \end{bmatrix} = \begin{bmatrix} a \\ 0 \end{bmatrix}.$$

Then

$$\operatorname{Im} T = \left\{ \begin{bmatrix} a \\ 0 \end{bmatrix} \,\middle|\, a \in F \right\}.$$

Note that

$$\begin{bmatrix} 1 \\ 2 \end{bmatrix} \notin \operatorname{Im} T, \qquad \begin{bmatrix} 0 \\ 1 \end{bmatrix} \notin \operatorname{Im} T,$$

and so on. ∎

9.4.4 EXAMPLE Let $T : F^2 \longrightarrow F^2$, where

$$T \begin{bmatrix} a \\ b \end{bmatrix} = \begin{bmatrix} 2a \\ 3b \end{bmatrix}.$$

Here

$$\operatorname{Im} T = \left\{ \begin{bmatrix} 2a \\ 3b \end{bmatrix} \,\middle|\, a, b \in F \right\}.$$

But since every $\begin{bmatrix} c \\ d \end{bmatrix}$ in F^2 can be rewritten as $\begin{bmatrix} 2(\frac{c}{2}) \\ 3(\frac{d}{3}) \end{bmatrix}$, it follows that each $\begin{bmatrix} c \\ d \end{bmatrix}$ is in $\text{Im } T$. Thus $\text{Im } T = F^2$. ∎

9.4.5 Property

Let $T : V \longrightarrow W$ be a linear mapping. Then $\text{Im } T$ is a subspace of W.

Proof Suppose $T(x)$, $T(y) \in \text{Im } T$. Then $T(x) - T(y) = T(x - y)$, since T is a linear mapping. Thus $T(x) - T(y) \in \text{Im } T$. Furthermore, $\alpha T(x) = T(\alpha x)$, $\alpha \in F$. So $\alpha T(x) \in \text{Im } T$. This proves that $\text{Im } T$ is a subspace of W.

We know that if T is a linear mapping, then $T(0) = 0$. But it is also possible that $T(x) = 0$ for some $x \neq 0$. For example, if $T : F^2 \longrightarrow F^2$, where

$$T \begin{bmatrix} a \\ b \end{bmatrix} = \begin{bmatrix} a \\ 0 \end{bmatrix},$$

then

$$T \begin{bmatrix} 0 \\ 1 \end{bmatrix} = \begin{bmatrix} 0 \\ 0 \end{bmatrix}.$$

The set of vectors x for which $T(x) = 0$ is quite an important set in the theory of vector spaces. To make our point about its importance, consider a linear mapping $T : F^3 \longrightarrow F^3$, where $T(x) = Ax$, A being some fixed 3×3 matrix. Then $T(x) = 0$ gives $Ax = 0$, which is a homogeneous LS. Thus the set of solutions of this LS is precisely the set of vectors x for which $T(x) = 0$. We call this set the *kernel* of T, as defined below.

9.4.6 DEFINITION

Kernel of a Linear Mapping

Let $T : V \to W$ be a linear mapping. The set

$$S = \{x \in V \mid T(x) = 0\}$$

is called the *kernel of* T, denoted by $\text{Ker } T$.

9.4.7 Property

If T is a linear mapping from V to W, then $\text{Ker } T$ is a subspace of V.

Proof Let $x, y \in \text{Ker } T$. Then $T(x) = 0 = T(y)$. This gives $T(x - y) = T(x) - T(y) = 0$, and so $x - y \in \text{Ker } T$. Furthermore, if $\alpha \in F$, then $T(\alpha x) = \alpha T(x) = \alpha(0) = 0$. Hence $\alpha x \in \text{Ker } T$. This proves that $\text{Ker } T$ is a subspace of V.

9.4.8
DEFINITION

One-to-One Mapping

A mapping f is called one-to-one if $f(x) = f(y)$ implies $x = y$, that is, distinct elements have distinct images.

For example, let

$$A = \begin{bmatrix} 1 & 1 & 3 \\ 2 & 2 & 4 \end{bmatrix},$$

and let $f : \mathbf{R}^3 \to \mathbf{R}^2$ be a linear mapping such that

$$f \begin{bmatrix} a \\ b \\ c \end{bmatrix} = A \begin{bmatrix} a \\ b \\ c \end{bmatrix}.$$

Then

$$f \begin{bmatrix} 1 \\ 0 \\ 0 \end{bmatrix} = \begin{bmatrix} 1 \\ 2 \end{bmatrix}, \quad \text{and} \quad f \begin{bmatrix} 0 \\ 1 \\ 0 \end{bmatrix} = \begin{bmatrix} 1 \\ 2 \end{bmatrix}.$$

Thus f is not a one-to-one mapping.

9.4.9 Property

If T is a linear mapping from V to W, then Ker T is zero if and only if T is one-to-one mapping.

Proof Assume first that T is one-to-one. Let x be in Ker T. Then $T(x) = 0$. Also by Property 9.4.1, $T(0) = 0$. Thus we get $x = 0$ because T is one-to-one. This proves that Ker $T = 0$.

Next assume that Ker $T = 0$. By Property 9.4.1(ii), $T(-x) = -T(x)$. Thus $T(x + (-y)) = T(x) + T(-y) = T(x) - T(y)$. Suppose now that $T(x) = T(y)$. Then we get $T(x + (-y)) = 0$. But then $x + (-y) = 0$, and so $x = y$, because Ker $T = 0$. This proves that $T(x) = T(y)$ implies that $x = y$. Hence T is one-to-one. This completes the proof.

9.4.10
DEFINITION

Onto Mapping

A mapping f from V to W is called an *onto* mapping if for each $w \in W$, there exists $v \subset V$ such that $f(v) = w$. In other words, Im $f = W$.

The mapping $f : \mathbf{R}^3 \to \mathbf{R}^2$ given by

$$f \begin{bmatrix} a \\ b \\ c \end{bmatrix} = \begin{bmatrix} a \\ b \end{bmatrix}$$

is clearly an onto mapping.

9.4.11
DEFINITION

Isomorphism

Let U, V be vector spaces over F. A linear mapping $f : U \rightarrow V$ is called an *isomorphism* from U to V if f is both one-to-one and onto. If there exists an isomorphism from U to V, then U is said to be isomorphic to V, written $U \cong V$.

9.4.12
DEFINITION

Composition of Mappings

If $f : U \rightarrow V$, and $g : V \rightarrow W$ are two mappings, then we define $gf : U \rightarrow W$ by $(gf)(x) = g(f(x))$ for all x in U.

The mapping gf is called the *composition* of f and g in this order.

9.4.13
DEFINITION

Identity Mapping

If $f : U \rightarrow U$ is a mapping from a set U to itself such that $f(x) = x$ for all x in U, then f is called the *identity mapping* on U.

9.4.14 Property

If f is an isomorphism from a vector space U to a vector space V, then there exists a one-to-one and onto linear mapping $g : V \rightarrow U$ such that $(fg)(x) = x$, for all x in V, and $(gf)(x) = x$, for all x in U. In other words, fg and gf are identity mappings on V and U, respectively.

The mapping g obtained in the above property is unique and is called the inverse of f, denoted by f^{-1}.

Proof This proof is left as an exercise.

9.4.15 Property

Let V be a vector space over F, and let dim $V = n$. Then $V \cong F^n$. (This shows that any finite-dimensional vector space can be looked on as F^n for some n.)

Proof Let $\{v_1, \ldots, v_n\}$ be a basis of V. Then each element $v \in V$ can be uniquely expressed as
$$v = \alpha_1 v_1 + \cdots + \alpha_n v_n.$$
Define the mapping $f : V \longrightarrow F^n$ given by

$$f(v) = \begin{bmatrix} \alpha_1 \\ \alpha_2 \\ \vdots \\ \alpha_n \end{bmatrix}.$$

It is a simple matter to verify the two properties for f to be a linear mapping.

Next, because the set of vectors $v_1, \ldots, v_n$ is a linearly independent set, it follows by using the definition of one-to-one mapping that f is one-to-one. Furthermore, given

any member $\begin{bmatrix} \alpha_1 \\ \alpha_2 \\ \vdots \\ \alpha_n \end{bmatrix}$ of F^n,

$$f(\alpha_1 v_1 + \cdots + \alpha_n v_n) = \begin{bmatrix} \alpha_1 \\ \alpha_2 \\ \vdots \\ \alpha_n \end{bmatrix}.$$

This proves that f is onto. Hence $V \cong F^n$, completing the proof.

9.4 Exercises

1. Find $\mathrm{Ker}\, f$ in each of the following cases. Also find $\dim \mathrm{Ker}\, f$.

   a. Exercise 2 of Section 9.3.
   b. Exercise 3 of Section 9.3.
   c. Exercises 8(a), (c) and (d) of Section 9.3.
   d. Exercises 9(a) and (c) of Section 9.3.
   e. Exercises 10(a) and (b) of Section 9.3.

2. Find the condition for which $(a, b, c) \in \mathrm{Im}\, f$, where f is the linear mapping in Exercise 2 of Section 9.3.

3. What are the conditions in order that $(a, b, c) \in \mathrm{Im}\, f$, where f is the linear mapping in Exercise 3 of Section 9.3?

4. In Exercises 8(a), 8(c), and 8(d) of Section 9.3, find $\mathrm{Im}\, T$ and $\dim \mathrm{Im}\, T$. Verify the well-known important result in linear algebra that

$$\dim\ V = \dim \mathrm{Ker}\ T + \dim \mathrm{Im}\, T,$$

where $T : V \longrightarrow W$ is a linear mapping. (Note: V in Exercise 8 of Section 9.3 is $\mathbf{R}^2$.)

5. For the linear mappings T in Exercises 10(a) and 10(b) of Section 9.3, verify the result that

$$\dim\ V = \dim \mathrm{Ker}\ T + \dim \mathrm{Im}\, T.$$

(Note: $\dim V$ in Exercises 10(a) and 10(b) of Section 9.3 is 4.)

6. (Drill 9.3) Let T be the linear transformation given by the matrix

$$A = \begin{bmatrix} 1 & 2 & 3 & -1 \\ 3 & 5 & 8 & -2 \\ 1 & 1 & 3 & 0 \end{bmatrix}.$$

a. Find a basis of the kernel of T.
b. Find a basis of the range of T.
c. Verify the rank-nullity theorem.

7. (Drill 9.4) Find the dimension of Ker T, where T is a linear transformation from $\mathbf{R}^2$ to $\mathbf{R}^3$ given by $T(x_1, x_2) = (x_1, x_1 + x_2, x_1 + x_2)$. Can you obtain the dimension of the range of T without actually computing the range?

9.5 Linear Mappings and Matrices

In this section, we illustrate through examples that there is a one-to-one correspondence between the set of $m \times n$ matrices and the set of linear mappings $T : V \longrightarrow W$, where $\dim V = n$ and $\dim W = m$. (Note that the matrix that corresponds to T is of size $m \times n$, not of size $n \times m$.) Our object here is to give the steps (or the procedure) for finding the matrix of a linear mapping. For this, we shall need bases of V and W. As you will notice, different bases will give rise to different matrices. (But such matrices A and B (say) do have a relation, namely, $B = QAP$ for some invertible matrices P and Q. We do not intend to discuss this in this elementary book.)

Let

$$A = \begin{bmatrix} a & b & c \\ p & q & r \end{bmatrix}$$

be a 2×3 matrix. Let us define a mapping $T : F^3 \longrightarrow F^2$ by

$$T \begin{bmatrix} x_1 \\ x_2 \\ x_3 \end{bmatrix} = \begin{bmatrix} a & b & c \\ p & q & r \end{bmatrix} \begin{bmatrix} x_1 \\ x_2 \\ x_3 \end{bmatrix}.$$

This is clearly a linear mapping. In particular,

$$T \begin{bmatrix} 1 \\ 0 \\ 0 \end{bmatrix} = \begin{bmatrix} a & b & c \\ p & q & r \end{bmatrix} \begin{bmatrix} 1 \\ 0 \\ 0 \end{bmatrix} = \begin{bmatrix} a \\ p \end{bmatrix}$$

$$= a \begin{bmatrix} 1 \\ 0 \end{bmatrix} + p \begin{bmatrix} 0 \\ 1 \end{bmatrix}.$$

Similarly,

$$T \begin{bmatrix} 0 \\ 1 \\ 0 \end{bmatrix} = b \begin{bmatrix} 1 \\ 0 \end{bmatrix} + q \begin{bmatrix} 0 \\ 1 \end{bmatrix},$$

$$T \begin{bmatrix} 0 \\ 0 \\ 1 \end{bmatrix} = c \begin{bmatrix} 1 \\ 0 \end{bmatrix} + r \begin{bmatrix} 0 \\ 1 \end{bmatrix}.$$

This shows that given a matrix A, we can produce a linear mapping T. We also notice that $T(e_1)$, where e_1 is a basis vector of the domain, has been expressed as a linear combination of the basis vectors of the range with the entries of the first column A as the coefficients. A similar statement holds for $T(e_2)$ and $T(e_3)$. This fact suggests to us the procedure for handling the converse question of finding the matrix of a linear mapping, as is illustrated in the following examples.

9.5.1 EXAMPLE Let $T : V \longrightarrow W$ be a linear mapping, and suppose that $\{e_1, e_2, e_3\}$ is a basis of V and $\{f_1, f_2\}$ is a basis of W. We construct below a matrix that we call the matrix of T relative to the given bases of V and W.

Step 1 Find $T(e_1)$, $T(e_2)$, and $T(e_3)$.

Step 2 Since $T(e_1)$, $T(e_2)$, and $T(e_3)$ belong to W, these can be expressed as linear combinations of basis vectors of W. So write

$$T(e_1) = af_1 + bf_2,$$
$$T(e_2) = cf_1 + df_2,$$
$$T(e_3) = pf_1 + qf_2.$$

Step 3 Set

$$A = \begin{bmatrix} a & c & p \\ b & d & q \end{bmatrix}$$

as the 2×3 matrix whose first column consists of the coefficients of f_1 and f_2 in the expression of $T(e_1)$, the second column consists of the coefficients of f_1 and f_2 in the expression of $T(e_2)$, and the third column consists of the coefficients of f_1 and f_2 in the expression of $T(e_3)$.

Then A is the desired matrix. ■

9.5.2 Remark If $T : V \longrightarrow W$ is a linear mapping where the dimensions of V and W are m and n, respectively, then by choosing bases of V and W, we can write down an $n \times m$ matrix of T as described above.

9.5.3 EXAMPLE Find the matrix of $T : F^2 \longrightarrow F^2$, where

$$T\begin{bmatrix} a \\ b \end{bmatrix} - \begin{bmatrix} 2a \\ 3b \end{bmatrix}$$

with respect to the standard basis of F^2, that is,

$$\left\{ e_1 = \begin{bmatrix} 1 \\ 0 \end{bmatrix}, \quad e_2 = \begin{bmatrix} 0 \\ 1 \end{bmatrix} \right\}.$$

(Here, we are taking the same basis both for the domain and the range of the function T.)

Step 1

$$T(e_1) = T \begin{bmatrix} 1 \\ 0 \end{bmatrix} = \begin{bmatrix} 2 \\ 0 \end{bmatrix},$$

$$T(e_2) = T \begin{bmatrix} 0 \\ 1 \end{bmatrix} = \begin{bmatrix} 0 \\ 3 \end{bmatrix}.$$

Step 2

$$\begin{bmatrix} 2 \\ 0 \end{bmatrix} = 2 \begin{bmatrix} 1 \\ 0 \end{bmatrix} + 0 \begin{bmatrix} 0 \\ 1 \end{bmatrix},$$

$$\begin{bmatrix} 0 \\ 3 \end{bmatrix} = 0 \begin{bmatrix} 1 \\ 0 \end{bmatrix} + 3 \begin{bmatrix} 0 \\ 3 \end{bmatrix}.$$

Step 3 The matrix of T is $\begin{bmatrix} 2 & 0 \\ 0 & 3 \end{bmatrix}$. ■

9.5.4 Important Remark

Observe that the matrix $\begin{bmatrix} 2 & 0 \\ 0 & 3 \end{bmatrix}$ behaves like the function T, since

$$T \begin{bmatrix} a \\ b \end{bmatrix} = \begin{bmatrix} 2 & 0 \\ 0 & 3 \end{bmatrix} \begin{bmatrix} a \\ b \end{bmatrix}.$$

9.5.5 EXAMPLES

a. Find the matrix of T given in Example 9.5.3 by choosing the basis of the domain as $\begin{bmatrix} 1 \\ 1 \end{bmatrix}$ and $\begin{bmatrix} 0 \\ 1 \end{bmatrix}$ and the basis of the range as $\begin{bmatrix} 1 \\ 0 \end{bmatrix}$ and $\begin{bmatrix} 2 \\ 3 \end{bmatrix}$.
Here,

$$e_1 = \begin{bmatrix} 1 \\ 1 \end{bmatrix}, \qquad e_2 = \begin{bmatrix} 0 \\ 1 \end{bmatrix}; \qquad f_1 = \begin{bmatrix} 1 \\ 0 \end{bmatrix}, \qquad f_2 = \begin{bmatrix} 2 \\ 3 \end{bmatrix}.$$

Step 1

$$T(e_1) = T \begin{bmatrix} 1 \\ 1 \end{bmatrix} = \begin{bmatrix} 2 \\ 3 \end{bmatrix},$$

$$T(e_2) = T \begin{bmatrix} 0 \\ 1 \end{bmatrix} = \begin{bmatrix} 0 \\ 3 \end{bmatrix}.$$

Step 2 We want to express $\begin{bmatrix} 2 \\ 3 \end{bmatrix}$ and $\begin{bmatrix} 0 \\ 3 \end{bmatrix}$ as linear combinations of $\begin{bmatrix} 1 \\ 0 \end{bmatrix}$ and $\begin{bmatrix} 2 \\ 3 \end{bmatrix}$. For this, we assume that

$$\begin{bmatrix} 2 \\ 3 \end{bmatrix} = \alpha \begin{bmatrix} 1 \\ 0 \end{bmatrix} + \beta \begin{bmatrix} 2 \\ 3 \end{bmatrix}$$

and compute α and β. Now, $2 = \alpha + 2\beta$; $3 = 3\beta$. Thus $\beta = 1, \alpha = 0$. This gives

$$\begin{bmatrix} 2 \\ 3 \end{bmatrix} = 0 \begin{bmatrix} 1 \\ 0 \end{bmatrix} + 1 \begin{bmatrix} 2 \\ 3 \end{bmatrix}. \tag{1}$$

Next we assume that

$$\begin{bmatrix} 0 \\ 3 \end{bmatrix} = a \begin{bmatrix} 1 \\ 0 \end{bmatrix} + b \begin{bmatrix} 2 \\ 3 \end{bmatrix}$$

and compute a and b. We have $0 = a + 2b$; $3 = 3b$. Thus $b = 1, a = -2$. Therefore

$$\begin{bmatrix} 0 \\ 3 \end{bmatrix} = -2 \begin{bmatrix} 1 \\ 0 \end{bmatrix} + 1 \begin{bmatrix} 2 \\ 3 \end{bmatrix}. \tag{2}$$

Step 3 From Equations (1) and (2), the matrix of T is $\begin{bmatrix} 0 & -2 \\ 1 & 1 \end{bmatrix}$.

b. Let $T : F^3 \longrightarrow F^3$ be a linear mapping given by

$$T \begin{bmatrix} x_1 \\ x_2 \\ x_3 \end{bmatrix} = \begin{bmatrix} 2x_1 \\ x_2 \\ -x_3 \end{bmatrix}.$$

Then choosing the standard basis

$$e_1 = \begin{bmatrix} 1 \\ 0 \\ 0 \end{bmatrix}, \qquad e_2 = \begin{bmatrix} 0 \\ 1 \\ 0 \end{bmatrix}, \qquad e_3 = \begin{bmatrix} 0 \\ 0 \\ 1 \end{bmatrix},$$

we obtain

$$T(e_1) = T \begin{bmatrix} 1 \\ 0 \\ 0 \end{bmatrix} = \begin{bmatrix} 2 \\ 0 \\ 0 \end{bmatrix} = 2 \begin{bmatrix} 1 \\ 0 \\ 0 \end{bmatrix} = 2e_1 + 0e_2 + 0e_3.$$

Similarly, $T(e_2) = 0e_1 + 1e_2 + 0e_3$, and $T(e_3) = 0e_1 + 0e_2 - e_3$. Thus the matrix of T is $\begin{bmatrix} 2 & 0 & 0 \\ 0 & 1 & 0 \\ 0 & 0 & -1 \end{bmatrix}$.

c. Let V be the vector space $\mathbf{R}^2$. Find the matrix of the differential operator D from V to V, with respect to the basis $\{\sin x, 2\cos x\}$

$$D(\sin x) = \cos x = 0 \ \sin x + \frac{1}{2}(2 \cos x),$$
$$D(2 \cos x) = -2 \ \sin x = -2 \sin x + 0 \cos x.$$

Thus the matrix of D is $\begin{bmatrix} 0 & -2 \\ \frac{1}{2} & 0 \end{bmatrix}$. ∎

9.5 Exercises

1. Find the matrix associated with the given linear mapping f:

   a. $f : \mathbf{R}^2 \longrightarrow \mathbf{R}^2$ given by $f(a, b) = (2a, 3b)$, and the bases of the domain and ranges are $\{(1, 2), (1, 3)\}$ and the standard basis, respectively.

   b. $f : \mathbf{R}^4 \longrightarrow \mathbf{R}^4$ given by $f(a, b, c, d) = (2a, 0, 0, c + d)$ with respect to the standard basis.

   c. $f : \mathbf{R}^n \longrightarrow \mathbf{R}^n$ given by $f(u) = \alpha u$, $u \in \mathbf{R}^n$ (α is some fixed element in $\mathbf{R}$) with respect to the standard basis.

2. (Drill 9.5) Let $T : \mathbf{R}^3 \to \mathbf{R}^3$ be a linear mapping given by $T(a, b, c) = (a + b, b + c, c + a)$.

   a. Find the matrix associated with T with respect to the standard basis.

   b. Find the matrix associated with T with respect to the basis $\{(1, 1, 0), (1, 0, 1), (0, 1, 1)\}$.

   c. Find a nonsingular matrix X such that $XB = AX$, where A is the matrix associated with the standard basis and B is the matrix associated with the nonstandard basis given in part (b).

3. Let $T : \mathbf{R}^4 \to \mathbf{R}^4$ be a linear mapping given by $T(a, b, c, d) = (a + b, 2a - c, c - d, d)$. Find the matrix associated with T

   a. with respect to the standard basis;

   b. with respect to the basis $\{(1, 0, 1, 2), (1, 1, -1, 0), (1, -2, -1, 0), (1, 0, 1, -1)\}$.

4. Let $V = \mathbf{R}^{2 \times 2}$. Find the matrix of the linear mapping $\phi : V \longrightarrow V$ given by

$$\phi(v) = \begin{bmatrix} 1 & 2 \\ 3 & -1 \end{bmatrix} v$$

   with respect to the standard basis

$$\left\{ \begin{bmatrix} 1 & 0 \\ 0 & 0 \end{bmatrix}, \begin{bmatrix} 0 & 1 \\ 0 & 0 \end{bmatrix}, \begin{bmatrix} 0 & 0 \\ 1 & 0 \end{bmatrix}, \begin{bmatrix} 0 & 0 \\ 0 & 1 \end{bmatrix} \right\}.$$

5. Let ϕ be a linear mapping on $\mathbf{R}^2$ defined by $\phi(x_1, x_2) = (-x_2, x_1)$.

   a. What is the matrix of ϕ with respect to the standard basis?

   b. What is the matrix of ϕ with respect to the basis (e_1, e_2), where $e_1 = (1, 2)$ and $e_2 = (-1, 1)$?

6. Let V be the vector space of all polynomials over $\mathbf{R}$ of degree less than or equal to 3. Let D be the differential operator, and let B be the basis of V consisting of the polynomials f_1, f_2, f_3, and f_4 defined by $f_i(x) = x^{i-1}$. Find the matrix of D with respect to B.

7. (Drill 9.6)

a. Let R_θ be the mapping of a plane onto itself given by rotating a given coordinate system counterclockwise by an angle θ. Show that the matrix of R_θ with respect to the standard basis is

$$\begin{bmatrix} \cos\theta & -\sin\theta \\ \sin\theta & \cos\theta \end{bmatrix}.$$

b. Compute the matrix of the linear map $R_\theta R_\phi$ with respect to the standard basis, and show that it is the matrix of $R_{\theta+\phi}$. Show that R_θ is invertible, and show that the matrix of R_θ^{-1} is the matrix associated with $R_{-\theta}$ with respect to the standard basis.

c. Find the new coordinates of the point $P = (1, 1)$ if the plane is first rotated by 60 degrees clockwise, then by 15 degrees counterclockwise, and then by 40 degrees clockwise. You must show the coordinates of the point after each rotation.

8. Let V be the vector space generated by a finite set B of linearly independent differentiable functions from $\mathbf{R}$ to $\mathbf{R}$, and let D denote the differential operator d/dx. Find the matrix of the linear mapping $D : V \longrightarrow V$ with respect to the basis B, where

a. $B = \{e^{-x}, e^{3x}\}$.   **b.** $B = \{e^x, xe^x\}$.

c. $B = \{1, x, x^2\}$.   **d.** $B = \{1, x, e^x, e^{2x}, xe^{2x}\}$.

e. $B = \{\sin x, \cos x\}$.

9. (Drill 9.7) The translation matrix in homogeneous coordinates is given by the matrix

$$T(a, b) = \begin{bmatrix} 1 & 0 & a \\ 0 & 1 & b \\ 0 & 0 & 1 \end{bmatrix},$$

where a and b are the translations along the x axis and y axis, respectively.

a. Suppose that the coordinates of the vertices of a rectangle are given by the matrix

$$A = \begin{bmatrix} -0.4 & 0 & 0.4 & -0.4 \\ -1 & 1 & -1 & 1 \end{bmatrix},$$

where the first row of A contains the x coordinates and the second row of A contains the corresponding y coordinates. Using Matlab, draw the rectangle.

b. Move the rectangle 0.2 unit along x axis and 0.3 unit along the y axis. Using the translation matrix, find the new coordinates of the rectangle, and plot the original rectangle and the new rectangle.

9.6 Chapter Review Questions and Projects

1. Find whether the following sets are vector spaces over the indicated field F. (Section 9.1)

a. The set of 2×2 matrices whose entries belong to F.

b. The set of 3×3 upper or lower triangular matrices with entries from F.

c. The set of matrices of the form $\begin{bmatrix} 1 & a \\ 0 & 1 \end{bmatrix}$, where a is any real number over $F = \mathbf{R}$.

d. The set of all polynomials with real coefficients over $F = \mathbf{R}$.

e. The set of all polynomials of degree one with coefficients belonging to rational numbers over $F = \mathbf{Q}$.

f. The set of solutions of the differential equation $(dy/dx) - y = a$, where a is a fixed real number over $F = \mathbf{R}$.

2. Check whether the following sets are linearly independent or linearly dependent over $\mathbf{R}$.

   a. $1 + x, 1 + x^2, x$.
   b. $x^2 + x + 1, 2x + 1, 2x^2 - 1$.
   c. $\sin 2x, \cos 2x$.
   d. $e^{-x}, e^x, \sin x$. (Section 9.1)

3. Find a basis of the following vector spaces.

   a. the vector space of all 2×2 matrices over the field of real numbers
   b. the vector space of all 3×3 matrices over the field of rational numbers
   c. the vector space of all 3×3 upper (and lower) triangular matrices over the field of complex numbers. (Section 9.1)

4. What is meant by a linear transformation of a vector space to another vector space? Give an example. (Section 9.3)

5. Find which of the following functions are linear transformations:

   a. $T : \mathbf{R}^2 \longrightarrow \mathbf{R}^3$, where $T(x_1, x_2) = (x_1 + x_2, x_1, 1)$.
   b. $T : \mathbf{R}^2 \longrightarrow \mathbf{R}^2$, where $T(x_1, x_2) = (x_1 + x_2, x_1 - x_2)$.
   c. $S : \mathbf{R}^2 \longrightarrow \mathbf{R}^2$, where $S(x_1, x_2)=(y_1, y_2)$, $y_1 = -2x_1 + x_2, y_2 = x_1 + 2x_2$. (Section 9.3)

6. Let S and T be the linear transformations from $\mathbf{R}^2$ to $\mathbf{R}^2$ given in Questions 5(b) and 5(c). Compute $S+T$ and ST, where $(S+T)(x) = S(x)+T(x)$ and $(ST)(x) = S(T(x))$, $x \in \mathbf{R}^2$. Check whether $S + T$ and ST are linear transformations or not. (Section 9.3)

7. Find the linear transformation $T : \mathbf{R}^3 \longrightarrow \mathbf{R}^2$ in each of the following cases:

   a. $T(1, 0, 1) = (1, -1), T(0, 1, 0) = (-1, 0), T(0, 0, 2) = (2, -1)$.
   b. $T(1, 0, 0) = (2, 3), T(1, 1, 0) = (-1, -2), T(1, 1, 1) = (0, 0)$. (Section 9.3)

8. Find the kernel of T in Questions 7(a) and 7(b). (Section 9.4)

9. Find the matrices of linear transformations S, T, and U from a vector space V to V with basis $\{u_1, u_2\}$:

   a. $S(u_1) = u_1, S(u_2) = -u_2$.
   b. $T(u_1) = u_1 + u_2, T(u_2) = -u_2 + 2u_1$.

c. $U(u_1) = -u_1 + 3u_2$, $U(u_2) = 0$. (Section 9.5)

10. Let D be a linear transformation from the vector space of all polynomials of degree less than 4 to itself given by $D(f) = f'$.

Find the matrix of D with respect to each of the following bases:

i. $\{1, x, x^2, x^3\}$,

ii. $\{1, 1 + x, 1 + x^2, x^3\}$. (Section 9.5)

Chapter Projects

1. In computer graphics, one must frequently use operations such as rotations and translations. Such a transformation $T : \mathbf{R}^n \longrightarrow \mathbf{R}^n$ is given by $T(x) = Ax + b$, where A is an $n \times n$ matrix and b is a vector in $\mathbf{R}^n$, and is known as an *affine transformation*. An affine transformation is a composition of a linear transformation S given by $S(x) = Ax$ followed by a translation (which is indeed not a linear transformation) that shifts the point given by a vector x to the new point $x + b$.

Both these operations, however, may be viewed as linear transformations that change the orientation or the position of the object. The matrix corresponding to a translation in two-dimensional space is a 3×3 matrix T given by

$$T = \begin{bmatrix} 1 & 0 & a \\ 0 & 1 & b \\ 0 & 0 & 1 \end{bmatrix},$$

where the object is moved by a units along x axis and b units along y axis. The third component, 1, is used as a dummy coordinate.

The rotation matrix to rotate an object counterclockwise (with respect to the origin) by an angle x is given by

$$R = \begin{bmatrix} \cos x & -\sin x \\ \sin x & \cos x \end{bmatrix}.$$

Given a list of points that make up a house in two-dimensional space, say $(1,1)$, $(1,2)$, $(1,3)$, $(1,4)$, $(2,1)$, $(2,4)$, $(3,1)$, $(3,2)$, $(3,3)$, $(3,4)$, $(3.5,1.5)$, $(3.5,3.5)$, $(4,2)$, $(4,2.5)$, and $(4,3)$.

a. Translate the house so that the bottom left point of the house coincides with the origin.

b. Rotate the house by 90 degrees counterclockwise.

c. What is the combined effect of both translation and rotation given by parts (a) and (b)?

2. Let S and T be linear mappings given by $S(x_1, x_2, \ldots, x_6) = (2x_1 + 3x_2, 3x_2 + 4x_3, 4x_3 + 5x_4, 5x_4 + 6x_5, 6x_5 + 7x_6, 7x_6 + 8x_1)$ and $T(x_1, x_2, \ldots, x_6) = (x_1 - x_2, x_2 - x_3, x_3 - x_4, x_4 - x_5, x_5 - x_6, x_6 - x_1)$, respectively.

a. Find the matrices of S and T with respect to (i) the standard basis of $\mathbf{R}^6$ and (ii) the basis $\{(1, 1, 1, 0, 0, 0), (1, 1, 0, 0, 0, 0), (1, 0, 0, 0, 0, 0), (0, 1, 0, 1, 0, 1), (0, 0, 0, 1, 0, 0), (0, 0, 0, 0, 1, 0)\}$.

b. Compute $S + T$ and ST.

c. Find whether the maps S and T are invertible onto?

d. Find T, T^2, T^3, T^4, and so on. What do you conjecture about any power of T?

e. Identifying T with its matrix, show that T, T^2, T^3, T^4, T^5, and T^6 are linearly dependent over the field of real numbers, and obtain a dependency relation.

Key Words

Field

Vector Space

Subspace

Linear Independence and Dependence

Basis

Zero Mapping (Function)

One-to-One Mapping (Function)

Onto Mapping

Linear Mapping (Transformation)

Inverse of a Linear Mapping

Isomorphism

Matrix of a Linear Mapping

Key Phrases

- Every n-dimensional vector space over a field F can be regarded as a vector space F^n of n-tuples.

- Every linear transformation of a vector space F^m to a vector space F^n can be identified (abstractly) with a suitable $n \times m$ matrix.

10 Determinants* (Revisited)

10.1 Permutations

10.2 Determinants

10.3 Cofactor Expansion

10.4 Adjoint of a Matrix

10.5 Cramer's Rule

10.6 Product Theorem of Determinants

Introduction

In this chapter, we give a systematic development of the theory of determinants as promised in *Chapter 6*. Our major tool in this development is the theory of permutations of a finite set.

10.1 Permutations

10.1.1 DEFINITION

Permutation
A one-to-one mapping of a set S onto itself is called a *permutation* of S.

For example, let $S = \{a, b, c\}$ be a set of three distinct elements. Then the mapping sending

$$a \to b,$$
$$b \to c,$$
$$c \to a$$

is a permutation of S, denoted by $\begin{bmatrix} a & b & c \\ b & c & a \end{bmatrix}$. In general, the symbol

$$\begin{bmatrix} a_1 & a_2 & \cdots & a_n \\ x_1 & x_2 & \cdots & x_n \end{bmatrix}$$

will denote the mapping that sends

$$a_1 \to x_1,$$
$$a_2 \to x_2,$$
$$\vdots$$
$$a_n \to x_n.$$

---

*Readers who are not interested in formal proofs may skip this chapter.

It is immaterial in which order the columns of

$$\begin{bmatrix} a_1 & a_2 & \cdots & a_n \\ x_1 & x_2 & \cdots & x_n \end{bmatrix}$$

are written. For example,

$$\begin{bmatrix} 1 & 2 & 3 \\ 2 & 3 & 1 \end{bmatrix} = \begin{bmatrix} 2 & 1 & 3 \\ 3 & 2 & 1 \end{bmatrix} = \begin{bmatrix} 3 & 2 & 1 \\ 1 & 3 & 2 \end{bmatrix},$$

and so on denote the same permutation.

It is easy to see that there are precisely six permutations of the set $\{1, 2, 3\}$. These will be denoted as follows throughout this chapter:

$$\sigma_1 = \begin{bmatrix} 1 & 2 & 3 \\ 1 & 2 & 3 \end{bmatrix}, \qquad \sigma_2 = \begin{bmatrix} 1 & 2 & 3 \\ 2 & 1 & 3 \end{bmatrix}, \qquad \sigma_3 = \begin{bmatrix} 1 & 2 & 3 \\ 3 & 2 & 1 \end{bmatrix},$$

$$\sigma_4 = \begin{bmatrix} 1 & 2 & 3 \\ 1 & 3 & 2 \end{bmatrix}, \qquad \sigma_5 = \begin{bmatrix} 1 & 2 & 3 \\ 2 & 3 & 1 \end{bmatrix}, \qquad \sigma_6 = \begin{bmatrix} 1 & 2 & 3 \\ 3 & 1 & 2 \end{bmatrix}.$$

The set of six permutations of $\{1, 2, 3\}$ is denoted by S_3. In general, S_n denotes the set of permutations of $\{1, 2, \ldots, n\}$, and it contains $n!$ elements.

We note the following facts about the permutations of $\{1, 2, 3\}$ as listed above:

$\sigma_1 = \begin{bmatrix} 1 & 2 & 3 \\ 1 & 2 & 3 \end{bmatrix}$ maps each number of the set into itself. It is called the *identity permutation i*.

$\sigma_2 = \begin{bmatrix} 1 & 2 & 3 \\ 2 & 1 & 3 \end{bmatrix}$ interchanges 1 and 2 and keeps 3 fixed (i.e., sends 3 to itself). It is called a *transposition* and is denoted by (1 2). Similarly, $\sigma_3 = (1\ 3)$ and $\sigma_4 = (2\ 3)$ are also transpositions in the sense defined below.

$\sigma_5 = \begin{bmatrix} 1 & 2 & 3 \\ 2 & 3 & 1 \end{bmatrix}$ is the mapping that sends $1 \to 2 \to 3 \to 1$. It is called a *cycle* and is denoted by (1 2 3).

$\sigma_6 = \begin{bmatrix} 1 & 2 & 3 \\ 3 & 1 & 2 \end{bmatrix}$ is the mapping that sends $1 \to 3 \to 2 \to 1$. It is also a cycle (in the sense defined below) and is denoted by (1 3 2).

10.1.2
DEFINITION

Transposition
Let a and b be two distinct elements of a set. Then a permutation σ that sends $a \to b \to a$ and keeps each of the remaining elements of S fixed is called a *transposition* and is denoted by $(a\ b)$.

Obviously, $(a\ b) = (b\ a)$.

10.1.3 DEFINITION	Cycle
	Let $a_1, a_2, \ldots, a_m$ be m distinct elements of a set S. If a permutation σ of S is such that it sends
	$$a_1 \rightarrow a_2 \rightarrow a_3 \rightarrow \cdots \rightarrow a_m \rightarrow a_1$$
	and keeps the remaining elements, if any, of S fixed, then σ is called a *cycle* of length m and is denoted by $(a_1 \ a_2 \cdots a_m)$.

Clearly, a transposition is a cycle of length 2.

10.1.4 DEFINITION	Product of Permutations
	Since permutations are mappings, we define the product of two permutations as the composition of two mappings.

For example, consider the product

$$\sigma_2 \cdot \sigma_3 = \begin{bmatrix} 1 & 2 & 3 \\ 2 & 1 & 3 \end{bmatrix} \begin{bmatrix} 1 & 2 & 3 \\ 3 & 2 & 1 \end{bmatrix}.$$

Now, σ_3 maps $1 \rightarrow 3$ and σ_2 maps $3 \rightarrow 3$, and so $\sigma_2\sigma_3$ maps $1 \rightarrow 3$. Thus

$$\sigma_2\sigma_3(1) = \sigma_2(\sigma_3(1)) = \sigma_2(3) = 3.$$

Similarly,

$$\sigma_2\sigma_3(2) = \sigma_2(\sigma_3(2)) = \sigma_2(2) = 1,$$
$$\sigma_2\sigma_3(3) = \sigma_2(\sigma_3(3)) = \sigma_2(1) = 2.$$

So $\sigma_2\sigma_3$ is the mapping that sends $1 \rightarrow 3, 2 \rightarrow 1, 3 \rightarrow 2$, that is,

$$\sigma_2\sigma_3 = \begin{bmatrix} 1 & 2 & 3 \\ 3 & 1 & 2 \end{bmatrix} = (1 \ 3 \ 2) = \sigma_6.$$

Similarly,

$$\sigma_5\sigma_6 = \begin{bmatrix} 1 & 2 & 3 \\ 2 & 3 & 1 \end{bmatrix} \begin{bmatrix} 1 & 2 & 3 \\ 3 & 1 & 2 \end{bmatrix} = \begin{bmatrix} 1 & 2 & 3 \\ 1 & 2 & 3 \end{bmatrix} = \sigma_1,$$

and so on.

Next we note that every permutation either is a transposition or can be expressed as a product of transpositions. For example,

$$(1 \quad 3 \quad 2) = (1 \quad 2)(1 \quad 3),$$

since

$$(1\ 2)(1\ 3) = \begin{bmatrix} 1 & 2 & 3 \\ 2 & 1 & 3 \end{bmatrix} \begin{bmatrix} 1 & 2 & 3 \\ 3 & 2 & 1 \end{bmatrix}$$

$$= \begin{bmatrix} 1 & 2 & 3 \\ 3 & 1 & 2 \end{bmatrix} = (1\ 3\ 2).$$

Similarly,

$$(1\ 2\ 3) = (1\ 3)(1\ 2).$$

If $(a\ b)$ is a transposition, then $(a\ b)(a\ b) = $ the identity permutation. For example,

$$(2\ 3)(2\ 3) = \begin{bmatrix} 1 & 2 & 3 \\ 1 & 3 & 2 \end{bmatrix} \begin{bmatrix} 1 & 2 & 3 \\ 1 & 3 & 2 \end{bmatrix} = \begin{bmatrix} 1 & 2 & 3 \\ 1 & 2 & 3 \end{bmatrix} = i.$$

Suppose a permutation σ has been expressed as a product of r transpositions:

$$\sigma = \sigma_1 \sigma_2 \cdots \sigma_r.$$

Since for any transposition $(a\ b)$, the product $(a\ b)(a\ b)$ is the identity permutation, σ can also be expressed as

$$\sigma = (a\ b)(a\ b)\sigma_1 \sigma_2 \cdots \sigma_r \qquad \text{(a product of } r+2 \text{ transpositions)}$$
$$= \sigma_1 \sigma_2 \cdots \sigma_r (a\ b)(a\ b)(b\ c)(b\ c) \qquad \text{(a product of } r+4 \text{ transpositions),}$$

and so on. Indeed, we have the following general theorem.

10.1.5 THEOREM

If a permutation σ is expressed as a product of r transpositions and also as a product of s transpositions, then r and s will differ by an even number; that is, if r is odd, then s is odd; and if r is even, then s is even. (The proof is omitted.)

10.1.6 DEFINITION

Even and Odd Permutations

If a permutation σ is a product of an even number of transpositions, then it is called an *even permutation*. If σ is a product of an odd number of transpositions, then it is called an *odd permutation*.

For example, in S_3, we have that σ_1 is even, $\sigma_2 = (1\ 2)$ is odd, $\sigma_3 = (1\ 3)$ is odd, $\sigma_4 = (2\ 3)$ is odd, $\sigma_5 = (1\ 2\ 3) = (1\ 3)(1\ 2)$ is even, and $\sigma_6 = (1\ 3\ 2) = (1\ 2)(1\ 3)$ is even.

10.1.7 DEFINITION

Inverse of a Permutation

Let $\sigma, \tau \in S_n$. If $\sigma\tau = $ identity permutation, then τ is called the *inverse* of σ. We write the inverse of σ as σ^{-1}. Obviously, if $\sigma\tau = i$, then $\tau\sigma = i$.

The inverse of any permutation can be easily written. For example, if

$$\sigma = \begin{bmatrix} 1 & 2 & 3 \\ 3 & 1 & 2 \end{bmatrix},$$

then

$$\sigma^{-1} = \begin{bmatrix} 3 & 1 & 2 \\ 1 & 2 & 3 \end{bmatrix} = \begin{bmatrix} 1 & 2 & 3 \\ 2 & 3 & 1 \end{bmatrix}.$$

Further, if $\sigma = (a\ b)$ is a transposition, then $\sigma^{-1} = (a\ b)$, since the product $(a\ b)(a\ b)$ maps $a \to b \to a$ and $b \to a \to b$, that is, $(a\ b)(a\ b) = i$. It can be shown that σ and σ^{-1} are either both even permutations or both odd permutations. In the example given above, $\sigma = (1\ 3\ 2) = (1\ 2)(1\ 3)$ is even and $\sigma^{-1} = (1\ 2\ 3) = (1\ 3)(1\ 2)$ is even. Indeed, if

$$\sigma = \eta_1 \eta_2 \cdots \eta_s,$$

a product of transpositions, then

$$\sigma^{-1} = \eta_s \cdots \eta_2 \eta_1,$$

since

$$\sigma\sigma^{-1} = (\eta_1 \eta_2 \cdots \eta_s)(\eta_s \cdots \eta_2 \eta_1) = i.$$

The following theorem will be used later.

10.1.8 THEOREM

If $T = \{\sigma^{-1} \mid \sigma \in S_n\}$, then $T = S_n$.

Proof If $\sigma \in S_n$, then $\sigma = (\sigma^{-1})^{-1} = \eta^{-1}$, where $\eta = \sigma^{-1} \in S_n$. Thus $\sigma \in T$. Thus $S_n \subset T$. Since trivially $S_n \supset T$, $T = S_n$.

10.2 Determinants

In the following, S_n denotes the set of permutations of the set $\{1, 2, \ldots, n\}$. In particular, S_3 is the set of permutations of $\{1, 2, 3\}$.

10.2.1 DEFINITION

Determinant of an $n \times n$ Matrix
Let $A = (a_{ij})$ be a square matrix of order n having entries in the given field. Then the *determinant* of A, denoted by det A, is given by

$$\det A = \sum_{\sigma \in S_n} \varepsilon_\sigma a_{1\sigma(1)} a_{2\sigma(2)} a_{3\sigma(3)} \cdots a_{n\sigma(n)},$$

where

$$\varepsilon_\sigma = \begin{cases} 1, & \text{if } \sigma \text{ is even,} \\ -1, & \text{if } \sigma \text{ is odd.} \end{cases}$$

Let A be a 3×3 matrix; then, since $S_3 = \{\sigma_1, \sigma_2, \sigma_3, \sigma_4, \sigma_5, \sigma_6\}$,

$$\det A = a_{1\sigma_1(1)}a_{2\sigma_1(2)}a_{3\sigma_1(3)}$$
$$- a_{1\sigma_2(1)}a_{2\sigma_2(2)}a_{3\sigma_2(3)}$$
$$- a_{1\sigma_3(1)}a_{2\sigma_3(2)}a_{3\sigma_3(3)}$$
$$- a_{1\sigma_4(1)}a_{2\sigma_4(2)}a_{3\sigma_4(3)}$$
$$+ a_{1\sigma_5(1)}a_{2\sigma_5(2)}a_{3\sigma_5(3)}$$
$$+ a_{1\sigma_6(1)}a_{2\sigma_6(2)}a_{3\sigma_6(3)}$$

(since $\varepsilon_{\sigma_1} = 1, \varepsilon_{\sigma_2} = -1, \varepsilon_{\sigma_3} = -1, \varepsilon_{\sigma_4} = -1, \varepsilon_{\sigma_5} = 1, \varepsilon_{\sigma_6} = 1$)

$$= a_{11}a_{22}a_{33} - a_{12}a_{21}a_{33} - a_{13}a_{22}a_{31}$$
$$- a_{11}a_{23}a_{32} + a_{12}a_{23}a_{31} + a_{13}a_{21}a_{32}.$$

10.2.2
Remarks

i. By the definition of the determinant and Theorem 10.1.8, it follows that we can also write

$$\det A = \sum_{\sigma \in S_n} \varepsilon_\sigma a_{1\sigma^{-1}(1)}a_{2\sigma^{-1}(2)} \cdots a_{n\sigma^{-1}(n)}.$$

ii. For each permutation $\sigma \in S_n$, there is a unique term in det A. For example, if

$$\sigma = \begin{bmatrix} 2 & 3 & 1 \\ 1 & 2 & 3 \end{bmatrix}$$

and $A = (a_{ij})$ is a 3×3 matrix, the term in det A corresponding to σ is $a_{21}a_{32}a_{13}$.

iii. det A is the sum of $n!$ terms corresponding to the $n!$ permutations in S_n.

iv. Each term of det A contains exactly one element from each row of A.

v. Each term of det A contains exactly one element from each column, since $\sigma(1)$, $\sigma(2), \ldots, \sigma(n)$ are n distinct numbers.

10.2.3
EXAMPLES

a.

$$A = \begin{bmatrix} -1 & 2 & 5 \\ 0 & 0 & 0 \\ 6 & 2 & 1 \end{bmatrix}.$$

By the remark following the definition of det A, each term of det A contains one element from each row of A; in particular, each term in det A contains as a factor an element from the second row, that is, 0. So each term of det A is 0. Hence det $A = 0$.

b.

$$A = \begin{bmatrix} -1 & 2 & 3 \\ 0 & 4 & 5 \\ 0 & 0 & 2 \end{bmatrix} = \begin{bmatrix} a_{11} & a_{12} & a_{13} \\ a_{21} & a_{22} & a_{23} \\ a_{31} & a_{32} & a_{33} \end{bmatrix}.$$

det A is the sum of the following six terms:

i. The permutation $\sigma_1 = \begin{bmatrix} 1 & 2 & 3 \\ 1 & 2 & 3 \end{bmatrix}$ gives the term $\varepsilon_{\sigma_1} a_{1\sigma_1(1)} \cdot a_{2\sigma_1(2)} \cdot$
$a_{3\sigma_1(3)} = 1 \cdot a_{11} \cdot a_{22} \cdot a_{33} = 1 - 1 \cdot 4 \cdot 22 = -8.$

ii. $\sigma_2 = \begin{bmatrix} 1 & 2 & 3 \\ 2 & 1 & 3 \end{bmatrix}$ gives the term $\varepsilon_{\sigma_2} a_{1\sigma_2(1)} \cdot a_{2\sigma_2(2)} \cdot a_{3\sigma_2(3)} = \varepsilon_{\sigma_2} a_{12} \cdot$
$a_{21} \cdot a_{33} = -1 \cdot 2 \cdot 0 \cdot 2 = 0.$

iii. $\sigma_3 = \begin{bmatrix} 1 & 2 & 3 \\ 3 & 2 & 1 \end{bmatrix}$ gives the term $\varepsilon_{\sigma_3} a_{1\sigma_3(1)} \cdot a_{2\sigma_3(2)} \cdot a_{3\sigma_3(3)} = -1 \cdot a_{13} \cdot$
$a_{22} \cdot a_{31} = -1 \cdot 3 \cdot 4 \cdot 0 = 0.$

iv. $\sigma_4 = \begin{bmatrix} 1 & 2 & 3 \\ 1 & 3 & 2 \end{bmatrix}$ gives the term $\varepsilon_{\sigma_4} a_{1\sigma_4(1)} \cdot a_{2\sigma_4(2)} \cdot a_{3\sigma_4(3)} = -1 \cdot a_{11} \cdot$
$a_{23} \cdot a_{32} = -1 \cdot -1 \cdot 5 \cdot 0 = 0.$

v. $\sigma_5 = \begin{bmatrix} 1 & 2 & 3 \\ 2 & 3 & 1 \end{bmatrix}$ gives the term $\varepsilon_{\sigma_5} a_{1\sigma_5(1)} \cdot a_{2\sigma_5(2)} \cdot a_{3\sigma_5(3)} = 1 \cdot a_{12} \cdot$
$a_{23} \cdot a_{31} = 1 \cdot 2 \cdot 5 \cdot 0 = 0.$

vi. $\sigma_6 = \begin{bmatrix} 1 & 2 & 3 \\ 3 & 1 & 2 \end{bmatrix}$ gives the term $\varepsilon_{\sigma_6} a_{1\sigma_6(1)} \cdot a_{2\sigma_6(2)} \cdot a_{3\sigma_6(3)} = 1 \cdot a_{13} \cdot$
$a_{21} \cdot a_{32} = 1 \cdot 3 \cdot 0 \cdot 0 = 0.$

c. If A is an upper (lower) triangular matrix, then det A is equal to the product of the diagonal elements. The proof follows easily from the definition of det A. ∎

**10.2.4
THEOREM**

Let A be a 3×3 matrix. Then det $A = \det A^T$.

Proof Let $A = (a_{ij})$. Then $A^T = (a'_{ij})$, where $a'_{ij} = a_{ji}$. By definition of the determinant,

$$\det A^T = \sum \varepsilon_\sigma a'_{1\sigma(1)} a'_{2\sigma(2)} a'_{3\sigma(3)}$$
$$= \sum \varepsilon_\sigma a_{\sigma(1)1} a_{\sigma(2)2} a_{\sigma(3)3}$$
$$= \sum \varepsilon_\sigma a_{\sigma(1)\sigma^{-1}(\sigma(1))} a_{\sigma(2)\sigma^{-1}(\sigma(2))} a_{\sigma(3)\sigma^{-1}(\sigma(3))}$$
$$= \sum \varepsilon_\sigma a_{1\sigma^{-1}(1)} a_{2\sigma^{-1}(2)} a_{3\sigma^{-1}(3)} = \det A,$$

since the set $\{\sigma(1), \sigma(2), \sigma(3)\}$ is the same as the set $\{1, 2, 3\}$, by Remark 10.2.2(i) following the definition of the determinant.

10.2.5 THEOREM

Let

$$A = \begin{bmatrix} a_{11} & a_{12} & a_{13} \\ a_{21} & a_{22} & a_{23} \\ a_{31} & a_{32} & a_{33} \end{bmatrix},$$

and let $B = (b_{ij})$ be the matrix obtained by interchanging row 2 and row 3 of A. Then $\det B = -\det A$. In particular, if two rows of a matrix A are identical, then $\det A = 0$.

Proof Let σ be a permutation in S_3. Then the term in $\det B$ corresponding to σ is

$$\varepsilon_\sigma b_{1\sigma(1)} b_{2\sigma(2)} b_{3\sigma(3)} = \varepsilon_\sigma a_{1\sigma(1)} a_{3\sigma(2)} a_{2\sigma(3)}$$

$$\text{(since } b_{2i} = a_{3i} \text{ for each } i \text{ and } b_{3j} = a_{2j} \text{ for each } j)$$

$$= \varepsilon_\sigma a_{1\sigma(1)} a_{2\sigma(3)} a_{3\sigma(2)},$$

which is the term of $\det A$ corresponding to the permutation σ', where

$$\sigma' = \begin{bmatrix} 1 & 2 & 3 \\ \sigma(1) & \sigma(2) & \sigma(3) \end{bmatrix}.$$

But σ' is the product of the transposition $(\sigma(2) \ \ \sigma(3))$ and σ, that is,

$$\sigma' = (\sigma(2) \ \ \sigma(3))\sigma.$$

So

$$\varepsilon_{\sigma'} = -\varepsilon_\sigma.$$

Thus each term in $\det B$ is a term of $\det A$ with opposite sign.

10.2.6 THEOREM

Let $A = (a_{ij})$ be a 3×3 matrix. Let $B = (b_{ij})$ be a 3×3 matrix obtained from A by multiplying one of its rows by α. Then

$$\det B = \alpha \ \det A.$$

Proof Suppose B is obtained from A by multiplying the second row A_2 of A by α. By definition,

$$\det B = \sum_\sigma \varepsilon_\sigma b_{1\sigma(1)} b_{2\sigma(2)} b_{3\sigma(3)}$$

$$= \sum_\sigma \varepsilon_\sigma a_{1\sigma(1)} \alpha a_{2\sigma(2)} a_{3\sigma(3)},$$

since $B_1 = A_1$, $B_2 = \alpha A_2$, and $B_3 = A_3$. So

$$\det B = \alpha \left[\sum_{\sigma} \varepsilon_{\sigma} a_{1\sigma(1)} a_{2\sigma(2)} a_{3\sigma(3)} \right] = \alpha \det A.$$

10.2.7 THEOREM

Let A be a 3×3 matrix, and let B be a matrix obtained from A by adding α times one row of A to another row of A. Then $\det B = \det A$.

Proof For definiteness, let B be obtained from A by adding α times the second row of A to the first row of A. Let

$$A = \begin{bmatrix} a_{11} & a_{12} & a_{13} \\ a_{21} & a_{22} & a_{23} \\ a_{31} & a_{32} & a_{33} \end{bmatrix}.$$

Then

$$B = \begin{bmatrix} a_{11} + \alpha a_{21} & a_{12} + \alpha a_{22} & a_{13} + \alpha a_{23} \\ a_{21} & a_{22} & a_{23} \\ a_{31} & a_{32} & a_{33} \end{bmatrix}.$$

We write $B = (b_{ij})$. Then

$$\det B = \sum_{\sigma} \varepsilon_{\sigma} b_{1\sigma(1)} b_{2\sigma(2)} b_{3\sigma(3)} = \sum_{\sigma} \varepsilon_{\sigma} (a_{1\sigma(1)} + \alpha a_{2\sigma(1)}) a_{2\sigma(2)} a_{3\sigma(3)},$$

since for any number $i = 1, 2, 3$, $b_{1i} = a_{1i} + \alpha a_{2i}$ and $b_{2i} = a_{2i}$, $b_{3i} = a_{3i}$. We can write this as

$$\sum_{\sigma} \varepsilon_{\sigma} a_{1\sigma(1)} a_{2\sigma(2)} a_{3\sigma(3)} + \alpha \sum_{\sigma} \varepsilon_{\sigma} a_{2\sigma(1)} a_{2\sigma(2)} a_{3\sigma(3)}.$$

The first sum is $\det A$; the second sum is

$$\det \begin{bmatrix} a_{21} & a_{22} & a_{23} \\ a_{21} & a_{22} & a_{23} \\ a_{31} & a_{32} & a_{33} \end{bmatrix}.$$

So

$$\det B = \det A + 0 = \det A.$$

Theorems 10.2.5, 10.2.6, and 10.2.7 can be put together in the following theorem.

**10.2.8
THEOREM**

Under elementary row operations, the determinant of a matrix A changes its value as given below:

$$\det A \xrightarrow{R_i \longleftrightarrow R_j} -\det A,$$

$$\det A \xrightarrow{\alpha R_i} \alpha \det A,$$

$$\det A \xrightarrow{R_i + \alpha R_j} \det A.$$

**10.2.9
Remarks**

i. Since $\det A = \det A^T$ (Theorem 10.2.4), the above theorem also holds for elementary column operations.

ii. Theorems 10.2.4 through 10.2.7 can be proved along similar lines for matrices of higher orders.

**10.2.10
EXAMPLE**

Let

$$A = \begin{bmatrix} 0 & 1 & -1 & 2 \\ 1 & 0 & 2 & 3 \\ -1 & 3 & 4 & 5 \\ 0 & 0 & 0 & 1 \end{bmatrix}$$

be a square matrix of order 4. To find $\det A$ from the definition, we first list the elements of S_4, that is, the permutations of the set $\{1, 2, 3, 4\}$. S_4 has $4! = 4 \cdot 3 \cdot 2 \cdot 1 = 24$ elements. They are

$$\begin{bmatrix} 1 & 2 & 3 & 4 \\ 1 & 2 & 3 & 4 \end{bmatrix}, \begin{bmatrix} 1 & 2 & 3 & 4 \\ 1 & 2 & 4 & 3 \end{bmatrix}, \begin{bmatrix} 1 & 2 & 3 & 4 \\ 1 & 3 & 2 & 4 \end{bmatrix}, \begin{bmatrix} 1 & 2 & 3 & 4 \\ 1 & 3 & 4 & 2 \end{bmatrix},$$

$$\begin{bmatrix} 1 & 2 & 3 & 4 \\ 1 & 4 & 2 & 3 \end{bmatrix}, \begin{bmatrix} 1 & 2 & 3 & 4 \\ 1 & 4 & 3 & 2 \end{bmatrix}, \begin{bmatrix} 1 & 2 & 3 & 4 \\ 2 & 1 & 3 & 4 \end{bmatrix}, \begin{bmatrix} 1 & 2 & 3 & 4 \\ 2 & 1 & 4 & 3 \end{bmatrix},$$

$$\begin{bmatrix} 1 & 2 & 3 & 4 \\ 2 & 3 & 4 & 1 \end{bmatrix}, \begin{bmatrix} 1 & 2 & 3 & 4 \\ 2 & 3 & 1 & 4 \end{bmatrix}, \begin{bmatrix} 1 & 2 & 3 & 4 \\ 2 & 4 & 1 & 3 \end{bmatrix}, \begin{bmatrix} 1 & 2 & 3 & 4 \\ 2 & 4 & 3 & 1 \end{bmatrix},$$

$$\begin{bmatrix} 1 & 2 & 3 & 4 \\ 3 & 4 & 1 & 2 \end{bmatrix}, \begin{bmatrix} 1 & 2 & 3 & 4 \\ 3 & 4 & 2 & 1 \end{bmatrix}, \begin{bmatrix} 1 & 2 & 3 & 4 \\ 3 & 1 & 2 & 4 \end{bmatrix}, \begin{bmatrix} 1 & 2 & 3 & 4 \\ 3 & 1 & 4 & 2 \end{bmatrix},$$

$$\begin{bmatrix} 1 & 2 & 3 & 4 \\ 3 & 2 & 1 & 4 \end{bmatrix}, \begin{bmatrix} 1 & 2 & 3 & 4 \\ 3 & 2 & 4 & 1 \end{bmatrix}, \begin{bmatrix} 1 & 2 & 3 & 4 \\ 4 & 1 & 2 & 3 \end{bmatrix}, \begin{bmatrix} 1 & 2 & 3 & 4 \\ 4 & 1 & 3 & 2 \end{bmatrix},$$

$$\begin{bmatrix} 1 & 2 & 3 & 4 \\ 4 & 2 & 1 & 3 \end{bmatrix}, \begin{bmatrix} 1 & 2 & 3 & 4 \\ 4 & 2 & 3 & 1 \end{bmatrix}, \begin{bmatrix} 1 & 2 & 3 & 4 \\ 4 & 3 & 1 & 2 \end{bmatrix}, \begin{bmatrix} 1 & 2 & 3 & 4 \\ 4 & 3 & 2 & 1 \end{bmatrix}.$$

We note that the $(1, 1)$, $(2, 2)$, $(4, 1)$, $(4, 2)$, $(4, 3)$ entries of A are all 0. So any permutation σ in S_4 for which $\sigma(1) = 1$ or $\sigma(2) = 2$ or $\sigma(4) = 1$ or $\sigma(4) = 2$ or

$\sigma(4) = 3$ will give a zero term in det A. We strike off all such permutations. So the only permutations that contribute to det A are

$$\sigma = \begin{bmatrix} 1 & 2 & 3 & 4 \\ 2 & 1 & 3 & 4 \end{bmatrix} = (1\ 2),$$

$$\sigma' = \begin{bmatrix} 1 & 2 & 3 & 4 \\ 2 & 3 & 1 & 4 \end{bmatrix} = (1\ 2\ 3) = (1\ 3)(1\ 2),$$

$$\sigma'' = \begin{bmatrix} 1 & 2 & 3 & 4 \\ 3 & 1 & 2 & 4 \end{bmatrix} = (1\ 3\ 2) = (1\ 2)(1\ 3).$$

Hence

$$\det A = -a_{12}a_{21}a_{33}a_{44} + a_{12}a_{23}a_{31}a_{44} + a_{13}a_{21}a_{32}a_{44}$$
$$= -1 \cdot 1 \cdot 4 \cdot 1 + 1 \cdot 2 \cdot (-1) \cdot 1 - 1 \cdot 1 \cdot 3 \cdot 1 = -4 - 2 - 3 = -9.$$

10.3 Cofactor Expansion

10.3.1 DEFINITION

Submatrix
Let A be an $n \times n$ matrix. If some rows or columns or both of A are deleted, the resulting array of scalars is called a *submatrix* of A. By convention, A is also considered as a submatrix of A.

For example, let

$$A = \begin{bmatrix} 2 & 1 & 5 \\ -1 & 0 & 2 \end{bmatrix}$$

be a 2×3 matrix. If we delete row 2 of A, we get the 1×3 submatrix $[2\ 1\ 5]$. By deleting column 3, we get the 2×2 matrix $\begin{bmatrix} 2 & 1 \\ -1 & 0 \end{bmatrix}$. By deleting row 1, column 2, and column 3, we get the 1×1 matrix $[-1]$.

10.3.2 DEFINITION

Cofactor
Let A be a square matrix, and let a_{pq} be any entry of A. Then the *cofactor* of a_{pq}, written as A_{pq}, is given by

$$A_{pq} = (-1)^{p+q} \det \text{ (the submatrix obtained by deleting}$$
$$\text{the } p\text{th row and } q\text{th column of } A).$$

For example, let

$$A = \begin{bmatrix} -2 & 3 & 4 \\ 5 & 6 & 2 \\ 1 & 0 & 5 \end{bmatrix}.$$

Then

$$A_{11} = (-1)^{1+1} \det \begin{bmatrix} 6 & 2 \\ 0 & 5 \end{bmatrix} = 30,$$

$$A_{31} = (-1)^{3+1} \det \begin{bmatrix} 3 & 4 \\ 6 & 2 \end{bmatrix} = 6 - 24 = -18,$$

$$A_{23} = (-1)^{2+3} \det \begin{bmatrix} -2 & 3 \\ 1 & 0 \end{bmatrix} = -(0 - 3) = 3,$$

and so on.

Let

$$A = (a_{ij})_{3\times 3}.$$

Then det A is the sum of six terms

$$\det A = \sum_{\sigma \in S_3} \varepsilon_\sigma a_{1\sigma(1)} a_{2\sigma(2)} a_{3\sigma(3)}.$$

By the Remark 10.2.2(iv) following the definition of the determinant function, each term of det A contains just one entry from the first row. Let us collect together all the terms of det A that contain a_{11}, then all the terms that contain a_{12}, and finally all the terms that contain a_{13}.

i. The terms of det A that contain a_{11} are given by those permutations σ for which $\sigma(1) = 1$. These permutations are

$$\sigma_1 = \begin{bmatrix} 1 & 2 & 3 \\ 1 & 2 & 3 \end{bmatrix}, \qquad \sigma_4 = \begin{bmatrix} 1 & 2 & 3 \\ 1 & 3 & 2 \end{bmatrix}.$$

So the sum of the terms given by σ_1 and σ_4 is

$$\varepsilon_{\sigma_1} a_{1\sigma_1(1)} a_{2\sigma_1(2)} a_{3\sigma_1(3)} + \varepsilon_{\sigma_4} a_{1\sigma_4(1)} a_{2\sigma_4(2)} a_{3\sigma_4(3)} = a_{11} a_{22} a_{33} - a_{11} a_{23} a_{32}$$
$$= a_{11}(a_{22} a_{33} - a_{23} a_{32})$$
$$= a_{11} A_{11}.$$

ii. Similarly, the terms containing a_{12} are given by those permutations σ for which $\sigma(1) = 2$. These permutations are

$$\sigma_2 = \begin{bmatrix} 1 & 2 & 3 \\ 2 & 1 & 3 \end{bmatrix}, \qquad \sigma_5 = \begin{bmatrix} 1 & 2 & 3 \\ 2 & 3 & 1 \end{bmatrix}.$$

So the sum of the terms in det A that contain a_{12} is

$$\varepsilon_{\sigma_2} a_{1\sigma_2(1)} a_{2\sigma_2(2)} a_{3\sigma_2(3)} \varepsilon_{\sigma_5} a_{1\sigma_5(1)} a_{2\sigma_5(2)} a_{3\sigma_5(3)} = -a_{12} a_{21} a_{33} + a_{12} a_{23} a_{31}$$
$$= -a_{12}(a_{21} a_{33} - a_{23} a_{31})$$
$$= a_{12} A_{12}.$$

iii. The terms in det A that contain a_{13} are given by those permutations σ for which $\sigma(1) = 3$, that is, by the permutations

$$\sigma_3 = \begin{bmatrix} 1 & 2 & 3 \\ 3 & 2 & 1 \end{bmatrix}, \qquad \sigma_6 = \begin{bmatrix} 1 & 2 & 3 \\ 3 & 1 & 2 \end{bmatrix}.$$

So the sum of such terms is

$$\varepsilon_{\sigma_3} a_{1\sigma_3(1)} a_{2\sigma_3(2)} a_{3\sigma_3(3)} + \varepsilon_{\sigma_6} a_{1\sigma_6(1)} a_{2\sigma_6(2)} a_{3\sigma_6(3)} = -a_{13}a_{22}a_{31} + a_{13}a_{21}a_{32}$$
$$= a_{13}(a_{21}a_{32} - a_{31}a_{22})$$
$$= a_{13}A_{13}.$$

Thus

$$\det A = a_{11}A_{11} + a_{12}A_{12} + a_{13}A_{13}. \tag{1}$$

This is called the expansion of det A by row 1. A similar procedure will give five similar results:

$$\det A = a_{21}A_{21} + a_{22}A_{22} + a_{23}A_{23} \qquad \text{(expansion by row 2)},$$
$$\det A = a_{31}A_{31} + a_{32}A_{32} + a_{33}A_{33} \qquad \text{(expansion by row 3)},$$
$$\det A = a_{11}A_{11} + a_{21}A_{21} + a_{31}A_{31} \qquad \text{(expansion by column 1)},$$
$$\det A = a_{12}A_{12} + a_{22}A_{22} + a_{32}A_{32} \qquad \text{(expansion by column 2)},$$
$$\det A = a_{13}A_{13} + a_{23}A_{23} + a_{33}A_{33} \qquad \text{(expansion by column 3)}.$$

We put these results in the following theorem.

10.3.3 THEOREM

Theorem of Cofactors

Let $A = (a_{ij})$ be a 3×3 matrix. Then

  i. det $A = a_{i1}A_{i1} + a_{i2}A_{i2} + a_{i3}A_{i3}$ (expansion by the ith row).

  ii. det $A = a_{1j}A_{1j} + a_{2j}A_{2j} + a_{3j}A_{3j}$ (expansion by the jth column).

    Further, if $i \neq j$, then

 iii. $a_{i1}A_{j1} + a_{i2}A_{j2} + a_{i3}A_{j3} = 0$,

 iv. $a_{1i}A_{1j} + a_{2i}A_{2j} + a_{3i}A_{3j} = 0$.

Proof We have proved (i) and (ii) above. We now prove (iii) and (iv). As before, let

$$A = \begin{bmatrix} a_{11} & a_{12} & a_{13} \\ a_{21} & a_{22} & a_{23} \\ a_{31} & a_{32} & a_{33} \end{bmatrix}.$$

Consider now the matrix

$$B = \begin{bmatrix} a_{11} & a_{12} & a_{13} \\ a_{11} & a_{12} & a_{13} \\ a_{31} & a_{32} & a_{33} \end{bmatrix}$$

obtained from A by replacing row 2 by row 1. Let B_{ij} denote the cofactor of the (i, j) entry of B. Then obviously,

$$B_{11} = A_{21}, \qquad B_{12} = A_{22}, \qquad B_{13} = A_{23}.$$

So

$$\det B = b_{11}B_{11} + b_{12}B_{12} + b_{13}B_{13}$$
$$= a_{11}A_{21} + a_{12}A_{22} + a_{13}A_{23}.$$

Since B has two equal rows, by Theorem 10.2.5, $\det B = 0$. Hence

$$A_{11}a_{21} + A_{12}a_{22} + A_{13}a_{23} = 0.$$

We can prove in a similar way the other cases.

10.3.4
Remark The above theorem can be proved on similar lines for matrices of higher order.

The cofactor theorem provides an easy method to evaluate the determinant of a matrix A. We first select a row or column that contains the largest number of zeros. Then we expand $\det A$ by that row or column. To produce a sufficient number of zeros in a selected row or column of A, we perform elementary row or column operations on A. We then proceed with the cofactor expansion of $\det A$, taking into account the fact that under a Type I operation, $\det A \rightarrow -\det A$; under a Type II operation, $\det A \rightarrow \alpha \det A$; and under a Type III operation $\det A$ does not change.

However, a simpler method to evaluate the determinant of a matrix with numerical entries is to reduce A to upper (or lower) triangular form by elementary row (or column) operations and use the result that the determinant of an upper or lower triangular matrix is the product of the diagonal entries. (See Example 10.2.3(c).)

10.3.5
EXAMPLE Let

$$A = \begin{bmatrix} 3 & -1 & 4 \\ 0 & 2 & 1 \\ 1 & -1 & -2 \end{bmatrix}.$$

Expanding by the first column, we get

$$\det A = 3(-1)^{1+1} \det \begin{bmatrix} 2 & 1 \\ -1 & -2 \end{bmatrix} + 1(-1)^{3+1} \det \begin{bmatrix} -1 & 4 \\ 2 & 1 \end{bmatrix}$$
$$= 3 \cdot 1 \cdot (-4 + 1) + 1 \cdot 1 \cdot (-1 - 8) = -9 - 9 = -18.$$

We can also find $\det A$ by first reducing A to an upper triangular matrix by elementary row operations:

$$\det A = -\det \begin{bmatrix} 1 & -1 & -2 \\ 0 & 2 & 1 \\ 3 & -1 & 4 \end{bmatrix} \quad \text{(negative sign due to interchange of row 1 and row 3)}$$

$$= -\det \begin{bmatrix} 1 & -1 & -2 \\ 0 & 2 & 1 \\ 0 & 2 & 10 \end{bmatrix} \quad \text{(by performing } R_3 - 3R_1\text{)}$$

$$= -\det \begin{bmatrix} 1 & -1 & -2 \\ 0 & 2 & 1 \\ 0 & 0 & 9 \end{bmatrix} \quad \text{(by performing } R_3 - 1R_2\text{)}$$

$$= (-1) \text{ product of the diagonal elements } = -18.$$

10.4 Adjoint of a Matrix

10.4.1 DEFINITION

Let $A = (a_{ij})$ be a square matrix of order n. Then

$$\text{adj}(A) = \begin{bmatrix} A_{11} & A_{12} & \dots & A_{1n} \\ A_{21} & A_{22} & \dots & A_{2n} \\ \vdots & \vdots & & \vdots \\ A_{n1} & A_{n2} & \dots & A_{nn} \end{bmatrix}^T$$

is called the *adjoint matrix* of A, or simply the *adjoint* of A. We write it as adj (A).

10.4.2 THEOREM

If A is an $n \times n$ nonsingular matrix, then

$$A^{-1} = \frac{1}{\det A} \text{ adj } (A).$$

Proof For simplicity, we can take A as a 3×3 matrix. Let $B = (b_{ij})_{3 \times 3}$ be the product A adj (A). Then

$$b_{11} = a_{11}A_{11} + a_{12}A_{12} + a_{13}A_{13} = \det A,$$
$$b_{12} = a_{11}A_{21} + a_{12}A_{22} + a_{13}A_{23} = 0,$$

and so on. Thus

$$A \text{ adj } (A) = \begin{bmatrix} \det A & 0 & \dots & 0 \\ 0 & \det A & \dots & 0 \\ \vdots & \vdots & \ddots & \vdots \\ 0 & 0 & \dots & \det A \end{bmatrix} = \det A \begin{bmatrix} 1 & 0 & \dots & 0 \\ 0 & 1 & \dots & 0 \\ \vdots & \vdots & \ddots & \vdots \\ 0 & 0 & \dots & 1 \end{bmatrix}.$$

So

$$\frac{1}{\det A} A \text{ adj } (A) = I,$$

that is,

$$A^{-1} = \frac{1}{\det A} \text{ adj } (A).$$

10.4.3
EXAMPLES

a. Let

$$A = \begin{bmatrix} 1 & 2 \\ -2 & 1 \end{bmatrix}.$$

Then

$$\det A = 5,$$
$$A_{11} = 1, \qquad A_{21} = -2,$$
$$A_{12} = 2, \qquad A_{22} = 1.$$

So

$$A^{-1} = \frac{1}{5} \begin{bmatrix} 1 & -2 \\ 2 & 1 \end{bmatrix} = \begin{bmatrix} \frac{1}{5} & -\frac{2}{5} \\ \frac{2}{5} & \frac{1}{5} \end{bmatrix}.$$

b. Let

$$A = \begin{bmatrix} 2 & 1 & 0 \\ 3 & 0 & 1 \\ 0 & 1 & 1 \end{bmatrix}.$$

Then

$$A_{11} = (-1)^{1+1} \det \begin{bmatrix} 0 & 1 \\ 1 & 1 \end{bmatrix} = -1, \qquad A_{12} = (-1)^{1+2} \det \begin{bmatrix} 3 & 1 \\ 0 & 1 \end{bmatrix} = -3,$$

$$A_{13} = (-1)^{1+3} \det \begin{bmatrix} 3 & 0 \\ 0 & 1 \end{bmatrix} = 3.$$

Similarly,

$$A_{21} = -1, \quad A_{22} = 2, \quad A_{23} = -2,$$
$$A_{31} = 1, \qquad A_{32} = -2, \quad A_{33} = -3.$$

So

$$\text{adj } (A) = \begin{bmatrix} -1 & -3 & 3 \\ -1 & 2 & -2 \\ 1 & -2 & -3 \end{bmatrix}^T = \begin{bmatrix} -1 & -1 & 1 \\ -3 & 2 & -2 \\ 3 & -2 & -3 \end{bmatrix}.$$

Also, expanding by the first row, we have $\det A = 2A_{11} + 1A_{12} = -2 - 3 = -5$.
So

$$A^{-1} = \frac{1}{-5} \begin{bmatrix} -1 & -1 & 1 \\ -3 & 2 & -2 \\ 3 & -2 & -3 \end{bmatrix} = \begin{bmatrix} \frac{1}{5} & \frac{1}{5} & -\frac{1}{5} \\ \frac{3}{5} & -\frac{2}{5} & \frac{2}{5} \\ -\frac{3}{5} & \frac{2}{5} & \frac{3}{5} \end{bmatrix}.$$

We remark that in general, the computation of the inverse of a matrix with numerical entries is simpler by the method of elementary row operations given in *Chapter 5*.

We emphasize once again the fact that the Gauss elimination method is the golden method to determine consistency of an LS, to solve a consistent LS, to find the basis of the null space and the basis of the range of a matrix, to find the rank of a matrix, to find the inverse of a nonsingular matrix, and to solve a host of related problems. We have discussed both elementary row and column operations. In performing Gauss elimination, it is enough if we stick to one method, either elementary row operations or elementary column operations. In fact, it will be confusing if we sometimes resort to row operations and sometimes to column operations.

Let A be a nonsingular $n \times n$ matrix. Then the unique solution of the LS

$$Ax = b$$

is given by $x = A^{-1}b$. As was remarked earlier, A^{-1} can be found by Gauss elimination (see Example 5.1.7(a)), and so the product $A^{-1}b$, which is an $n \times 1$ matrix, gives the unique solution of the LS. Sometimes Cramer's rule, given in the following theorem, is found to be handy for solving an LS.

10.5 Cramer's Rule

10.5.1 THEOREM

Cramer's Rule

Let $A = [A^{(1)}, A^{(2)}, \ldots, A^{(n)}]$ be a nonsingular $n \times n$ matrix. Then the solution of the LS

$$Ax = b$$

is given by $x_i = (\det B_i)/(\det A), i = 1, 2, \ldots, n$, where $B_i = (A^{(1)}, \ldots, A^{(i-1)}, b, A^{(i+1)}, \ldots, A^{(n)})$, the matrix obtained by replacing the ith column of A by b.

Proof We give the proof for a 3×3 matrix. The LS $Ax = b$ can be written as

$$[A^{(1)} \; A^{(2)} \; A^{(3)}] \begin{bmatrix} x_1 \\ x_2 \\ x_3 \end{bmatrix} = b,$$

that is, $x_1 A^{(1)} + x_2 A^{(2)} + x_3 A^{(3)} = b$. Since A is nonsingular, the LS has a unique solution. This means that there are unique scalars x_1, x_2, and x_3 such that

$$x_1 A^{(1)} + x_2 A^{(2)} + x_3 A^{(3)} = b. \tag{1}$$

So

$$x_1 \det A = x_1 \det [A^{(1)} \; A^{(2)} \; A^{(3)}]$$
$$= \det [x_1 A^{(1)} \; A^{(2)} \; A^{(3)}] \quad \text{(by Theorem 10.2.6)}$$
$$= \det [b - x_2 A^{(2)} - x_3 A^{(3)} \; A^{(2)} \; A^{(3)}] \quad \text{(by using Equation (1))}$$
$$= \det [b \; A^{(2)} \; A^{(3)}] \quad \text{(by Theorem 10.2.7).}$$

Hence $x_1 = (\det\ B_1)/(\det\ A)$. Similarly, $x_2 = (\det\ B_2)/(\det\ A)$, and $x_3 = (\det\ B_3)/(\det\ A)$.

10.5.2
EXAMPLES

Consider the LS

$$
\begin{aligned}
2x_1 - x_2 + x_3 &= 3, \\
x_1 + x_2 &= 3, \\
x_2 - x_3 &= 1.
\end{aligned}
$$

This can be written as

$$
\begin{bmatrix} 2 & -1 & 1 \\ 1 & 1 & 0 \\ 0 & 1 & -1 \end{bmatrix} \begin{bmatrix} x_1 \\ x_2 \\ x_3 \end{bmatrix} = \begin{bmatrix} 3 \\ 3 \\ 1 \end{bmatrix}.
$$

Then

$$
\det A = \det \begin{bmatrix} 2 & -1 & 0 \\ 1 & 1 & 1 \\ 0 & 1 & 0 \end{bmatrix} = 1 \cdot (-1)^{2+3} \cdot 2 = -2,
$$

(by expanding by the third column),

$$
\det\ B_1 = \det \begin{bmatrix} 3 & -1 & 1 \\ 3 & 1 & 0 \\ 1 & 1 & -1 \end{bmatrix} = \det \begin{bmatrix} 3 & -1 & 0 \\ 3 & 1 & 1 \\ 1 & 1 & 0 \end{bmatrix} = -4,
$$

$$
\det\ B_2 = \det \begin{bmatrix} 2 & 3 & 0 \\ 1 & 3 & 1 \\ 0 & 1 & 0 \end{bmatrix} = -2,
$$

$$
\det\ B_3 = \det \begin{bmatrix} 2 & -1 & 3 \\ 1 & 1 & 3 \\ 0 & 1 & 1 \end{bmatrix} = 0.
$$

So

$$
x_1 = \frac{-4}{-2} = 2, \qquad x_2 = \frac{-2}{-2} = 1, \qquad x_3 = 0.
$$

10.6 Product Theorem of Determinants

In this section, we prove the product theorem for determinants and other theorems based on it. We recall that an elementary row operation on a matrix A is equivalent to premultiplying A by an elementary matrix, while an elementary column operation on A is equivalent to postmultiplying A by an elementary matrix. There are three types of elementary matrices of a given order n:

$E_{(q,p)}$ = the matrix obtained by interchanging the pth and qth rows or the pth and the qth columns of the identity matrix I_n.

$E_{\alpha(p)}$ = the matrix obtained by multiplying the pth row or the pth column of I_n by a nonzero scalar α.

$E_{(p)+\alpha(q)}$ = the matrix obtained by adding α times the qth row of I_n to the pth row of I_n or by adding α times the pth column of I_n to the qth column of I_n.

10.6.1 THEOREM

a. $\det E_{(p,q)} = -1$.
b. $\det E_{\alpha(p)} = \alpha$.
c. $\det E_{(p)+\alpha(q)} = 1$.

Proof

a. $E_{(p,q)}$ is the matrix obtained by interchanging two rows of I_n, and so $\det E_{p,q} = -\det I_n = -1$ (Theorem 10.2.5).

b. $E_{\alpha(p)}$ is the matrix obtained by multiplying the pth row of I_n by α, so $\det E_{\alpha(p)} = \alpha \det I_n = \alpha$ (Theorem 10.2.6).

c. $E_{(p)+\alpha(q)}$ is the matrix obtained by adding α times the qth row of I_n to the pth row of I_n, and so $\det E_{(p)+\alpha(q)} = \det I = 1$ (Theorem 10.2.7).

10.6.2 THEOREM

If A is an $n \times n$ matrix and E is an elementary $n \times n$ matrix, then

a. $\det(EA) = \det E \cdot \det A$.
b. $\det(AE) = \det A \cdot \det E$.

Proof (a) and (b) are similar. Let us prove (a).

Case (i)
Let $E = E_{p,q}$. Then the product EA is equal to the matrix obtained by interchanging the pth and the qth rows of A. So by Theorem 10.2.5, $\det(EA) = -\det A = \det E \cdot \det A$, since $\det E = -1$ by the previous theorem.

Case (ii)
Let $E = E_{\alpha(p)}$. Then EA is equal to the matrix obtained by multiplying the pth row of A by α. So by Theorem 10.2.6, $\det(EA) = \alpha \cdot \det A = \det E \cdot \det A$, since $\det E = \alpha$ by the previous theorem.

Case (iii)
Let $E = E_{(p)+\alpha(q)}$. Then EA is equal to the matrix obtained by adding α times the qth row of A to the pth row of A, so by Theorem 10.2.7, $\det(EA) = \det A = \det E \cdot \det A$, since $\det E = 1$ by the previous theorem.

10.6.3 COROLLARY

$$\det(E_p \cdots E_1 A) = \det E_p \cdots \det E_1 \det A;$$
$$\det(A E_1 \cdots E_p) = \det A \cdot \det E_1 \cdots \det E_p.$$

Proof We can prove the assertions easily by induction on p.

10.6.4 COROLLARY

If A is an $n \times n$ singular matrix, then $\det A = 0$.

Proof By elementary row operations, A can be reduced to a row echelon matrix whose last row must be a zero row. That is, there exist elementary matrices E_1, $E_2, \ldots, E_p$ such that $E_p \cdots E_2 E_1 A$ is a matrix whose last row is zero. So

$$\det(E_p \cdots E_2 E_1 A) = 0,$$

that is,

$$\det E_p \cdots \det E_1 \cdot \det A = 0,$$

by the previous corollary. But by Theorem 10.6.1, the determinant of an elementary matrix is a nonzero scalar. Hence

$$\det \ A = 0.$$

10.6.5 THEOREM

Product Theorem of Determinants
Let A and B be $n \times n$ matrices. Then

$$\det(AB) = \det A \cdot \det B.$$

Proof Suppose A is noninvertible. Now, if AB were invertible, then there would exist a matrix C such that $(AB)C = I$. But by associativity of matrix multiplication, we would then have $A(BC) = I$. This means that A must be invertible, which is not true. Thus AB is not invertible. By the above corollary, $\det A = 0$, and $\det(AB) = 0$. Therefore $\det(AB) = (\det A)(\det B)$.

Next, suppose that A is invertible. Then A is a product of elementary matrices:

$$A = E_1 E_2 \cdots E_p.$$

(Why?) So

$$
\begin{aligned}
\det(AB) &= \det(E_1 \cdots E_p B) \\
&= \det E_1 \cdots \det E_p \cdot \det B \qquad \text{(by Corollary 1)} \\
&= \det(E_1) \cdots \det(E_p) \cdot \det I \cdot \det B \qquad \text{(since} \ \det I = 1) \\
&= \det(E_1 \cdots E_p I) \cdot \det \ B \\
&= \det \ A \cdot \det \ B.
\end{aligned}
$$

10.6 Exercises

1. If A is an invertible matrix, prove that

$$\det(A^{-1}) = \frac{1}{\det A}.$$

2. If A is a square matrix such that $A = A^2$, prove that $\det A = 0$ or 1.

3. If A is a square matrix such that $A^k = 0$ for some positive integer k, prove that $\det A = 0$.

4. If A is a square matrix such that $A^2 = I$, prove that $\det A = \pm 1$.

5. If A is a square matrix, prove that $\det(P^{-1}AP) = \det A$, where P is any invertible matrix.

6. Prove that the characteristic polynomial of a square matrix A is the same as that of $P^{-1}AP$, where P is any invertible matrix.

Key Words

Permutation
Transposition
Cycle
Determinant

Cofactor
Adjoint
Cramer's Rule

Key Phrases

■ The determinant of the product of matrices is the product of determinants.

Answers and Hints to Selected Exercises

CHAPTER 1

1.2 Exercises

1. $x_1 = 26$, $x_2 = 12$, $x_3 = 9$.

2. $x_1 = 2$, $x_2 = 8$, $x_3 = 21$.

3. Inconsistent.

4. $x = \frac{59}{28}$, $y = -\frac{3}{7}$, $z = \frac{1}{8}$.

5. Infinitely many solutions.

6. a. $a \neq \frac{5}{2}$.

   b. $a = 10$.

   c. $a \neq -1$.

7. $x_1 = 1$, $x_2 = 2$, $x_3 = 2$.

8. Yes.

9. Inconsistent.

10. $b_1 = -9$.

1.3 Exercises

1. a. Homogeneous and the only solution is $x_1 = 0 = x_2$.

   b. Nonhomogeneous.

   c. Nonhomogeneous.

   d. Homogeneous and has a nontrivial solution (Theorem 1.3.3).

2. a. $x_1 = 0$, $x_2 = t$, $x_3 = t$.

   b. $x_1 = t_1 - t_2$, $x_2 = t_2$, $x_3 = t_1$.

   c. $x_3 = t_1$, $x_2 = t_2$, $x_1 = -t_1 - t_2$.

   d. $x_1 = t$, $x_2 = 0$, $x_3 = t$, $x_4 = 0$.

3. a.
$$\left. \begin{array}{r} x_2 + 2x_3 = 0 \\ x_1 - x_2 + 3x_3 = 0 \\ 2x_1 - x_2 + 8x_3 = 0 \end{array} \right\} \xrightarrow{R_1 \longleftrightarrow R_2}$$

$$\left. \begin{array}{r} x_1 - x_2 + 3x_3 = 0 \\ x_2 + 2x_3 = 0 \\ 2x_1 - x_2 + 8x_3 = 0 \end{array} \right\} \xrightarrow{R_3 - 2R_1}$$

$$\left. \begin{array}{r} x_1 - x_2 + 3x_3 = 0 \\ x_2 + 2x_3 = 0 \\ x_2 + 2x_3 = 0 \end{array} \right\} \xrightarrow{R_3 + (-1)R_2}$$

$$\left. \begin{array}{r} x_1 - x_2 + 3x_3 = 0 \\ x_2 + 2x_3 = 0 \\ 0 = 0 \end{array} \right\}$$

Put $x_3 = t$. So $x_2 = -2t$, $x_1 = -5t$, where t is arbitrary.

   b. No nontrivial solution.

   c. $x_1 = t_2 - 2t_1$, $x_2 = -4t_2 - 5t_1$, $x_3 = t_1$, $x_4 = t_2$.

   d. No nontrivial solution.

   e. $x_2 = t_2$, $x_1 = -t_1 - t_2$, $x_3 = t_1$.

4. a. has a nontrivial solution.

   b. has a nontrivial solution.

   c. has a nontrivial solution.

   d. has a nontrivial solution.

5. No nontrivial solution.

6. $x_1 = -t$, $x_2 = 0$, $x_3 = t$, $x_4 = 0$, $x_5 = 0$.

7. a. $c = -\frac{15}{2}$.

   b. $c = \frac{2}{3}$.

   c. $c = \frac{1}{5}$.

8. $x_1 = -t$, $x_2 = 0$, $x_3 = t$, $x_4 = 0$.

9. $x_1 = \frac{1}{2}t$, $x_2 = \frac{5}{2}t$, $x_3 = t$.

10. $x_1 = t_1$, $x_2 = t_2$, $x_3 = -t_1$, $x_4 = -t_2$.

11. $x_1 = 3t$, $x_2 = t$, $x_3 = -t$.

1.4 Exercises

1. a. Coefficient matrix $= \begin{bmatrix} 2 & -3 & 1 \\ 5 & 0 & 6 \\ 1 & -1 & 1 \end{bmatrix}$, augmented matrix $= \begin{bmatrix} 2 & -3 & 1 & 0 \\ 5 & 0 & 6 & 1 \\ 1 & -1 & 1 & 5 \end{bmatrix}$.

b. Coefficient matrix $= \begin{bmatrix} 5 & 6 & 7 & 8 \end{bmatrix}$,

augmented matrix $= \begin{bmatrix} 5 & 6 & 7 & 8 & 9 \end{bmatrix}$.

c. Coefficient matrix $= \begin{bmatrix} -1 & 2 & 6 \\ 1 & -1 & 7 \end{bmatrix}$,

augmented matrix $= \begin{bmatrix} -1 & 2 & 6 & 0 \\ 1 & -1 & 7 & 0 \end{bmatrix}$.

d. Coefficient matrix $= \begin{bmatrix} 2 & 3 & 4 & 7 \\ 5 & 6 & 7 & 8 \\ 0 & 1 & 0 & -6 \end{bmatrix}$,

augmented matrix $= \begin{bmatrix} 2 & 3 & 4 & 7 & 1 \\ 5 & 6 & 7 & 8 & 2 \\ 0 & 1 & 0 & -6 & 0 \end{bmatrix}$.

2. a. $x_1 + 2x_2 + 3x_3 = 4$,
$5x_1 + 6x_2 + 7x_3 = 8$.

b. $x_2 = 2$,
$x_1 = 3$,
$2x_1 + x_2 = 8$.

c. $-x_1 - 2x_2 + 3x_3 = 4$,
$5x_2 + x_3 = 1$,
$7x_1 + 6x_2 + 5x_3 = 3$.

d. $2x_1 + 4x_2 + 6x_3 + 8x_4 = 10$,
$x_1 + 3x_2 + 5x_3 + 7x_4 = 9$,
$-x_1 - 2x_2 = 1$,
$x_1 + 6x_4 = 5$.

The coefficient matrices for (a), (b), (c)

and (d) are $\begin{bmatrix} 1 & 2 & 3 \\ 5 & 6 & 7 \end{bmatrix}$, $\begin{bmatrix} 0 & 1 \\ 1 & 0 \\ 2 & 1 \end{bmatrix}$,

$\begin{bmatrix} -1 & -2 & 3 \\ 0 & 5 & 1 \\ 7 & 6 & 5 \end{bmatrix}$, and $\begin{bmatrix} 2 & 4 & 6 & 8 \\ 1 & 3 & 5 & 7 \\ -1 & -2 & 0 & 0 \\ 1 & 0 & 0 & 6 \end{bmatrix}$,

respectively.

3. a. $x^2 = 5x - 6$. So $x = 2, 3$.

b. $x = y + 1$, $y = 2$, $z = z$. So $x = 3$,
and z is arbitrary.

c. $y = 4 - x$, $x = x^2$, $x = 1$. So $y = 3$
and $x = 1$.

d. $x = y = z = 1$.

e. No solution.

5. a.

b. $\begin{bmatrix} 0 & 1 & 0 & 0 \\ 0 & 0 & 1 & 1 \\ 0 & 0 & 1 & 2 \\ 0 & 0 & 0 & 0 \end{bmatrix}$.

6. a.

b. $\begin{bmatrix} 1 & 1 & 0 & 0 & 0 \\ 0 & 0 & 0 & 0 & 0 \\ 0 & 0 & 0 & 1 & 0 \\ 0 & 0 & 0 & 0 & 2 \\ 0 & 1 & 0 & 0 & 0 \end{bmatrix}$.

7. a. , undirected.

b. , undirected.

1.5 Exercises

1. a. $\begin{bmatrix} 1 & 2 & -3 & 1 \\ 0 & 1 & -2 & 2 \\ 0 & 2 & -4 & 4 \end{bmatrix} \xrightarrow{R_3 - 2R_2} \begin{bmatrix} 1 & 2 & -3 & 1 \\ 0 & 1 & -2 & 2 \\ 0 & 0 & 0 & 0 \end{bmatrix}$.

The transformed LS is

$$x_1 + 2x_2 - 3x_3 = 1,$$
$$x_2 - 2x_3 = 2.$$

Put $x_3 = t$. So $x_2 = 2 + 2t$, and $x_1 = -3 - t$.

b. $\begin{bmatrix} 2 & 1 & 0 & 1 & 2 \\ 3 & 3 & 3 & 5 & 4 \\ 3 & -3 & 0 & -2 & 3 \end{bmatrix} \xrightarrow{\frac{1}{2}R_1}$

$$\begin{bmatrix} 1 & \frac{1}{2} & 0 & \frac{1}{2} & 1 \\ 3 & 3 & 3 & 5 & 4 \\ 3 & -3 & 0 & -2 & 3 \end{bmatrix} \begin{array}{c} R_2-3R_1 \\ \longrightarrow \\ R_3-3R_1 \end{array}$$

$$\begin{bmatrix} 1 & \frac{1}{2} & 0 & \frac{1}{2} & 1 \\ 0 & \frac{3}{2} & 3 & \frac{7}{2} & 1 \\ 0 & -\frac{9}{2} & 0 & -\frac{7}{2} & 0 \end{bmatrix} \begin{array}{c} R_3+3R_2 \\ \longrightarrow \end{array}$$

$$\begin{bmatrix} 1 & \frac{1}{2} & 0 & \frac{1}{2} & 1 \\ 0 & \frac{3}{2} & 3 & \frac{7}{2} & 1 \\ 0 & 0 & 9 & 7 & 3 \end{bmatrix}.$$

The transformed LS is

$$x + \frac{1}{2}y + \frac{1}{2}w = 1,$$
$$\frac{3}{2}y + 3z + \frac{7}{2}w = 1,$$
$$9z + 7w = 3.$$

Thus the system has infinitely many solutions. We may choose $w = t$, t arbitrary and find x, y, and z in terms of t.

2. $x_1 = -2t, x_2 = t, x_3 = 0, x_4 = 0.$

3. a. $x = \frac{17}{5} - 2t, y = t, z = -\frac{3}{5}.$
  b. $x_1 = -66, x_2 = 27, x_3 = 6, x_4 = 4.$
  c. $x_1 = -1 - \frac{3}{5}t, x_2 = 1 + \frac{1}{5}t, x_3 = t.$
  d. $x_1 = 2, x_2 = -3, x_3 = 1, x_4 = -1.$

5. Suppose x, y, z, and w ounces of foods A, B, C, and D, respectively, are used. Then

$$20x + 10y + 10z + 15w = 70,$$
$$5x + 5y + 10z + 15w = 35,$$
$$5x + 15y + 5z + 10w = 35,$$
$$8x + 10y + 10z + 20w = 50.$$

This gives $x = \frac{60}{29}, y = \frac{23}{29}, z = \frac{18}{29}, w = \frac{28}{29}.$

6. Suppose x, y, and z are the numbers of hours put in by I, II, and III, respectively. Then

$$x + y + z = 500,$$
$$8x + 6y + 5z = 2800.$$

$$\begin{bmatrix} 1 & 1 & 1 & 500 \\ 8 & 6 & 5 & 2800 \end{bmatrix} \begin{array}{c} R_2-8R_1 \\ \longrightarrow \end{array}$$

$$\begin{bmatrix} 1 & 1 & 1 & 500 \\ 0 & -2 & -3 & -1200 \end{bmatrix} \begin{array}{c} -R_2 \\ \longrightarrow \end{array}$$

$$\begin{bmatrix} 1 & 1 & 1 & 500 \\ 0 & 2 & 3 & 1200 \end{bmatrix}.$$

This gives

$$x + y + z = 500,$$
$$2y + 3z = 1200.$$

Put

$$z = t,$$
$$y = 600 - \frac{3}{2}t \geq 0,$$
$$x = \frac{t}{2} - 100 \geq 0.$$

Thus $200 \leq t \leq 400$. Considering $z \geq x$ and $z \geq y$, it follows that $t \geq 240$. Hence $240 \leq t \leq 400$.

7. Suppose the numbers of products produced of types I, II, and III are x, y, and z thousand, respectively. Then

$$30x + 20y + 30z = 350,$$
$$10x + 10y + 30z = 150,$$
$$5x + 10y + 5z = 100.$$

This gives

$$x = \frac{55}{8}, \qquad y = \frac{25}{4}, \qquad z = \frac{5}{8}.$$

8. $u = 500, t = 800, y = 2000 - x$, and $z = 100 + x$, where $x \leq 2000$.

  a. $x = 0, y = 2000, z = 100, t = 800, u = 500.$
  b. Not possible.

1.7 Review Questions and Project

4. c. i. $\begin{bmatrix} 1 & 0 & -1 & 0 \\ 0 & 1 & 3 & 0 \\ 0 & 0 & -3 & 5 \end{bmatrix}, \begin{bmatrix} 1 & 0 & 0 & -\frac{5}{3} \\ 0 & 1 & 0 & 5 \\ 0 & 0 & 1 & -\frac{5}{3} \end{bmatrix}.$

  ii. $\begin{bmatrix} 0 & 0 & 0 \\ 0 & 1 & 2 \\ 0 & 0 & 1 \end{bmatrix}, \begin{bmatrix} 0 & 0 & 0 \\ 0 & 1 & 0 \\ 0 & 0 & 1 \end{bmatrix}.$

5. c. $x_1 = -4t$, $x_2 = t$, $x_3 = 3t$.

6. a. Inconsistent.

   b. $x_1 = 70 - t$, $x_2 = 48$, $x_3 = t$, $x_4 = 8$.

   c. $x_1 = \left(-\frac{2}{7}\right) - 2t$, $x_2 = t$, $x_3 = \frac{-17}{7}$,
   $x_4 = \frac{-5}{14}$.

   d. $x_1 = \frac{26}{5}$, $x_2 = \frac{73}{5}$, $x_3 = \frac{54}{5}$, $x_4 = \frac{28}{5}$.

Project

a. $x_1 + x_2 - x_3 - x_4 = 10$, $-x_1 + x_2 - x_5 = -20$, $-x_2 + x_6 = -10$, $-2x_2 + x_3 + x_4 = 0$, $x_2 + x_5 = 30$, and $x_6 = 10$.

b. $x_1 = 30$, $x_2 = 20$, $x_3 = 40 - t$, $x_4 = t$, $x_5 = 10$, $x_6 = 10$.

c. Because $x_3 = 40 - t$ must be greater than or equal to zero.

d. $x_1 + x_2 = 14$, $x_1 - x_3 - x_4 + x_5 = -45$, $x_3 - x_6 + x_7 = -20$, $x_2 = 10$, $x_4 - x_5 - x_8 = 10$, and $x_6 - x_7 + x_8 = 59$.

Solution: $x_1 = 4$, $x_2 = 10$, $x_3 = 39 - t_2$, $x_4 = 10 + t_1 + t_2$, $x_5 = t_1$, $x_6 = 59 - t_3 + t_2$, $x_7 = t_2$, $x_8 = t_3$.

CHAPTER 2

2.1 Exercises

1. a. $A - 2B = \begin{bmatrix} -2 & -17 & -2 \\ 5 & 13 & -3 \\ -3 & -2 & -1 \end{bmatrix}$,

$6A + 7B = \begin{bmatrix} 26 & 69 & 26 \\ 11 & -36 & 39 \\ 20 & 26 & 13 \end{bmatrix}$.

b. $X = B - A - 10I$.

2. a. $X = \frac{1}{3}(B - 4A) = \frac{1}{3}\begin{bmatrix} -4 & -7 & -12 \\ 6 & 3 & -1 \\ 2 & -4 & -5 \end{bmatrix}$.

b. $X = \frac{1}{5}(B - A) = \frac{1}{5}\begin{bmatrix} -1 & -1 & -3 \\ 6 & 6 & -1 \\ -1 & -1 & -2 \end{bmatrix}$.

c. $X = \frac{1}{7}\begin{bmatrix} 2 & 1 & 6 \\ -18 & -19 & 3 \\ 4 & 2 & 5 \end{bmatrix}$.

3. i. $\begin{bmatrix} 3 & 19 & 5 \\ 13 & 17 & 20 \end{bmatrix}$,

ii. $\begin{bmatrix} 6 & 21 & -9 \\ 1 & -2 & -3 \end{bmatrix}$,

iii. $\begin{bmatrix} -2 & 7 & 1 \\ 3 & 6 & 7 \end{bmatrix}$.

4. a. $X = \begin{bmatrix} -3 & -3 \\ 1 & -5 \end{bmatrix}$.

b. $X = \begin{bmatrix} -1 & -2 & -3 \\ -2 & 1 & -5 \end{bmatrix}$.

c. $X = B$.

5. a. $\alpha = 3$.

b. $\alpha = 3$.

2.2 Exercises

1. a. -2.   **b.** 23.   **c.** -3.

2. a. 8.   **b.** 4.

3. a. $x = -2$, $y = -4$.

   b. $x = \frac{13}{7}$, $y = \frac{3}{7}$.

   c. $x = -1$, $y = \frac{9}{8}$, $z = \frac{11}{8}$.

   d. $x = \frac{t+1}{2}$, $y = \frac{5t+3}{2}$, $z = t$, where t is a parameter.

4. $AB = \begin{bmatrix} 30 & 38 & 10 \\ 15 & 21 & 16 \\ -21 & -27 & -9 \end{bmatrix}$,

$BA = \begin{bmatrix} 6 & 0 & 5 \\ 30 & 36 & 11 \\ -1 & -2 & 0 \end{bmatrix}$.

5. $CD = \begin{bmatrix} 21 & 24 & 27 \\ 47 & 54 & 61 \end{bmatrix}$; DC undefined.

6. $A = \begin{bmatrix} 0 & 1 \\ 0 & 0 \end{bmatrix}$, $B = \begin{bmatrix} 0 & 2 \\ 0 & 0 \end{bmatrix}$.

9. a. $AX = b$, where $A = \begin{bmatrix} 1 & 2 & 3 \\ 5 & 1 & 6 \end{bmatrix}$,

$b = \begin{bmatrix} 1 \\ 2 \end{bmatrix}$, $X = \begin{bmatrix} x \\ y \\ z \end{bmatrix}$.

b. $Ax = b$, where $A = \begin{bmatrix} 1 & -2 & 3 & 4 \\ 0 & 2 & -1 & 1 \\ 5 & 6 & 7 & 8 \end{bmatrix}$,

$b = \begin{bmatrix} 1 \\ 2 \\ 9 \end{bmatrix}$, $X = \begin{bmatrix} x_1 \\ x_2 \\ x_3 \\ x_4 \end{bmatrix}$.

c. $Ax = b$, where $A = \begin{bmatrix} 2 & 3 & -5 \\ 1 & 1 & -7 \end{bmatrix}$,

$b = \begin{bmatrix} 1 \\ 6 \end{bmatrix}$, $X = \begin{bmatrix} u \\ v \\ w \end{bmatrix}$.

11. $A = \begin{bmatrix} 1 & 1 & 1 \\ 1 & 0 & 0 \\ 0 & 1 & 0 \end{bmatrix}$.

12. $A = \begin{bmatrix} 1 & 2 \\ 1 & 0 \end{bmatrix}$, $a_4 = 5$.

13. $A = \begin{bmatrix} 1 & 1 \\ 1 & 0 \end{bmatrix}$.

14. $P(n) = A^n P(0)$, where $P(n) = \begin{bmatrix} A_n \\ B_n \end{bmatrix}$,

$A = \begin{bmatrix} 0.6 & 0.7 \\ 0.1 & 1.2 \end{bmatrix}$, and

$P(0) = \begin{bmatrix} A_0 \\ B_0 \end{bmatrix} = \begin{bmatrix} 50{,}000 \\ 100{,}000 \end{bmatrix}$.

15. i. $S_5 = 91{,}458$ (rounded), $F_5 = 119{,}760$ (rounded), $S_6 = 111{,}926$ (rounded), $F_6 = 141{,}883$ (rounded).

ii. $S_5 = 77{,}946$ (rounded), $F_5 = 79{,}412$ (rounded), $S_6 = 86{,}329$ (rounded), $F_6 = 79{,}706$ (rounded).

16. Biweekly budget vector is $\begin{bmatrix} 86 \\ 98 \\ 76 \\ 79 \\ 78 \end{bmatrix}$.

2.3 Exercises

3. $A = \begin{bmatrix} 1 & 2 & 3 \\ 2 & 4 & 5 \\ 3 & 5 & 6 \end{bmatrix}$, $A = \begin{bmatrix} a & h & g \\ h & b & f \\ g & f & c \end{bmatrix}$.

4. $A = \begin{bmatrix} -1 & 1 \\ 1 & 1 \end{bmatrix}$.

5. $A = \begin{bmatrix} 5 & 3 \\ 3 & 7 \end{bmatrix}$.

6. $A = \begin{bmatrix} a & h & g \\ h & b & f \\ g & f & c \end{bmatrix}$.

7. c. $A = \begin{bmatrix} 1 & 1 \\ 1 & 0 \end{bmatrix}$, $B = \begin{bmatrix} 1 & 1 \\ 1 & 1 \end{bmatrix}$.

9. $A = \begin{bmatrix} 1 & 0 \\ 1 & 1 \end{bmatrix}$, $B = \begin{bmatrix} 1 & 1 \\ 0 & 1 \end{bmatrix}$.

2.5 Review Questions and Projects

9. A^2 and $A + B$ are symmetric. AB need not be symmetric.

10. $A = \begin{bmatrix} 0 & 1 & 0 & 1 \\ 0 & 0 & 0 & 1 \\ 0 & 1 & 0 & 1 \\ 0 & 0 & 0 & 1 \end{bmatrix}$, $A^2 = \begin{bmatrix} 0 & 0 & 0 & 2 \\ 0 & 0 & 0 & 1 \\ 0 & 0 & 0 & 2 \\ 0 & 0 & 0 & 1 \end{bmatrix}$,

$A^3 = \begin{bmatrix} 0 & 0 & 0 & 2 \\ 0 & 0 & 0 & 1 \\ 0 & 0 & 0 & 2 \\ 0 & 0 & 0 & 1 \end{bmatrix}$.

$(2, 3)$ and $(1, 3)$ entries of A^2 (A^3) give, respectively, the number of paths from 2 to 3 and 1 to 3 of length 2 (3).

Projects

1. a. Every worker can contaminate every other worker via a chain of at most $n - 1$ workers.
b. $n = 6$.
c. Worker 2 has the least potential.
d. If the jth column has the largest column sum then the jth worker has the greatest potential to get contaminated.

2. a. $p^{(i)} = p_i$.
c. starting with higher amount would result in higher probability in this situation.
d. $(6, 11)$ entry of T^{200} is 0.5 and so Bill has 50% chance of accumulating \$10 in 200 steps.
e. 60% chance of going to movie after 200 steps.

CHAPTER 3

3.1 Exercises

1. a. $R_3 = -3R_1 + R_2$, where R_1, R_2, and R_3 are first, second, and third rows, respectively.

b. $x_1 = -\frac{1}{19}$, $x_2 = 0$, $x_4 = \frac{14}{19}$.

c. $\alpha_2 = -1$, $\alpha_3 = -2$, $\alpha_4 = 2$.

3. a. $x_1 = 1 + 6t$, $x_2 = -\frac{24}{5}t$, $x_3 = -\frac{13}{5}t$, $x_4 = t$.

b. $x_1 = 2$, $x_2 = 1$, $x_3 = -1$, $x_4 = 3$.

3.2 Exercises

1. Let $x = [a\ b\ 0]$, $y = [a'\ b'\ 0] \in W$. Then $x + y = [a + a'\ b + b'\ 0] \in W$. If $\alpha \in F$, then $\alpha x = [\alpha a\ \alpha b\ 0] \in W$. Thus W is a subspace.

2. Let $x = \begin{bmatrix} a \\ b \\ c \end{bmatrix}$, $y = \begin{bmatrix} a' \\ b' \\ c' \end{bmatrix} \in W$. Then $a + b + c = 0$, and $a' + b' + c' = 0$. Thus

$$x + y = \begin{bmatrix} a + a' \\ b + b' \\ c + c' \end{bmatrix} \in W, \text{ since } a + a' + b + b' +$$

$c + c' = 0$. Also, if $\alpha \in F$, then $\alpha x = \begin{bmatrix} \alpha a \\ \alpha b \\ \alpha c \end{bmatrix}$

$\in W$, since $\alpha a + \alpha b + \alpha c = \alpha(a + b + c) = 0$. Hence W is a subspace.

3. Let $x = \begin{bmatrix} x_1 \\ x_2 \\ x_3 \end{bmatrix}$, $y = \begin{bmatrix} y_1 \\ y_2 \\ y_3 \end{bmatrix} \in W$. Then $x_3 = 2x_1 - x_2$, and $y_3 = 2y_1 - y_2$. So $x + y = \begin{bmatrix} x_1 + y_1 \\ x_2 + y_2 \\ x_3 + y_3 \end{bmatrix} \in W$, since $(x_3 + y_3) = 2x_1 - x_2 + 2y_1 - y_2 = 2(x_1 + y_1) - (x_2 + y_2)$.

If $\alpha \in F$, then $\alpha x = \begin{bmatrix} \alpha x_1 \\ \alpha x_2 \\ \alpha x_3 \end{bmatrix} \in W$, since $\alpha x_3 = \alpha(2x_1 - x_2) = 2(\alpha x_1) - (\alpha x_2)$.

4. Since for each $x = \begin{bmatrix} x_1 \\ x_2 \end{bmatrix} \in W$, $x_2 = 1$, the zero vector $\begin{bmatrix} 0 \\ 0 \end{bmatrix}$ is not in W. Hence W cannot be a subspace.

5. W is not a subspace.

6. If x, $y \in W$, then $Ax = 0 = Ay$. So $A(x + y) = Ax + Ay = 0$. Thus $x + y \in W$. Furthermore, if $\alpha \in F$, then $A(\alpha x) = \alpha(Ax) = 0$. So $\alpha x \in W$. Hence W is a subspace.

7. a. No. **b.** No. **c.** Yes.

8. Since $\sqrt{2} \begin{bmatrix} 1 \\ 1 \\ 1 \end{bmatrix} = \begin{bmatrix} \sqrt{2} \\ \sqrt{2} \\ \sqrt{2} \end{bmatrix} \notin W$, but $\begin{bmatrix} 1 \\ 1 \\ 1 \end{bmatrix} \in$

W and $\sqrt{2} \in \mathbf{R}$, W is not a subspace.

12. We want to check whether $(1\ 2) = \alpha(1\ 0) + \beta(1\ 1) + \gamma(-1\ 0)$ for some α, β, and γ—in other words, whether $(1\ 2) = (\alpha + \beta - \gamma\ \beta)$, that is, whether

$$1 = \alpha + \beta - \gamma,$$
$$2 = \beta,$$

has a solution. Clearly, $\beta = 2$, $\alpha = -1$, $\gamma = 0$ is one of the solutions. Thus $(1\ 2)$ belongs to the subspace spanned by $(1\ 0)$, $(1\ 1)$, and $(-1\ 0)$. Similarly, $(1\ 2)$ belongs to the subspace spanned by $(1\ 0)$ and $(1\ 1)$. But $(1\ 2)$ does not belong to the subspace spanned by $(1\ 0)$, since $(1\ 2) = \alpha(1\ 0)$ is not possible for any α.

13. No. Hint:

$$\alpha + 4\beta + 3\gamma = 1,$$
$$3\alpha + 0\beta + \gamma = 1,$$
$$4\alpha + \beta + 2\gamma = 1,$$

is an inconsistent linear system.

14. We want to check whether

$$(2\ 0\ 4\ -2) = \alpha(0\ 2\ 1\ -1) + \beta(1\ -1\ 1\ 0) + \gamma(2\ 1\ 0\ -2)$$

for some α, β, and γ—in other words, whether the LS

$$2 = 0\alpha + \beta + 2\gamma,$$
$$0 = 2\alpha - \beta + \gamma,$$
$$4 = \alpha + \beta + 0\gamma,$$
$$-2 = -\alpha + 0\beta - 2\gamma,$$

is consistent. We perform elementary row operations on the augmented matrix:

$$\begin{bmatrix} 0 & 1 & 2 & 2 \\ 2 & -1 & 1 & 0 \\ 1 & 1 & 0 & 4 \\ -1 & 0 & -2 & -2 \end{bmatrix} \xrightarrow{R_1 \longleftrightarrow R_3}$$

$$\begin{bmatrix} 1 & 1 & 0 & 4 \\ 2 & -1 & 1 & 0 \\ 0 & 1 & 2 & 2 \\ -1 & 0 & -2 & -2 \end{bmatrix} \xrightarrow[R_4+R_1]{R_2-2R_1}$$

$$\begin{bmatrix} 1 & 1 & 0 & 4 \\ 0 & -3 & 1 & -8 \\ 0 & 1 & 2 & 2 \\ 0 & 1 & -2 & 2 \end{bmatrix} \xrightarrow{R_2 \longleftrightarrow R_3}$$

$$\begin{bmatrix} 1 & 1 & 0 & 4 \\ 0 & 1 & 2 & 2 \\ 0 & -3 & 1 & -8 \\ 0 & 1 & -2 & 2 \end{bmatrix} \xrightarrow[R_4-R_2]{R_3+3R_2}$$

$$\begin{bmatrix} 1 & 1 & 0 & 4 \\ 0 & 1 & 2 & 2 \\ 0 & 0 & 7 & -2 \\ 0 & 0 & -4 & 0 \end{bmatrix} \xrightarrow{R_3 \longleftrightarrow R_4}$$

$$\begin{bmatrix} 1 & 1 & 0 & 4 \\ 0 & 1 & 2 & 2 \\ 0 & 0 & -4 & 0 \\ 0 & 0 & 7 & -2 \end{bmatrix} \xrightarrow{-\frac{1}{4}R_3}$$

$$\begin{bmatrix} 1 & 1 & 0 & 4 \\ 0 & 1 & 2 & 2 \\ 0 & 0 & 1 & 0 \\ 0 & 0 & 7 & -2 \end{bmatrix} \xrightarrow{R_4-7R_3} \begin{bmatrix} 1 & 1 & 0 & 4 \\ 0 & 1 & 2 & 2 \\ 0 & 0 & 1 & 0 \\ 0 & 0 & 0 & -2 \end{bmatrix}.$$

Thus the LS reduces to

$$\alpha + \beta + 0\gamma = 4,$$
$$\beta + 2\gamma = 2,$$
$$\gamma = 0,$$
$$0 = -2.$$

Therefore the LS is inconsistent. Hence $[2 \ 0 \ 4 \ -2]$ does not belong to the subspace of the exercise.

15. a. $[1 \ 0 \ 2 \ 3], [1 \ -1 \ 1 \ 5].$

b. $\begin{bmatrix} 0 \\ 1 \\ -1 \end{bmatrix}.$

d. No.

e. $\begin{bmatrix} 1 \\ 0 \\ -1 \end{bmatrix}$ does not belong to column space of

A, but $\begin{bmatrix} 0 \\ 0 \\ 0 \end{bmatrix}$ does belong to any subspace.

f. No.

3.3 Exercises

1. Suppose

$$\alpha \begin{bmatrix} 2 \\ 6 \\ -2 \end{bmatrix} + \beta \begin{bmatrix} 3 \\ 1 \\ 2 \end{bmatrix} + \gamma \begin{bmatrix} 8 \\ 16 \\ -3 \end{bmatrix} = \begin{bmatrix} 0 \\ 0 \\ 0 \end{bmatrix}.$$

This gives

$$2\alpha + 3\beta + 8\gamma = 0,$$
$$6\alpha + \beta + 16\gamma = 0,$$
$$-2\alpha + 2\beta - 3\gamma = 0.$$

This coefficient matrix is

$$\begin{bmatrix} 2 & 3 & 8 \\ 6 & 1 & 16 \\ -2 & 2 & -3 \end{bmatrix} \xrightarrow[R_3+R_1]{R_2-3R_1} \begin{bmatrix} 2 & 3 & 8 \\ 0 & -8 & -8 \\ 0 & 5 & 5 \end{bmatrix}$$

$$\xrightarrow{-\frac{1}{8}R_2} \begin{bmatrix} 2 & 3 & 8 \\ 0 & 1 & 1 \\ 0 & 5 & 5 \end{bmatrix} \xrightarrow{R_3-5R_2} \begin{bmatrix} 2 & 3 & 8 \\ 0 & 1 & 1 \\ 0 & 0 & 0 \end{bmatrix}.$$

So

$$2\alpha + 3\beta + 8\gamma = 0,$$
$$\beta + \gamma = 0.$$

Choose $\gamma = 1$. Then $\beta = -1, \alpha = -\frac{3}{2}$. Therefore $-\frac{5}{2} \begin{bmatrix} 2 \\ 6 \\ -2 \end{bmatrix} - \begin{bmatrix} 3 \\ 1 \\ 2 \end{bmatrix} + \begin{bmatrix} 8 \\ 16 \\ -3 \end{bmatrix} = \begin{bmatrix} 0 \\ 0 \\ 0 \end{bmatrix}$

is a dependence relation, showing that the given list of vectors is linearly dependent.

2. Suppose

$$\alpha \begin{bmatrix} 4 \\ 5 \\ 1 \end{bmatrix} + \beta \begin{bmatrix} 3 \\ 0 \\ 2 \end{bmatrix} + \gamma \begin{bmatrix} a \\ 10 \\ 9 \end{bmatrix} = \begin{bmatrix} 0 \\ 0 \\ 0 \end{bmatrix}.$$

So

$$\begin{aligned} 4\alpha + 3\beta + a\gamma &= 0, \\ 5\alpha + 10\gamma &= 0, \quad (1) \\ \alpha + 2\beta + 9\gamma &= 0. \end{aligned}$$

Reducing $\begin{bmatrix} 4 & 3 & a \\ 5 & 0 & 10 \\ 1 & 2 & 9 \end{bmatrix}$ into echelon form

$\begin{bmatrix} 1 & 2 & 9 \\ 0 & 1 & \frac{7}{2} \\ 0 & 0 & a - \frac{37}{2} \end{bmatrix}$, it follows that the LS (1)

has a nontrivial solution if $a = \frac{37}{2}$, in which case the given list of vectors is linearly dependent.

3. Suppose $\alpha\,[2\ 1\ 1\ 1] + \beta\,[3\ -2\ 1\ 0] + \gamma\,[a\ -1\ 2\ 0] = 0$. This gives

$$\begin{aligned} 2\alpha + 3\beta + a\gamma &= 0, \\ \alpha - 2\beta - \gamma &= 0, \\ \alpha + \beta + 2\gamma &= 0, \\ \alpha &= 0, \end{aligned}$$

and so

$$\begin{aligned} 3\beta + a\gamma &= 0, \\ -2\beta - \gamma &= 0, \\ \beta + 2\gamma &= 0. \end{aligned}$$

The last two equations give $\beta = 0 = \gamma$. Hence the given list of vectors is linearly independent for all a.

4. a, any nonzero number and $b \neq \frac{2}{3}$.

5. No. For example, $\begin{bmatrix} 1 \\ 2 \end{bmatrix}$ is not a linear combination of $\begin{bmatrix} 1 \\ 1 \end{bmatrix}$.

6. Yes. First, the given vectors are linearly independent. Second, if $\begin{bmatrix} a \\ b \end{bmatrix} \in \mathbf{R}^2$, we can find α, β such that $\begin{bmatrix} a \\ b \end{bmatrix} = \alpha \begin{bmatrix} 1 \\ 1 \end{bmatrix} + \beta \begin{bmatrix} 2 \\ 3 \end{bmatrix}$.

7. No, the given list is linearly dependent.

8. No, $\begin{bmatrix} 1 \\ 0 \end{bmatrix}, \begin{bmatrix} 0 \\ 1 \end{bmatrix}$ form a basis of $\mathbf{R}^2$.

9. The given vectors form a basis.

10. The given list of vectors is linearly independent and so is a basis of the subspace spanned by this list.

12. a. 1. **b.** 1. **c.** 1. **d.** 2.

13. a. 0. **b.** 0. **c.** 0. **d.** 1.
　　e. 1. **f.** 2. **g.** 1.

14. a. 2. **b.** 2.

15. a. 2. **b.** 2.

16. a. $\begin{bmatrix} -1 \\ 0 \\ 1 \\ -1 \\ -1 \end{bmatrix}, \begin{bmatrix} 2 \\ 2 \\ -1 \\ 2 \\ 2 \end{bmatrix}, \begin{bmatrix} 3 \\ 3 \\ 3 \\ 3 \\ 3 \end{bmatrix}.$

b. $\begin{bmatrix} -1 & 2 & 3 & 1 & 11 \\ 0 & 2 & 3 & 4 & 8 \\ 1 & -2 & 3 & 1 & 1 \\ -1 & 2 & 3 & 1 & 11 \\ -1 & 2 & 3 & 1 & 11 \end{bmatrix} \begin{matrix} \\ \\ R_3+R_1 \\ R_4-R_1 \\ \xrightarrow{} \\ R_5-R_1 \end{matrix}$

$\begin{bmatrix} -1 & 2 & 3 & 1 & 11 \\ 0 & 2 & 3 & 4 & 8 \\ 0 & 0 & 6 & 0 & 12 \\ 0 & 0 & 0 & 0 & 0 \\ 0 & 0 & 0 & 0 & 0 \end{bmatrix}.$ So $[-1\ 2\ 3\ 1\ 11]$,

$[0\ 2\ 3\ 4\ 8], [0\ 0\ 6\ 0\ 12]$ is a basis of the row space of A.

17. a. The dimension of the subspace spanned by rows = 2, 4, respectively. The dimension of the subspace spanned by columns is also =2, 4, respectively.

b. The dimension of the subspace spanned by the rows of a matrix is equal to that of the subspace spanned by the columns of the matrix.

3.5 Review Questions and Project

7. a. No.   **b.** No.

10. $\mathbf{R}^2$.

11. No.

  a. (2 1 0), (−1 3 1), (0 7 3).
  b. (0 0 1 1), (−1 −1 2 3), (1 0 −2 1).

12. i. {(1 0), (0 1} and {(1 0), (1 1)}.

  ii. {(1 0 0), (0 1 0), (0 0 1)} and
    {(1 0 0), (1 1 0), (1 1 1)}.

  Any two bases of a vector space has the same number of elements.

Project

a. Mix B, Mix C, Mix D, Mix E.

b. Unique.

c. 2.5 packets of Mix B, 1.5 packet of Mix C, and 1 packet of Mix D.

d. $7.58.

e. The new mixture cannot be made.

f. Linear independence, linear dependence, basis.

CHAPTER 4

4.1 Exercises

1. a. $\begin{bmatrix} 1 & 2 \\ 3 & 4 \end{bmatrix} \xrightarrow{R_2-3R_1} \begin{bmatrix} 1 & 2 \\ 0 & -2 \end{bmatrix}$. Basis
(row space): {[1 2], [0 − 2]}

$\begin{bmatrix} 1 & 3 \\ 2 & 4 \end{bmatrix} \xrightarrow{R_2-2R_1} \begin{bmatrix} 1 & 3 \\ 0 & -2 \end{bmatrix}$. Basis

(column space): $\left\{ \begin{bmatrix} 1 \\ 3 \end{bmatrix}, \begin{bmatrix} 0 \\ -2 \end{bmatrix} \right\}$.

  Rank $A = 2$.

b. Row space basis: {[1 0 1], [0 2 −2]};

column space basis: $\left\{ \begin{bmatrix} 1 \\ 3 \end{bmatrix}, \begin{bmatrix} 0 \\ 2 \end{bmatrix} \right\}$.

  Rank $A = 2$.

c. $\begin{bmatrix} 2 & 3 \\ 4 & 5 \\ 6 & 8 \end{bmatrix} \xrightarrow[R_3-3R_1]{R_2-2R_1} \begin{bmatrix} 2 & 3 \\ 0 & -1 \\ 0 & -1 \end{bmatrix}$

$\xrightarrow{R_3 - R_2} \begin{bmatrix} 2 & 3 \\ 0 & -1 \\ 0 & 0 \end{bmatrix}$.

Basis (row space): {[2 3], [0 −1]}.

$\begin{bmatrix} 2 & 4 & 6 \\ 3 & 5 & 8 \end{bmatrix} \xrightarrow{R_2-R_1} \begin{bmatrix} 2 & 4 & 6 \\ 1 & 1 & 2 \end{bmatrix}$

$\xrightarrow{R_1 \longleftrightarrow R_2} \begin{bmatrix} 1 & 1 & 2 \\ 2 & 4 & 6 \end{bmatrix} \xrightarrow{R_2-2R_1}$

$\begin{bmatrix} 1 & 1 & 2 \\ 0 & 2 & 2 \end{bmatrix}$.

Basis (column space):

$\left\{ \begin{bmatrix} 1 \\ 1 \\ 2 \end{bmatrix}, \begin{bmatrix} 0 \\ 2 \\ 2 \end{bmatrix} \right\}$.

Rank $A = 2$.

d. Row space basis:

  {[1 2 0], [0 5 4], [0 0 − 1]};
column space basis:

$\left\{ \begin{bmatrix} 1 \\ -1 \\ 0 \end{bmatrix}, \begin{bmatrix} 0 \\ 5 \\ 4 \end{bmatrix}, \begin{bmatrix} 0 \\ 0 \\ -1 \end{bmatrix} \right\}$.

Rank $A = 3$.

e. Basis of row space:

  {[5 6 7], [0 5 6], [0 0 5]};
basis of column space:

$\left\{ \begin{bmatrix} 5 \\ 0 \\ 0 \end{bmatrix}, \begin{bmatrix} 6 \\ 5 \\ 0 \end{bmatrix}, \begin{bmatrix} 7 \\ 6 \\ 5 \end{bmatrix} \right\}$.

Rank $A = 3$.

2. a. $\begin{bmatrix} 2 & 3 & 4 & 5 \\ 0 & 1 & 5 & 6 \\ 0 & 0 & 7 & 8 \\ 0 & 0 & 5 & 3 \end{bmatrix} \xrightarrow{R_4-\frac{5}{7}R_3}$

$\begin{bmatrix} 2 & 3 & 4 & 5 \\ 0 & 1 & 5 & 6 \\ 0 & 0 & 7 & 8 \\ 0 & 0 & 0 & -\frac{19}{7} \end{bmatrix}$.

For aesthetic reasons only, perform $-\frac{7}{19}R_4$

to obtain $\begin{bmatrix} 2 & 3 & 4 & 5 \\ 0 & 1 & 5 & 6 \\ 0 & 0 & 7 & 8 \\ 0 & 0 & 0 & 1 \end{bmatrix}$.

Basis (row space):

$$\{[2\ 3\ 4\ 5], [0\ 1\ 5\ 6],$$
$$[0\ 0\ 7\ 8], [0\ 0\ 0\ 1]\}.$$

Considering the transpose of the given matrix, a basis of the column space is

$$\left\{ \begin{bmatrix} 2 \\ 0 \\ 0 \\ 0 \end{bmatrix}, \begin{bmatrix} 0 \\ 1 \\ 0 \\ 0 \end{bmatrix}, \begin{bmatrix} 0 \\ 0 \\ 7 \\ 5 \end{bmatrix}, \begin{bmatrix} 0 \\ 0 \\ 0 \\ 1 \end{bmatrix} \right\}.$$

b.
$$\begin{bmatrix} 1 & 2 & 3 & 4 \\ 2 & 4 & 6 & 8 \\ 3 & 5 & 7 & 9 \\ 4 & 6 & 8 & 10 \end{bmatrix} \begin{matrix} R_2 - 2R_1 \\ \longrightarrow \\ R_3 - 3R_1 \\ R_4 - 4R_1 \end{matrix}$$

$$\begin{bmatrix} 1 & 2 & 3 & 4 \\ 0 & 0 & 0 & 0 \\ 0 & -1 & -2 & -3 \\ 0 & -2 & -4 & -6 \end{bmatrix} \begin{matrix} R_2 \longleftrightarrow R_4 \end{matrix}$$

$$\begin{bmatrix} 1 & 2 & 3 & 4 \\ 0 & -2 & -4 & -6 \\ 0 & -1 & -2 & -3 \\ 0 & 0 & 0 & 0 \end{bmatrix} \begin{matrix} R_3 - \frac{1}{2}R_2 \\ \longrightarrow \end{matrix}$$

$$\begin{bmatrix} 1 & 2 & 3 & 4 \\ 0 & -2 & -4 & -6 \\ 0 & 0 & 0 & 0 \\ 0 & 0 & 0 & 0 \end{bmatrix}.$$

For aesthetic reasons only, perform $-\frac{1}{2}R_2$ to transform the last matrix to

$$\begin{bmatrix} 1 & 2 & 3 & 4 \\ 0 & 1 & 2 & 3 \\ 0 & 0 & 0 & 0 \\ 0 & 0 & 0 & 0 \end{bmatrix}.$$ So the basis (row space) is

$$\{[1\ 2\ 3\ 4], [0\ 1\ 2\ 3]\}.$$

For the column space, we consider the transpose of the given matrix and perform elementary row operations on that matrix to reduce it to row echelon form. Then the transpose of each of the nonzero rows forms a basis of the column space:

Basis (column space):
$$\left\{ \begin{bmatrix} 1 \\ 2 \\ 3 \\ 4 \end{bmatrix}, \begin{bmatrix} 0 \\ 0 \\ -1 \\ -2 \end{bmatrix} \right\}.$$

3. a. Write $A = \begin{bmatrix} 1 & 1 & 1 \\ 2 & 2 & 2 \\ 0 & 0 & 0 \end{bmatrix}$, and find a basis of the row space of A. Just by inspection or by reducing A to $\begin{bmatrix} 1 & 1 & 1 \\ 0 & 0 & 0 \\ 0 & 0 & 0 \end{bmatrix}$, we see that

$[1\ 1\ 1]$ is a basis.

b. $\begin{bmatrix} 3 & 2 & 1 \\ 4 & 3 & 2 \\ 1 & 1 & 1 \end{bmatrix} \begin{matrix} R_1 \longleftrightarrow R_3 \\ \longrightarrow \end{matrix} \begin{bmatrix} 1 & 1 & 1 \\ 4 & 3 & 2 \\ 3 & 2 & 1 \end{bmatrix}$

$\begin{matrix} R_2 - 4R_1 \\ \longrightarrow \\ R_3 - 3R_1 \end{matrix} \begin{bmatrix} 1 & 1 & 1 \\ 0 & -1 & -2 \\ 0 & -1 & -2 \end{bmatrix} \begin{matrix} R_3 + (-1)R_2 \\ \longrightarrow \end{matrix}$

$\begin{bmatrix} 1 & 1 & 1 \\ 0 & -1 & -2 \\ 0 & 0 & 0 \end{bmatrix}$, and so $\{[1\ 1\ 1],$

$[0\ -1\ -2]\}$ is a basis.

c. $\left\{ \begin{bmatrix} 2 \\ 7 \\ 0 \end{bmatrix}, \begin{bmatrix} 3 \\ 5 \\ 1 \end{bmatrix}, \begin{bmatrix} 1 \\ 0 \\ 0 \end{bmatrix} \right\}$ is a basis.

4. a. If $x = \begin{bmatrix} x_1 \\ x_2 \\ x_3 \end{bmatrix}$ is a solution, then $x_3 = x_1 +$

x_2, and so $x = \begin{bmatrix} x_1 \\ x_2 \\ x_1 + x_2 \end{bmatrix} = x_1 \begin{bmatrix} 1 \\ 0 \\ 1 \end{bmatrix} +$

$x_2 \begin{bmatrix} 0 \\ 1 \\ 1 \end{bmatrix}$, where x_1, x_2 are arbitrary. Thus

$\begin{bmatrix} 1 \\ 0 \\ 1 \end{bmatrix}$ and $\begin{bmatrix} 0 \\ 1 \\ 1 \end{bmatrix}$ form a basis.

b. $x = \begin{bmatrix} t \\ -3t \\ t \end{bmatrix} = t \begin{bmatrix} 1 \\ -3 \\ 1 \end{bmatrix}$ is a solution,

where t is arbitrary, yielding that $\begin{bmatrix} 1 \\ -3 \\ 1 \end{bmatrix}$ is

a basis.

c. This LS has only the trivial solution $x_1 = 0 = x_2 = x_3$. Thus the solution space is the zero space, and by convention, the basis of the zero space is the empty set.

d.
$$\begin{bmatrix} 1 & 2 & 0 & 3 \\ 2 & 5 & 2 & 1 \\ 1 & 3 & 1 & -1 \end{bmatrix} \xrightarrow[R_3-R_1]{R_2-2R_1}$$

$$\begin{bmatrix} 1 & 2 & 0 & 3 \\ 0 & 1 & 2 & -5 \\ 0 & 1 & 1 & -4 \end{bmatrix} \xrightarrow{R_3-R_2}$$

$$\begin{bmatrix} 1 & 2 & 0 & 3 \\ 0 & 1 & 2 & 5 \\ 0 & 0 & -1 & 1 \end{bmatrix}.$$

This gives

$$x_1 + 2x_2 + 0x_3 + 3x_4 = 0,$$
$$x_2 + 2x_3 + 5x_4 = 0,$$
$$-x_3 + x_4 = 0.$$

So we put $x_4 = t$ and get $x_3 = t$, $x_2 = 3t$,

$$x_1 = -9t. \text{ Thus } x = \begin{bmatrix} -9t \\ 3t \\ t \\ t \end{bmatrix} = t \begin{bmatrix} -9 \\ 3 \\ 1 \\ 1 \end{bmatrix},$$

where t is arbitrary, is a solution. Hence

$$\begin{bmatrix} -9 \\ 3 \\ 1 \\ 1 \end{bmatrix} \text{ is a basis.}$$

5. $\begin{bmatrix} 1 & 2 & 0 \\ 2 & 1 & 3 \end{bmatrix} \xrightarrow{R_2-2R_1} \begin{bmatrix} 1 & 2 & 0 \\ 0 & -5 & 3 \end{bmatrix}$ gives

$$x_1 + 2x_2 + 0x_3 = 0,$$
$$-5x_2 + 3x_3 = 0,$$

and so we put $x_3 = t$ and obtain $x_2 = \frac{3}{5}t$,

$$x_1 = -\frac{6}{5}t. \text{ Thus } x = \begin{bmatrix} -\frac{6}{5}t \\ \frac{3}{5}t \\ t \end{bmatrix} = \frac{t}{5}\begin{bmatrix} -6 \\ 3 \\ 5 \end{bmatrix}$$

is a solution, where t is arbitrary. This shows

that $\begin{bmatrix} -6 \\ 3 \\ 5 \end{bmatrix}$ is a basis of the solution space.

6. a. $\begin{bmatrix} 1 & 5 \\ 2 & 6 \end{bmatrix} \xrightarrow{R_2-2R_1} \begin{bmatrix} 1 & 5 \\ 0 & -4 \end{bmatrix}$. Thus the LS $Ax = 0$ transforms to the equivalent LS

$$x_1 + 5x_2 = 0,$$
$$-4x_2 = 0,$$

where $x = \begin{bmatrix} x_1 \\ x_2 \end{bmatrix}$. This yields $x_2 = 0 = x_1$. The solution space is $\{0\}$, and a basis of the zero subspace by convention is $\emptyset$, the empty set.

b. $\begin{bmatrix} 2 & 3 & 5 \\ 1 & 0 & 1 \end{bmatrix} \xrightarrow{R_1 \longleftrightarrow R_2} \begin{bmatrix} 1 & 0 & 1 \\ 2 & 3 & 5 \end{bmatrix}$

$\xrightarrow{R_2-2R_1} \begin{bmatrix} 1 & 0 & 1 \\ 0 & 3 & 3 \end{bmatrix}$. The LS $Ax = 0$ transforms to

$$x_1 + x_3 = 0,$$
$$3x_2 + 3x_3 = 0.$$

Since the last equation has two unknowns, we put $x_3 = t$ arbitrary. Then $x_2 = -t$ and $x_1 = -t$. This gives $x = \begin{bmatrix} x_1 \\ x_2 \\ x_3 \end{bmatrix} = \begin{bmatrix} -t \\ -t \\ t \end{bmatrix} =$

$t \begin{bmatrix} -1 \\ -1 \\ 1 \end{bmatrix}$, showing that $\left\{ \begin{bmatrix} -1 \\ -1 \\ 1 \end{bmatrix} \right\}$ is a basis

of the solution space.

c. $\emptyset$ (the empty set).

d. The LS is $x_1 + 2x_2 + 3x_3 + 4x_4 = 0$. So put $x_4 = t_1$, $x_3 = t_2$, $x_2 = t_3$. Then $x_1 = -2t_3 - 3t_2 - 4t_1$. Therefore

$$x = \begin{bmatrix} -2t_3 - 3t_2 - 4t_1 \\ t_3 \\ t_2 \\ t_1 \end{bmatrix}, \text{ where } t_1,$$

t_2, t_3 are arbitrary numbers. Therefore a basis of the solution space is

$$\left\{ \begin{bmatrix} -2 \\ 1 \\ 0 \\ 0 \end{bmatrix}, \begin{bmatrix} -3 \\ 0 \\ 1 \\ 0 \end{bmatrix}, \begin{bmatrix} -4 \\ 0 \\ 0 \\ 1 \end{bmatrix} \right\}.$$

4.2 Exercises

1. a. If $A = [1\ 1\ 1]$, the question is to find a basis of the subspace consisting of the solutions of the LS $Ax = 0$. If $x = \begin{bmatrix} x_1 \\ x_2 \\ x_3 \end{bmatrix}$ is a solution, then $x_1 + x_2 + x_3 = 0$, and so $x = \begin{bmatrix} x_1 \\ x_2 \\ -x_1 - x_2 \end{bmatrix} = x_1 \begin{bmatrix} 1 \\ 0 \\ -1 \end{bmatrix} + x_2 \begin{bmatrix} 0 \\ 1 \\ -1 \end{bmatrix}$.

Thus the null space of A is spanned by
$$\begin{bmatrix} 1 \\ 0 \\ -1 \end{bmatrix} \text{ and } \begin{bmatrix} 0 \\ 1 \\ -1 \end{bmatrix}.$$

b. $\emptyset$.

c. $\begin{bmatrix} 0 & 1 & 1 \\ 1 & 0 & 0 \end{bmatrix} \longrightarrow \begin{bmatrix} 1 & 0 & 0 \\ 0 & 1 & 1 \end{bmatrix}$. The null space consists of $x = \begin{bmatrix} x_1 \\ x_2 \\ x_3 \end{bmatrix}$, where

$$x_1 = 0,$$
$$x_2 + x_3 = 0.$$

Putting $x_3 = t$, where t is arbitrary, $x = \begin{bmatrix} 0 \\ -t \\ t \end{bmatrix} = t \begin{bmatrix} 0 \\ -1 \\ 1 \end{bmatrix}$, and so $\begin{bmatrix} 0 \\ -1 \\ 1 \end{bmatrix}$ is a basis.

d. $\begin{bmatrix} 2 & 5 \\ 3 & 6 \\ 4 & 7 \end{bmatrix} \xrightarrow[R_3 - 2R_1]{R_2 - \frac{3}{2}R_1} \begin{bmatrix} 2 & 5 \\ 0 & -\frac{3}{2} \\ 0 & -3 \end{bmatrix} \xrightarrow[-\frac{1}{3}R_3]{-\frac{2}{3}R_2}$

$\begin{bmatrix} 2 & 5 \\ 0 & 1 \\ 0 & 1 \end{bmatrix} \xrightarrow{R_3 - R_2} \begin{bmatrix} 2 & 5 \\ 0 & 1 \\ 0 & 0 \end{bmatrix}$.

If $x = \begin{bmatrix} x_1 \\ x_2 \end{bmatrix}$ is in the null space of the given matrix, then

$$2x_1 + 5x_2 = 0,$$
$$x_2 = 0,$$

and so $x_1 = 0 = x_2$. Thus the empty set is the basis.

e. 0.

2. $\begin{bmatrix} 1 & 1 & -2 & 1 \\ 2 & -1 & 0 & 1 \end{bmatrix} \xrightarrow{R_2 - 2R_1}$

$\begin{bmatrix} 1 & 1 & -2 & 1 \\ 0 & -3 & 4 & -1 \end{bmatrix}$.

So the given LS is transformed to

$$x_1 + x_2 - 2x_3 + x_4 = 0,$$
$$-3x_2 + 4x_3 - x_4 = 0.$$

Since the last equation has three unknowns, put $x_4 = t_1$, and $x_3 = t_2$. So $x_2 = (-t_1 + 4t_2)/3$ and $x_1 = (2t_2 - 2t_1)/3$. Thus if $x = \begin{bmatrix} x_1 \\ x_2 \\ x_3 \\ x_4 \end{bmatrix}$ is a solution, then

$$x = \begin{bmatrix} (2t_2 - 2t_1)/3 \\ (-t_1 + 4t_2)/3 \\ t_2 \\ t_1 \end{bmatrix}$$

$$= t_1 \begin{bmatrix} -\frac{2}{3} \\ -\frac{1}{3} \\ 0 \\ 1 \end{bmatrix} + t_2 \begin{bmatrix} \frac{2}{3} \\ \frac{4}{3} \\ 1 \\ 0 \end{bmatrix}.$$

This proves that the solution space is spanned by $\begin{bmatrix} -\frac{2}{3} \\ -\frac{1}{3} \\ 0 \\ 1 \end{bmatrix}$ and $\begin{bmatrix} \frac{2}{3} \\ \frac{4}{3} \\ 1 \\ 0 \end{bmatrix}$, which are indeed linearly independent, showing that the dimension of the solution space is 2. Alternatively, since the rank of the coefficient matrix is 2, by Theorem 4.2.2, the dimension of the solution space is $4 - 2 = 2$.

3. a. Yes. **b.** Yes. **c.** Yes.

4. a. $\begin{bmatrix} 1 \\ -3 \\ -2 \\ 1 \end{bmatrix}$.

b. $\begin{bmatrix} -1 \\ -1 \\ 1 \\ 1 \\ 0 \end{bmatrix}, \begin{bmatrix} -1 \\ 0 \\ 0 \\ 0 \\ 1 \end{bmatrix}.$

c. $\begin{bmatrix} 2 \\ -3 \\ 0 \\ 1 \\ 0 \end{bmatrix}, \begin{bmatrix} 1 \\ -2 \\ 1 \\ 0 \\ 0 \end{bmatrix}.$

5. a. rank $A = 3$, nullity $A = 1$
  b. rank $A = 3$, nullity $A = 1$
  c. rank $A = $ nullity $A = 2$

4.3 Exercises

1. a. $I = \begin{bmatrix} 1 & 0 & 0 \\ 0 & 1 & 0 \\ 0 & 0 & 1 \end{bmatrix} \xrightarrow{2R_1} \begin{bmatrix} 2 & 0 & 0 \\ 0 & 1 & 0 \\ 0 & 0 & 1 \end{bmatrix} = E$,

say.

$A = \begin{bmatrix} 5 & 6 & 7 \\ 1 & 1 & 1 \\ 0 & 1 & 1 \end{bmatrix} \xrightarrow{2R_1} \begin{bmatrix} 10 & 12 & 14 \\ 1 & 1 & 1 \\ 0 & 1 & 1 \end{bmatrix} =$

B, say.

Then $B = EA$ by actual multiplication of E and A. Parts (b) and (c) can be shown in a similar manner.

2. b. Since A is a 2×3 matrix and performing elementary column operations is the same as postmultiplying by a suitable elementary matrix E, E must be a 3×3 matrix. So start with the 3×3 matrix I.

$I = \begin{bmatrix} 1 & 0 & 0 \\ 0 & 1 & 0 \\ 0 & 0 & 1 \end{bmatrix} \xrightarrow{C_3+2C_1} \begin{bmatrix} 1 & 0 & 2 \\ 0 & 1 & 0 \\ 0 & 0 & 1 \end{bmatrix} = E$,

$A = \begin{bmatrix} 2 & 3 & 7 \\ 1 & 0 & 0 \end{bmatrix} \xrightarrow{C_3+2C_1} \begin{bmatrix} 2 & 3 & 11 \\ 1 & 0 & 2 \end{bmatrix} = B$.

Then $B = AE$ by actual multiplication of A and E.

4. a. $P = E_{(3)+2(2)} \quad E_{(3)-1(1)} \quad E_{(2)-2(1)}.$

b. $P = E_{\frac{1}{17}(3)} \quad E_{(3)+1(1)} \quad E_{(2)-3(1)}.$

c. $P = E_{(3)-2(2)} \quad E_{(3)-(1)} \quad E_{(2)-2(1)}.$

5. a. $Q = E_{(1)-1(2)} \quad E_{(1)-1(3)} \quad E_{(1)-1(4)}$
$E_{(2)+1(3)} \quad E_{(3,4)}.$

b. $Q = E_{(1)-1(2)} \quad E_{(1)-2(3)} \quad E_{(1)-6(4)}$
$E_{(2)+7(3)} \quad E_{(2)+3(4)} \quad E_{(3)-2}.$

c. $Q = E_{(1)-3(2)} \quad E_{(1)+1(3)} \quad E_{(1)-1(4)}$
$E_{(2)+3(3)} \quad E_{(2)+3(4)} \quad E_{(3)-1(4)}.$

4.5 Review Questions and Project

1. a. i. 1;   **ii.** 3;   **iii.** 2.
2. i. 3;   **ii.** 0;   **iii.** 2.

Project

a. Rank $A = 8$. Ranks of augmented matrices formed by the six budget vectors, in order are 8, 9, 8, 9, 9, and 8.
b. First, Third, and Sixth budget vectors.
c. 8, 9, 8, 9, 9, and 8.
d. Corresponding to the first budget vector: $8, $7, $10, $9, $7, $7, $9, $11;
Corresponding to the third budget vector: $10, $7, $8, $7, $11, $7, $11, $9;
Corresponding to the sixth budget vector: $6, $9, $6, $7, $9, $7, $7, $9.
In other cases, it is not possible.
e. Nullity $A = 2$.

CHAPTER 5

5.1 Exercises

1. a. $\begin{bmatrix} 1 & 1 \\ 2 & 2 \end{bmatrix} \xrightarrow{R_2-2R_1} \begin{bmatrix} 1 & 1 \\ 0 & 0 \end{bmatrix}$, showing that A^{-1} does not exist.

b. $\begin{bmatrix} 1 & 0 \\ 0 & 1 \end{bmatrix} \xrightarrow{(-1)R_1} \begin{bmatrix} 1 & 0 \\ 0 & 1 \end{bmatrix}$, showing that A^{-1}, exists, and to find A^{-1}, we do as suggested in the remark preceding the examples and obtain $A^{-1} = \begin{bmatrix} -1 & 0 \\ 0 & 1 \end{bmatrix}$.

c. We can always check the existence of A^{-1} and also compute A^{-1} (in case it exists) simultaneously:

$$\left[\begin{array}{cc|cc} 2 & 3 & 1 & 0 \\ 1 & 1 & 0 & 1 \end{array}\right] \xrightarrow{R_1 \longleftrightarrow R_2}$$

$$\left[\begin{array}{cc|cc} 1 & 1 & 0 & 1 \\ 2 & 3 & 1 & 0 \end{array}\right] \xrightarrow{R_2 - 2R_1}$$

$$\left[\begin{array}{cc|cc} 1 & 1 & 0 & 1 \\ 0 & 1 & 1 & -2 \end{array}\right]$$

$$\xrightarrow{R_1 - R_2} \left[\begin{array}{cc|cc} 1 & 0 & -1 & 3 \\ 0 & 1 & 1 & -2 \end{array}\right], \text{ showing that}$$

A^{-1} exists and is $\begin{bmatrix} -1 & 3 \\ 1 & -2 \end{bmatrix}$.

d. A^{-1} does not exist.

e. $A^{-1} = -\frac{1}{5}\begin{bmatrix} -7 & 3 & 1 \\ 2 & 2 & -1 \\ 2 & -3 & -1 \end{bmatrix}$.

f. $A^{-1} = \frac{1}{3}\begin{bmatrix} -1 & 14 & 3 \\ 1 & -2 & 0 \\ 1 & 1 & 0 \end{bmatrix}$.

2. a. $\begin{bmatrix} -2 & 4 & 3 \\ -4 & 7 & 5 \\ -1 & 2 & 1 \end{bmatrix}$;

b. $\begin{bmatrix} -3 & 4 & 3 & -1 \\ -\frac{5}{3} & \frac{7}{3} & 2 & -1 \\ \frac{7}{3} & -\frac{8}{3} & -2 & 1 \\ 1 & -1 & -1 & 0 \end{bmatrix}$;

c. itself.

3. a. $x = A^{-1}\begin{bmatrix} 1 \\ 1 \\ 1 \end{bmatrix} = \begin{bmatrix} -2 & 4 & 3 \\ -4 & 7 & 5 \\ -1 & 2 & 1 \end{bmatrix}\begin{bmatrix} 1 \\ 1 \\ 1 \end{bmatrix} =$

$\begin{bmatrix} 5 \\ 8 \\ 2 \end{bmatrix}$; thus $x_1 = 5, x_2 = 8, x_3 = 2$.

b. $x = A^{-1}\begin{bmatrix} -1 \\ 1 \\ 0 \\ 1 \end{bmatrix} = \begin{bmatrix} 6 \\ 3 \\ -4 \\ -2 \end{bmatrix}$, if A is the

matrix in Exercise 2(b).

$$x = A^{-1}\begin{bmatrix} -1 \\ 1 \\ 0 \\ 1 \end{bmatrix} = \frac{1}{6}\begin{bmatrix} 3 \\ -7 \\ -1 \\ -7 \end{bmatrix}, \text{ if } A \text{ is the}$$

matrix in Exercise 2(c).

4. a. $x_1 = 1, x_2 = -2, x_3 = 4$.

b. $x_1 = 0, x_2 = 0, x_3 = 0$.

c. $x_1 = 8/3, x_2 = -13/3, x_3 = 1/3, x_4 = 3$.

5. $A^{-1} = A^2 - 2A$.

6. No.

8. a. $a \neq \frac{8}{3}$.

b. $a \neq 1$.

9. b. $x = A^{-1}\begin{bmatrix} 1 \\ 2 \\ 3 \\ 4 \end{bmatrix}$.

5.3 Exercises

1. Since the rank of the matrix is 1, we can choose $F = 1$ and $G = [1 \ 2 \ 3]$.

2. Since the rank of the matrix is 2, we can choose $F = I$, the 2×2 identity matrix, and $G = \begin{bmatrix} 1 & 2 \\ 3 & 4 \end{bmatrix}$.

3. $A = \begin{bmatrix} 1 & 2 & 4 \\ 3 & 0 & 1 \\ 1 & -4 & -7 \end{bmatrix} \xrightarrow[R_3 - 1R_1]{R_2 - 3R_1} \begin{bmatrix} 1 & 2 & 4 \\ 0 & -6 & -11 \\ 0 & -6 & -11 \end{bmatrix}$

$$\xrightarrow{R_3 + (-1)R_2} \begin{bmatrix} 1 & 2 & 4 \\ 0 & -6 & -11 \\ 0 & 0 & 0 \end{bmatrix} = \begin{bmatrix} G \\ 0 \end{bmatrix},$$

where $G = \begin{bmatrix} 1 & 2 & 4 \\ 0 & -6 & -11 \end{bmatrix}$.

To find F, we proceed as in Step 2 given after Theorem 5.3.3.

$$I = \begin{bmatrix} 1 & 0 & 0 \\ 0 & 1 & 0 \\ 0 & 0 & 1 \end{bmatrix} \xrightarrow{R_3 - 1R_2} \begin{bmatrix} 1 & 0 & 0 \\ 0 & 1 & 0 \\ 0 & -1 & 1 \end{bmatrix}$$

$$\xrightarrow[R_3 + 1R_1]{R_2 + 3R_1} \begin{bmatrix} 1 & 0 & 0 \\ 3 & 1 & 0 \\ 1 & -1 & 1 \end{bmatrix}.$$

Since rank $A = 2$, $F = \begin{bmatrix} 1 & 0 \\ 3 & 1 \\ 1 & -1 \end{bmatrix}$.

4. $A = \begin{bmatrix} 1 & 2 & 5 \\ 6 & 0 & 1 \end{bmatrix} \xrightarrow{R_2 - 6R_1} \begin{bmatrix} 1 & 2 & 5 \\ 0 & -12 & -29 \end{bmatrix} = G.$

We may find F by following the steps in the text. However, since rank $A = 2$, we may choose $F = I$, the 2×2 identity matrix, and $G = A$.

5. $F = A, G = I.$

5.4 Exercises

1. a. $A = \begin{bmatrix} 1 & 2 \\ 2 & 3 \end{bmatrix} \xrightarrow{R_2 - 2R_1} \begin{bmatrix} 1 & 2 \\ 0 & -1 \end{bmatrix} = U.$

To find L, we proceed as in Step 2 given after Theorem 5.4.1.

$I = \begin{bmatrix} 1 & 0 \\ 0 & 1 \end{bmatrix} \xrightarrow{R_2 + 2R_1} \begin{bmatrix} 1 & 0 \\ 2 & 1 \end{bmatrix} = L.$

b. Here, we need to perform an interchange of rows to obtain the desired decomposition. Interchanging of rows, if any, must be performed at the beginning. The result is $PA = LI$, where $P = \begin{bmatrix} 1 & 0 & 0 \\ 1 & 1 & 0 \\ 1 & 0 & 1 \end{bmatrix}$ and

$L = \begin{bmatrix} 1 & 0 & 0 \\ 1 & 2 & 0 \\ 1 & 0 & 3 \end{bmatrix}.$

c. $A = \begin{bmatrix} 2 & 3 & 4 \\ 0 & 1 & 3 \\ 1 & 1 & 1 \end{bmatrix} \xrightarrow{R_3 - \frac{1}{2}R_1}$

$\begin{bmatrix} 2 & 3 & 4 \\ 0 & 1 & 3 \\ 0 & -\frac{1}{2} & -1 \end{bmatrix} \xrightarrow{R_3 + \frac{1}{2}R_2}$

$\begin{bmatrix} 2 & 3 & 4 \\ 0 & 1 & 3 \\ 0 & 0 & \frac{1}{2} \end{bmatrix} = U.$ To find L, we perform elementary row operations on I as explained in Step 2.

$I = \begin{bmatrix} 1 & 0 & 0 \\ 0 & 1 & 0 \\ 0 & 0 & 1 \end{bmatrix} \xrightarrow{R_3 - \frac{1}{2}R_2} \begin{bmatrix} 1 & 0 & 0 \\ 0 & 1 & 0 \\ 0 & -\frac{1}{2} & 1 \end{bmatrix}$

$\xrightarrow{R_3 + \frac{1}{2}R_1} \begin{bmatrix} 1 & 0 & 0 \\ 0 & 1 & 0 \\ \frac{1}{2} & -\frac{1}{2} & 1 \end{bmatrix} = L.$

d. $A = LU, L = \begin{bmatrix} 1 & 0 & 0 \\ 1 & 1 & 0 \\ 4 & \frac{5}{2} & 1 \end{bmatrix}$, and $U =$

$\begin{bmatrix} 1 & 0 & 2 \\ 0 & 2 & -1 \\ 0 & 0 & -\frac{11}{2} \end{bmatrix}.$

e. As in part (b) we need to interchange rows. Here the result is $PA = LU$, where

$P = \begin{bmatrix} 0 & 1 & 0 \\ 1 & 0 & 0 \\ 0 & 0 & 1 \end{bmatrix}, \quad L = \begin{bmatrix} 1 & 0 & 0 \\ 0 & 1 & 0 \\ 1 & 1 & 1 \end{bmatrix},$

$U = \begin{bmatrix} 1 & 0 & 0 \\ 0 & 1 & 1 \\ 0 & 0 & 0 \end{bmatrix}.$

f. $L = \begin{bmatrix} 1 & 0 & 0 \\ 2 & 5 & 0 \\ 3 & 1 & 1 \end{bmatrix}, U = \begin{bmatrix} 1 & 2 & 3 & 0 \\ 0 & 1 & 0 & 2 \\ 0 & 0 & 1 & 5 \end{bmatrix}.$

g. $A = \begin{bmatrix} 1 & 1 & 1 & 7 \\ 6 & 12 & 18 & 0 \\ 3 & 9 & 6 & 10 \end{bmatrix} \xrightarrow[R_3 - 3R_1]{R_2 - 6R_1}$

$\begin{bmatrix} 1 & 1 & 1 & 7 \\ 0 & 6 & 12 & -42 \\ 0 & 6 & 3 & -11 \end{bmatrix} \xrightarrow{\frac{1}{6}R_2}$

$\begin{bmatrix} 1 & 1 & 1 & 7 \\ 0 & 1 & 2 & -7 \\ 0 & 6 & 3 & 11 \end{bmatrix} \xrightarrow{R_3 - 6R_2}$

$\begin{bmatrix} 1 & 1 & 1 & 7 \\ 0 & 1 & 2 & -7 \\ 0 & 0 & -9 & 31 \end{bmatrix} = U,$

$I = \begin{bmatrix} 1 & 0 & 0 \\ 0 & 1 & 0 \\ 0 & 0 & 1 \end{bmatrix} \xrightarrow{R_3 + 6R_2}$

$$\begin{bmatrix} 1 & 0 & 0 \\ 0 & 1 & 0 \\ 0 & 6 & 1 \end{bmatrix} \xrightarrow{6R_2} \begin{bmatrix} 1 & 0 & 0 \\ 0 & 6 & 0 \\ 0 & 6 & 1 \end{bmatrix}$$

$$\xrightarrow[R_3+3R_1]{R_2+6R_1} \begin{bmatrix} 1 & 0 & 0 \\ 6 & 6 & 0 \\ 3 & 6 & 1 \end{bmatrix} = L.$$

Observe that $A = LU$.

2. a. $A = \begin{bmatrix} 1 & 2 & 3 \\ 2 & 9 & 6 \\ 3 & 2 & 10 \end{bmatrix} \xrightarrow[R_2-2R_1]{R_3-3R_1}$

$$\begin{bmatrix} 1 & 2 & 3 \\ 0 & 5 & 0 \\ 0 & -4 & 1 \end{bmatrix} \xrightarrow{R_3+\frac{4}{5}R_2}$$

$$\begin{bmatrix} 1 & 2 & 3 \\ 0 & 5 & 0 \\ 0 & 0 & 1 \end{bmatrix} = U.$$

Next, we find L.

$$I = \begin{bmatrix} 1 & 0 & 0 \\ 0 & 1 & 0 \\ 0 & 0 & 1 \end{bmatrix} \xrightarrow{R_3-\frac{4}{5}R_2}$$

$$\begin{bmatrix} 1 & 0 & 0 \\ 0 & 1 & 0 \\ 0 & -\frac{4}{5} & 1 \end{bmatrix} \xrightarrow[R_3+3R_1]{R_2+2R_1}$$

$$\begin{bmatrix} 1 & 0 & 0 \\ 2 & 1 & 0 \\ 3 & -\frac{4}{5} & 1 \end{bmatrix} = L.$$

Thus $Ax = \begin{bmatrix} 0 \\ 10 \\ 7 \end{bmatrix}$ can be rewritten as

$LUx = \begin{bmatrix} 0 \\ 10 \\ 7 \end{bmatrix}$ or $Ly = \begin{bmatrix} 0 \\ 10 \\ 7 \end{bmatrix}$, where

$y = Ux$. But then $y = \begin{bmatrix} 0 \\ 10 \\ 15 \end{bmatrix}$, and from

$Ux = y$, we get $x = \begin{bmatrix} -49 \\ 2 \\ 15 \end{bmatrix}$.

b. $x_1 = 1, x_2 = 2, x_3 = -1$.

c. $x_1 = 1, x_2 = 0, x_3 = 1$.

3. Interchange the second and third rows to obtain the LU-decomposition. LU-decomposition of

the transformed matrix is $\begin{bmatrix} 1 & 0 & 0 & 0 \\ 1 & 1 & 0 & 0 \\ \frac{3}{2} & 0 & 1 & 0 \\ 2 & -3 & 10 & 1 \end{bmatrix}$

$\times \begin{bmatrix} 2 & 4 & 3 & 2 \\ 0 & 1 & -1 & -5 \\ 0 & 0 & \frac{1}{2} & -1 \\ 0 & 0 & 0 & 5 \end{bmatrix}$. The solution vector is

$$\begin{bmatrix} \frac{966}{5} \\ -\frac{404}{5} \\ -\frac{54}{5} \\ -\frac{77}{5} \end{bmatrix}.$$

5.6 Review Questions and Projects

2. a. $A = \begin{bmatrix} 1 & 1 \\ 0 & 1 \end{bmatrix}, B = \begin{bmatrix} -1 & 2 \\ 0 & -1 \end{bmatrix}$

b. $A = \begin{bmatrix} 1 & 2 \\ 2 & 4 \end{bmatrix}, B = \begin{bmatrix} 2 & 3 \\ 6 & 9 \end{bmatrix}$

3. No; yes; yes.

4. $m = n$.

5. $\begin{bmatrix} 2 & -1 & 0 \\ -1 & 2 & -1 \\ 0 & -1 & 1 \end{bmatrix}$, does not exist,

$$\begin{bmatrix} -1 & 1 & 0 & 0 \\ \frac{4}{3} & -1 & 0 & 0 \\ 0 & 0 & 6 & -5 \\ 0 & 0 & -7 & 6 \end{bmatrix}.$$

6. a. Full row rank. $F = I, G = A$.

b. $F = \begin{bmatrix} 1 & 0 \\ 2 & 1 \\ 1 & 2 \end{bmatrix}, G = \begin{bmatrix} 1 & -2 & 3 \\ 0 & -1 & -5 \end{bmatrix}$

c. $F = \begin{bmatrix} 1 & 0 \\ 2 & 1 \\ 3 & 1 \end{bmatrix}, G = \begin{bmatrix} 1 & 2 & -1 & 1 \\ 0 & 0 & -1 & -2 \end{bmatrix}$

d. Full row rank. $F = I, G = A$.

7. a.
$$\begin{bmatrix} 1 & 0 & 0 \\ -\frac{1}{2} & 1 & 0 \\ 0 & -\frac{2}{3} & 1 \end{bmatrix} \begin{bmatrix} -2 & 1 & 0 \\ 0 & -\frac{3}{2} & 1 \\ 0 & 0 & -\frac{4}{3} \end{bmatrix}$$

$$= \begin{bmatrix} 1 & 0 & 0 \\ -\frac{1}{2} & 1 & 0 \\ 0 & -\frac{2}{3} & 1 \end{bmatrix} \begin{bmatrix} -2 & 0 & 0 \\ 0 & -\frac{3}{2} & 0 \\ 0 & 0 & -\frac{4}{3} \end{bmatrix}$$

$$\times \begin{bmatrix} 1 & -\frac{1}{2} & 0 \\ 0 & 1 & -\frac{2}{3} \\ 0 & 0 & 1 \end{bmatrix}.$$

b.
$$\begin{bmatrix} 1 & 0 & 0 \\ 4 & 1 & 0 \\ 0 & -1 & 1 \end{bmatrix} \begin{bmatrix} 1 & 4 & 0 \\ 0 & -4 & 4 \\ 0 & 0 & 4 \end{bmatrix} =$$

$$\begin{bmatrix} 1 & 0 & 0 \\ 4 & 1 & 0 \\ 0 & -1 & 1 \end{bmatrix} \begin{bmatrix} 1 & 0 & 0 \\ 0 & -4 & 0 \\ 0 & 0 & 4 \end{bmatrix}$$

$$\times \begin{bmatrix} 1 & 4 & 0 \\ 0 & 1 & -1 \\ 0 & 0 & 1 \end{bmatrix}.$$

c.
$$\begin{bmatrix} 1 & 0 & 0 & 0 \\ 2 & 1 & 0 & 0 \\ 0 & -1 & 1 & 0 \\ 0 & 0 & 1 & 1 \end{bmatrix} \begin{bmatrix} 1 & 2 & 0 & 0 \\ 0 & -1 & 1 & 0 \\ 0 & 0 & 3 & 3 \\ 0 & 0 & 0 & 1 \end{bmatrix}$$

$$= \begin{bmatrix} 1 & 0 & 0 & 0 \\ 2 & 1 & 0 & 0 \\ 0 & -1 & 1 & 0 \\ 0 & 0 & 1 & 1 \end{bmatrix} \begin{bmatrix} 1 & 0 & 0 & 0 \\ 0 & -1 & 0 & 0 \\ 0 & 0 & 3 & 0 \\ 0 & 0 & 0 & 1 \end{bmatrix}$$

$$\times \begin{bmatrix} 1 & 2 & 0 & 0 \\ 0 & 1 & -1 & 0 \\ 0 & 0 & 1 & 1 \\ 0 & 0 & 0 & 1 \end{bmatrix}.$$

8. $\begin{bmatrix} 0 \\ \frac{1}{4} \\ -\frac{1}{2} \end{bmatrix}, \begin{bmatrix} 0 \\ \frac{1}{4} \\ -\frac{3}{4} \end{bmatrix}$

Projects

1. L is the lower triangular matrix with all entries equal to 1.

2. Scrambled message:

GONOOWASTNRTHE

CHAPTER 6

6.1 Exercises

1. 24.   **2.** 220.   **3.** 625. **4.** 1080.
5. 1.   **6.** $abcd$.   **7.** -3. **8.** -3.
9. -9.   **10.** 24 .   **11.** $abcd$.

6.2 Exercises

1. $\det \begin{bmatrix} 3 & 0 & 2 \\ -1 & 5 & 0 \\ 1 & 9 & 6 \end{bmatrix} = -\det \begin{bmatrix} 1 & 9 & 6 \\ -1 & 5 & 0 \\ 3 & 0 & 2 \end{bmatrix}$

(by $R_1 \longleftrightarrow R_3$) $= -\det \begin{bmatrix} 1 & 9 & 6 \\ 0 & 14 & 6 \\ 0 & -27 & -16 \end{bmatrix}$

(by $R_2 + 1R_1$ and $R_3 - 3R_1$) $=$

$-\det \begin{bmatrix} 1 & 9 & 6 \\ 0 & 14 & 6 \\ 0 & 0 & -\frac{62}{14} \end{bmatrix}$ (by $R_3 + \frac{27}{14}R_2$) $= 62$

(by multiplying the diagonal entries).

2. $\det \begin{bmatrix} 1 & 2 & -1 \\ 0 & 1 & 0 \\ 2 & 6 & 0 \end{bmatrix} = \det \begin{bmatrix} 1 & 2 & -1 \\ 0 & 1 & 0 \\ 0 & 2 & 2 \end{bmatrix}$ (by

$R_3 - 2R_1) = \det \begin{bmatrix} 1 & 2 & -1 \\ 0 & 1 & 0 \\ 0 & 0 & 2 \end{bmatrix}$ (by $R_3 -$

$2R_2) = 2.$

3. 30.

4. -2.

5. 0.

9. -72.

10. $\det(A - xI) = 0$ gives $x^3 - 5x^2 - 4x + 20 = 0$. Solving, we get $x = 5, 2, -2$.

6.3 Exercises

1. a. Since $\det A - 0$, A^{-1} does not exist.

b. $\det A = -1$. $A^{-1} = \frac{1}{-1} \begin{bmatrix} 1 & 0 \\ 0 & -1 \end{bmatrix}^T$

$= \begin{bmatrix} -1 & 0 \\ 0 & 1 \end{bmatrix}.$

c. $A^{-1} = -1 \begin{bmatrix} 1 & -1 \\ -3 & 2 \end{bmatrix}^T = \begin{bmatrix} -1 & 3 \\ 1 & -2 \end{bmatrix}$.

d. det $A = 0$. So A^{-1} does not exist.

e. det $A = -5$.

$$A^{-1} = -\frac{1}{5} \begin{bmatrix} -7 & 2 & 2 \\ 3 & 2 & -3 \\ 1 & -1 & -1 \end{bmatrix}^T$$

$$= \frac{1}{5} \begin{bmatrix} 7 & -3 & -1 \\ -2 & -2 & 1 \\ -2 & 3 & 1 \end{bmatrix}.$$

f. $A^{-1} = \frac{1}{3} \begin{bmatrix} -1 & 14 & 3 \\ 1 & -2 & 0 \\ 1 & 1 & 0 \end{bmatrix}$.

g. $\frac{1}{9} \begin{bmatrix} 1 & 4 & 2 & -3 \\ 8 & -4 & 7 & -6 \\ -5 & -2 & -1 & 6 \\ -3 & -3 & -6 & 9 \end{bmatrix}$.

h. $\begin{bmatrix} -1 & 3 & -\frac{3}{2} & -1 \\ -2 & 3 & -1 & -1 \\ -2 & 4 & -\frac{3}{2} & -2 \\ 1 & -2 & 1 & 1 \end{bmatrix}$.

6.4 Exercises

1. $x = 1$, $y = -4$, $z = 5$.

2. $x = -\frac{1}{3}$, $y = \frac{17}{9}$, $z = -\frac{34}{9}$.

3. $x_1 = -\frac{1}{15}$, $x_2 = -\frac{13}{15}$, $x_3 = \frac{1}{3}$.

4. $x = 1$, $y = 2$, $z = -2$.

5. $x = 2$, $y = -2$, $z = 1$.

6. $x_1 = 0$, $x_2 = 1$, $x_3 = 0$, $x_4 = 1$.

6.5 Review Questions and Projects

9. a. -15.

b. 3.

10. $x = -1, 3$.

11. a. -440.

b. $16i$.

12. a. $\begin{bmatrix} -2 & 1 \\ \frac{3}{2} & -\frac{1}{2} \end{bmatrix}$

b. does not exist.

c. $\begin{bmatrix} -\frac{13}{3} & \frac{7}{3} & \frac{2}{3} \\ \frac{5}{3} & -\frac{2}{3} & -\frac{1}{3} \\ \frac{1}{3} & -\frac{1}{3} & \frac{1}{3} \end{bmatrix}$.

13. $a_1 a_2' \neq a_2 a_1'$.

Projects

1. a. 2^{n-1}

b. $n!$

2. b. The inequalities are equalities if A is diagonal.

CHAPTER 7

7.2 Exercises

1. a. $2, 3$; $\begin{bmatrix} 2 \\ 1 \end{bmatrix}$, $\begin{bmatrix} 1 \\ 1 \end{bmatrix}$.

b. $2, 3, 1$; $\begin{bmatrix} 1 \\ 0 \\ 0 \end{bmatrix}$, $\begin{bmatrix} 1 \\ 1 \\ 0 \end{bmatrix}$, $\begin{bmatrix} 1 \\ -1 \\ 2 \end{bmatrix}$.

c. $1, 2, 7$; $\begin{bmatrix} 1 \\ 0 \\ 0 \end{bmatrix}$, $\begin{bmatrix} 3 \\ 1 \\ 0 \end{bmatrix}$, $\begin{bmatrix} 9 \\ -2 \\ 10 \end{bmatrix}$.

d. $1, 1, 2$; $\begin{bmatrix} 1 \\ 0 \\ 0 \end{bmatrix}$, $\begin{bmatrix} 0 \\ 1 \\ 1 \end{bmatrix}$, $\begin{bmatrix} 3 \\ 2 \\ 3 \end{bmatrix}$.

e. $3, 3, 3, -5$; $\begin{bmatrix} -1 \\ 0 \\ 0 \\ 1 \end{bmatrix}$, $\begin{bmatrix} -1 \\ 0 \\ 1 \\ 0 \end{bmatrix}$,

$\begin{bmatrix} -1 \\ 1 \\ 0 \\ 0 \end{bmatrix}$, $\begin{bmatrix} 1 \\ 1 \\ 1 \\ 1 \end{bmatrix}$.

2. a. $4, -1$; $\begin{bmatrix} 1 \\ 1 \end{bmatrix}$, $\begin{bmatrix} -2 \\ 3 \end{bmatrix}$.

b. $3, -1, -1$; $\begin{bmatrix} 2 \\ 1 \\ 2 \end{bmatrix}$, $\begin{bmatrix} -2 \\ 1 \\ 2 \end{bmatrix}$.

c. $1, 2, 3$; $\begin{bmatrix} 1 \\ 2 \\ -1 \end{bmatrix}, \begin{bmatrix} -1 \\ 1 \\ -1 \end{bmatrix}, \begin{bmatrix} 1 \\ 0 \\ 1 \end{bmatrix}.$

d. $0, 1, 1, 2$; $\begin{bmatrix} 1 \\ -1 \\ 0 \\ 0 \end{bmatrix}, \begin{bmatrix} 1 \\ 0 \\ 0 \\ 0 \end{bmatrix},$

$\begin{bmatrix} 0 \\ 0 \\ 1 \\ -1 \end{bmatrix}, \begin{bmatrix} 3 \\ 1 \\ 0 \\ 2 \end{bmatrix}.$

e. $4, 4, 1, 1$; $\begin{bmatrix} 1 \\ 0 \\ 0 \\ 0 \end{bmatrix}, \begin{bmatrix} -2 \\ -3 \\ 9 \\ 0 \end{bmatrix}, \begin{bmatrix} -5 \\ -3 \\ 0 \\ 9 \end{bmatrix}.$

f. $0, 1, 3$; $\begin{bmatrix} 1 \\ 1 \\ 1 \end{bmatrix}, \begin{bmatrix} 1 \\ 0 \\ -1 \end{bmatrix}, \begin{bmatrix} 1 \\ -2 \\ 1 \end{bmatrix}.$

3. If λ is an eigenvalue, then $Ax = \lambda x$, where $x \neq 0$. This implies that $A^2 x = \lambda A x = \lambda^2 x$, and so $\lambda x = \lambda^2 x$, since $A = A^2$ and $Ax = \lambda x$. Thus $(\lambda - \lambda^2)x = 0$, showing that $\lambda = \lambda^2$ because $x \neq 0$. This proves that $\lambda = 0, 1$.

4. If λ is an eigenvalue, then $Ax = \lambda x$, where $x \neq 0$. This implies that $A^2 x = \lambda A x = \lambda^2 x$, and so $\lambda^2 x = 0$, since $A^2 = 0$. But then $\lambda^2 = 0$, proving that $\lambda = 0$ is the only distinct eigenvalue.

9. The eigenvalues of A are $1, 2$, and 3; those of A^2 are $1, 4$, and 9; those of A^3 are $1, 8$, and 27; those of A^4 are $1, 16$, and 81, and those of A^{-1} are $1, \frac{1}{2}$, and $\frac{1}{3}$. The eigenvalues of A^n are the nth powers of the eigenvalues of A, and the eigenvalues of A^{-1} are the reciprocals of the eigenvalues of A.

7.4 Exercises

1. a. $C_A(x) = \det \begin{bmatrix} 1-x & 2 \\ 3 & 4-x \end{bmatrix} = (x-1)$
$\times (x-4) - 6 = x^2 - 5x - 2$. Thus by the Cayley-Hamilton theorem, $A^2 - 5A - 2I = 0$, and so $I = \frac{1}{2}A^2 - \frac{5}{2}A$. Thus

$A^{-1} = \frac{1}{2}A - \frac{5}{2}I = \frac{1}{2}\begin{bmatrix} 1 & 2 \\ 3 & 4 \end{bmatrix} - \frac{5}{2}\begin{bmatrix} 1 & 0 \\ 0 & 1 \end{bmatrix}$

$= \begin{bmatrix} -2 & 1 \\ \frac{3}{2} & -\frac{1}{2} \end{bmatrix}.$

b. $\begin{bmatrix} 1 & 1 \\ 1 & 0 \end{bmatrix}.$

c. $\frac{1}{12}\begin{bmatrix} -4 & -4 & 8 \\ -4 & 2 & 2 \\ 6 & 3 & -3 \end{bmatrix}.$

d. $C_A(x) = -(x^3 - 2x^2 + x)$. A is not invertible.

e. $C_A(x) = -(x^3 + x^2 - 5x - 8)$.

$A^{-1} = \frac{1}{8}\begin{bmatrix} 0 & 1 & 1 \\ 8 & 0 & 0 \\ 0 & 3 & -5 \end{bmatrix}.$

2. $C_A(x) = \det \begin{bmatrix} -3-x & -4 \\ 2 & 3-x \end{bmatrix} = x^2 - 1.$

Thus the eigenvalues are ± 1. Write $x^{100} = (x^2 - 1)q(x) + (a_0 + a_1 x)$ and put $x = 1, -1$ to obtain

$$1 = a_0 + a_1,$$
$$1 = a_0 - a_1.$$

Therefore $a_0 = 1$, $a_1 = 0$, and so $A^{100} = I$. For $A = \begin{bmatrix} 2 & -1 \\ 3 & -2 \end{bmatrix}$, we get similarly $A^{100} = I$.

3. $C_A(x) = \det \begin{bmatrix} -x-2 & 4 & 3 \\ 0 & -x & 0 \\ -1 & 5 & -x+2 \end{bmatrix}$

$= x(x^2 - 1)$. Thus the eigenvalues are $0, 1$, and -1. Write $x^{520} + 3x^{70} - 7 = x(x^2 - 1)q(x) + (a_0 + a_1 x + a_2 x^2)$ and put $x = 0, 1, -1$ in succession to obtain

$$a_0 = -7,$$
$$a_0 + a_1 + a_2 = -3,$$
$$a_0 - a_1 + a_2 = -3.$$

Therefore $a_1 = 0$, $a_2 = 4$, and so

$$A^{520} + 3A^{70} - 7I = -7I + 4A^2$$

$$= \begin{bmatrix} -3 & 28 & 0 \\ 0 & -7 & 0 \\ 0 & 24 & -3 \end{bmatrix}.$$

4. Eigenvalues: $2, -2, 2$; $A^{212} = aI + bA + cA^2$, where $a = -210(2)^{211}$, $b = 0$, $c = 53(2)^{211}$.

5. Eigenvalues: $1, 1, 1$;

$$A^{700} = \begin{bmatrix} 1401 & -1,955,800 & 2800 \\ 0 & 1 & 0 \\ -700 & 976,500 & -1399 \end{bmatrix}.$$

6. Eigenvalues: $6, -6$; $A^{15} = 6^{14}A$.

7. 0.

8. $a_k = \begin{cases} \dfrac{\alpha^k + \alpha^{-k}}{\alpha + \alpha^{-1}}, & \text{if } k \text{ is odd,} \\ \dfrac{\alpha^k - \alpha^{-k}}{\alpha + \alpha^{-1}}, & \text{if } k \text{ is even,} \end{cases}$

where $\alpha = \frac{\sqrt{5}+1}{2}$.

7.5 Exercises

1. By Property 7.5.2, the product of the eigen-

values is $\det \begin{bmatrix} 2 & 0 & 1 \\ 1 & 0 & -1 \\ 1 & 5 & 3 \end{bmatrix} = 15$. By Prop-

erty 7.5.1, the sum of the eigenvalues $= 5$.

2. By Property 7.5.1, the sum of the eigenvalues is $2 + a + 1$, the sum of the diagonal entries. Hence $a = 0$.

3. $a = \frac{3}{2}$.

4. $a = 1, b = -1$; $a = -1, b = 1$.

5. $a + b = 5$ and $ab - 5 = 1$, that is, $ab = 6$. So $a = 3, b = 2$; $a = 2, b = 3$.

7. a. $P = \begin{bmatrix} 1 & 1 \\ \frac{1}{3} & 1 \end{bmatrix}$, $P^{-1}AP = \begin{bmatrix} -1 & 0 \\ 0 & 1 \end{bmatrix}$.

b. $P = \begin{bmatrix} 1 & -1 & 1 \\ 2 & 1 & 0 \\ -1 & -1 & 1 \end{bmatrix}$,

$P^{-1}AP = \begin{bmatrix} 1 & 0 & 0 \\ 0 & 2 & 0 \\ 0 & 0 & 3 \end{bmatrix}.$

c. $P = \begin{bmatrix} 1 & 1 & 1 \\ 1 & 1 & 0 \\ 1 & 0 & 1 \end{bmatrix}$,

$P^{-1}AP = \begin{bmatrix} 1 & 0 & 0 \\ 0 & 3 & 0 \\ 0 & 0 & 2 \end{bmatrix}.$

d. $P = \begin{bmatrix} i & -i & 0 \\ 0 & 0 & 1 \\ 1 & 1 & 1 \end{bmatrix}$,

$P^{-1}AP = \begin{bmatrix} 1+i & 0 & 0 \\ 0 & 1-i & 0 \\ 0 & 0 & 1 \end{bmatrix}.$

8. $\text{Eig}(A) = 7, 3, 3$; $x = \begin{bmatrix} 1 \\ 0 \\ 1 \end{bmatrix}$,

$y = \begin{bmatrix} 1 \\ 0 \\ -1 \end{bmatrix}, z = \begin{bmatrix} -1 \\ 4 \\ -1 \end{bmatrix}.$

9. $\text{Eig}(A) = 3, 3, 3, -5$; $x = \begin{bmatrix} -1 \\ 0 \\ 0 \\ 1 \end{bmatrix}$,

$y = \begin{bmatrix} -1 \\ 0 \\ 1 \\ 0 \end{bmatrix}, z = \begin{bmatrix} -1 \\ 1 \\ 0 \\ 0 \end{bmatrix}, t = \begin{bmatrix} 1 \\ 1 \\ 1 \\ 1 \end{bmatrix}.$

13. $\det(A) = -85x + 15$. Since the product of the eigenvalues is 355, we get $-85x + 15 = 355$. Thus $x = 4$.

7.7 Review Questions and Projects

2. Yes; No; No; Yes.

3. $\begin{bmatrix} 3 \\ 4 \end{bmatrix}$ is an eigenvector.

6. a. $\left\{ \begin{bmatrix} 1 \\ -2 \\ 0 \end{bmatrix}, \begin{bmatrix} 0 \\ 0 \\ 1 \end{bmatrix}, \begin{bmatrix} 1 \\ 2 \\ 0 \end{bmatrix} \right\}.$

b. $\left\{ \begin{bmatrix} -1 \\ -1 \\ 1 \end{bmatrix}, \begin{bmatrix} -3 \\ -1 \\ 1 \end{bmatrix} \right\}.$

c. $\left\{ \begin{bmatrix} -1 \\ 1 \\ 0 \end{bmatrix}, \begin{bmatrix} -1 \\ 0 \\ 1 \end{bmatrix}, \begin{bmatrix} 2 \\ -1 \\ 1 \end{bmatrix} \right\}$.

7. $1, 1, 2^{-100}, 3^{-100}$.

8. (a) and (c) are diagonalizable.

9. i. $\begin{bmatrix} -2 & -1 & 0 \\ 1 & 0 & 1 \\ 1 & 1 & 0 \end{bmatrix}$.

ii. $\begin{bmatrix} 1 & 1 & 1 \\ 0 & 5 & 3 \\ 0 & 0 & 3 \end{bmatrix}$.

10. i. $\begin{bmatrix} -1 & \frac{1}{4} & \frac{1}{4} \\ 0 & \frac{1}{4} & \frac{1}{4} \\ 0 & 0 & \frac{1}{2} \end{bmatrix}$.

ii. $\begin{bmatrix} -2 & 1 \\ \frac{3}{2} & \frac{-1}{2} \end{bmatrix}$.

11. $(1 + 5^9 + 5^{49})A$.

Projects

1. a. Eigenvalues: 0.7, 1.2.

Eigenvectors: $v_1 = \begin{bmatrix} -0.9992 \\ -0.0400 \end{bmatrix}$, $v_2 = \begin{bmatrix} -0.0001 \\ -1 \end{bmatrix}$.

b. $\begin{pmatrix} S_0 \\ F_0 \end{pmatrix} = (-1.0004)v_1 + (-4.9600)v_2$.

c. $P_5 = 10^5 \begin{bmatrix} 0.0169 \\ 1.2349 \end{bmatrix}$, $P_{10} = 10^5 \begin{bmatrix} 0.0031 \\ 3.0712 \end{bmatrix}$,

$P_{13} = 10^5 \begin{bmatrix} 0.0014 \\ 5.3068 \end{bmatrix}$, $P_{14} = 10^5 \begin{bmatrix} 0.0012 \\ 6.3682 \end{bmatrix}$,

$P_{15} = 10^5 \begin{bmatrix} 0.0011 \\ 7.6418 \end{bmatrix}$, $P_{20} = 10^6 \begin{bmatrix} 0.0002 \\ 1.9015 \end{bmatrix}$.

d. Neither type would be extinct.

2. a. Rank $A^{25} = 2$, Rank $A^{26} = 1$.

b. Eigenvalue $= 1$, Eigenvector $= \begin{bmatrix} 0.3846 \\ 0.1538 \\ 0.4615 \end{bmatrix}$

c. Each column of the limit matrix is the equilibrium vector of A.

d. $\begin{bmatrix} 19/30 \\ 143/600 \\ 229/600 \end{bmatrix}, \begin{bmatrix} 5/13 \\ 2327/15125 \\ 6983/15130 \end{bmatrix}$.

CHAPTER 8

8.1 Exercises

1. a. No.   **b.** No.   **c.** No.   **d.** Yes.
   e. Yes.

2. $x + 2y + 3 = 0$ and $x + 1 = 0$. Therefore $x = -1, y = -1$.

3. $4x + 3y = 0$ and $-2x + 1 + 6y = 0$. Thus $x = \frac{1}{10}, y = -\frac{2}{15}$.

4. $a + c + 2d = 0, a + b - c = 0, a - 2b - c = 0$. These equations yield $a = -d, b = 0, c = -d$. Therefore for any value in the set $\{-d, 0, -d, d\}$, where d is arbitrary, the vectors are mutually orthogonal.

5. $2a + 1 + 1 = 0, 2b + 3 - c = 0$, and $ab + 3 - c = 0$. These equations yield $a = -1, b = 0, c = 3$.

6. $2a + b - 1 = 0, a + 3 - c = 0, 2 + 3b + c = 0$. Solving, we get $a = \frac{8}{5}, b = \frac{-11}{5}, c = \frac{23}{5}$.

7. a. $\frac{1}{\sqrt{6}} \begin{bmatrix} 1 \\ 2 \\ 1 \end{bmatrix}, \frac{1}{\sqrt{3}} \begin{bmatrix} 1 \\ -1 \\ 1 \end{bmatrix}, \frac{1}{\sqrt{2}} \begin{bmatrix} 1 \\ 0 \\ -1 \end{bmatrix}$.

b. $\frac{1}{\sqrt{2}} \begin{bmatrix} 1 \\ 1 \\ 0 \end{bmatrix}, \frac{1}{\sqrt{6}} \begin{bmatrix} -1 \\ 1 \\ 2 \end{bmatrix}$.

c. $\begin{bmatrix} 0 \\ 0 \\ 1 \\ 0 \end{bmatrix}, \frac{1}{\sqrt{2}} \begin{bmatrix} 1 \\ 0 \\ 0 \\ 1 \end{bmatrix}, \begin{bmatrix} 0 \\ 1 \\ 0 \\ 0 \end{bmatrix}$.

d. $\frac{1}{\sqrt{6}} \begin{bmatrix} 1 & 0 & 1 & 2 \end{bmatrix}, \frac{1}{\sqrt{3}} \begin{bmatrix} 1 & 1 & -1 & 0 \end{bmatrix}$,
   $\frac{1}{\sqrt{6}} \begin{bmatrix} 1 & -2 & -1 & 0 \end{bmatrix}$.

8. Let $\alpha_1 u_1 + \cdots + \alpha_n u_n = 0$, where $u_1, \ldots, u_n$ are orthogonal (column) vectors and $\alpha_i \in \mathbf{R}$. Multiplying both sides by u_1^T, we get

$\alpha_1 u_1 u_1^T = 0$, and so $\alpha_1 = 0$, since $u_1 u_1^T \neq 0$. (Note: $u_1 u_1^T = 0 \iff u_1 = 0$.) Similarly, $\alpha_2 = 0 = \cdots = \alpha_n$.

10. Write $U = [\, C_1 \ C_2 \ \cdots \ C_n \,]$, where $C_1, C_2,$

$\ldots, C_n$ are columns of U and $x = \begin{bmatrix} x_1 \\ x_2 \\ \vdots \\ x_n \end{bmatrix}$,

where $x_i \in \mathbf{R}$. Then $Ux = C_1 x_1 + C_2 x_2 + \cdots + C_n x_n$, and so

$$\begin{aligned} \|Ux\|^2 &= (C_1 x_1 + \cdots + C_n x_n) \\ &\quad \times (C_1 x_1 + \cdots + C_n x_n)^T \\ &= (C_1 x_1 + \cdots + C_n x_n) \\ &\quad \times (C_1^T x_1 + \cdots + C_n^T x_n) \\ &= x_1^2 + \cdots + x_n^2, \end{aligned}$$

by Exercise 9.

8.2 Exercises

1. a. $\frac{1}{\sqrt{2}} \begin{bmatrix} 1 \\ 0 \\ 1 \end{bmatrix}$, $\frac{1}{\sqrt{3}} \begin{bmatrix} -1 \\ 1 \\ 1 \end{bmatrix}$, $\frac{1}{\sqrt{6}} \begin{bmatrix} -1 \\ -2 \\ 1 \end{bmatrix}$;

eigenvalues: $-6, -6, 6$.

b. $\begin{bmatrix} 1 \\ 0 \\ 0 \\ 0 \end{bmatrix}$, $\frac{1}{\sqrt{2}} \begin{bmatrix} 0 \\ 1 \\ 0 \\ 1 \end{bmatrix}$, $\frac{1}{\sqrt{6}} \begin{bmatrix} 0 \\ -1 \\ 2 \\ 1 \end{bmatrix}$, $\frac{1}{\sqrt{3}} \begin{bmatrix} 0 \\ -1 \\ -1 \\ 1 \end{bmatrix}$;

eigenvalues: $-3, 2, 3, 3$.

c. $\frac{1}{\sqrt{2}} \begin{bmatrix} 0 \\ -1 \\ 0 \\ 1 \end{bmatrix}$, $\frac{1}{\sqrt{6}} \begin{bmatrix} 0 \\ -1 \\ 2 \\ -1 \end{bmatrix}$, $\begin{bmatrix} 1 \\ 0 \\ 0 \\ 0 \end{bmatrix}$, $\frac{1}{\sqrt{3}} \begin{bmatrix} 0 \\ 1 \\ 1 \\ 1 \end{bmatrix}$;

eigenvalues: $5, 5, 6, 8$.

d. $\frac{1}{\sqrt{2}} \begin{bmatrix} -1 \\ 1 \\ 0 \end{bmatrix}$, $\begin{bmatrix} 0 \\ 0 \\ 1 \end{bmatrix}$, $\frac{1}{\sqrt{2}} \begin{bmatrix} 1 \\ 1 \\ 0 \end{bmatrix}$;

eigenvalues: $0, 2, 2$.

2. a. $P = [\, u_1 \ u_2 \ u_3 \,]; u_1 = \frac{1}{\sqrt{2}} \begin{bmatrix} -1 \\ 1 \\ 0 \end{bmatrix}$,

$u_2 = \frac{1}{\sqrt{6}} \begin{bmatrix} -1 \\ -1 \\ 2 \end{bmatrix}$, $u_3 = \frac{1}{\sqrt{3}} \begin{bmatrix} 1 \\ 1 \\ 1 \end{bmatrix}$,

$P^T A P = \begin{bmatrix} 0 & 0 & 0 \\ 0 & 0 & 0 \\ 0 & 0 & 3 \end{bmatrix}$.

b. $A^{100} = P \begin{bmatrix} 0 & 0 & 0 \\ 0 & 0 & 0 \\ 0 & 0 & 3^{100} \end{bmatrix} P^T$,

$\sqrt{A} = P \begin{bmatrix} 0 & 0 & 0 \\ 0 & 0 & 0 \\ 0 & 0 & \sqrt{3} \end{bmatrix} P^T$.

3. i. Eigenvalues: $-4, 2, 2$, $\quad \frac{1}{\sqrt{3}} \begin{bmatrix} -1 \\ 1 \\ 1 \end{bmatrix}$,

$\frac{1}{\sqrt{2}} \begin{bmatrix} 1 \\ 0 \\ 1 \end{bmatrix}$, $\frac{1}{\sqrt{6}} \begin{bmatrix} 1 \\ 2 \\ -1 \end{bmatrix}$.

ii. Eigenvalues: $-1, 3$, $\quad \frac{1}{\sqrt{2}} \begin{bmatrix} 1 \\ -1 \end{bmatrix}$,

$\frac{1}{\sqrt{2}} \begin{bmatrix} 1 \\ 1 \end{bmatrix}$.

iii. Eigenvalues: $-4, 2, 2$, $\quad \frac{1}{\sqrt{2}} \begin{bmatrix} 1 \\ 1 \\ 0 \end{bmatrix}$,

$\frac{1}{\sqrt{2}} \begin{bmatrix} -1 \\ 1 \\ 0 \end{bmatrix}$, $\begin{bmatrix} 0 \\ 0 \\ 1 \end{bmatrix}$.

iv. Eigenvalues: $-2, 0, 6$, $\quad \frac{1}{\sqrt{2}} \begin{bmatrix} -1 \\ 0 \\ 1 \end{bmatrix}$,

$\frac{1}{\sqrt{3}} \begin{bmatrix} 1 \\ -1 \\ 1 \end{bmatrix}$, $\frac{1}{\sqrt{6}} \begin{bmatrix} 1 \\ 2 \\ 1 \end{bmatrix}$.

4. a. i. $0, 10$;
   ii. $8, 2, 1$.

b. i. $\left\{ \frac{1}{\sqrt{5}} \begin{bmatrix} -2 \\ 1 \end{bmatrix}, \frac{1}{\sqrt{5}} \begin{bmatrix} 1 \\ 2 \end{bmatrix} \right\}$

ii. $\left\{ \frac{1}{\sqrt{2}} \begin{bmatrix} 1 \\ 1 \\ 0 \end{bmatrix}, \frac{1}{\sqrt{2}} \begin{bmatrix} -1 \\ 1 \\ 0 \end{bmatrix}, \begin{bmatrix} 0 \\ 0 \\ 1 \end{bmatrix} \right\}$.

5. a. $V = [v_1 \ v_2]$, where $v_1 = \frac{1}{\sqrt{2}} \begin{bmatrix} 1 \\ 1 \end{bmatrix}$,

$v_2 = \frac{1}{\sqrt{2}} \begin{bmatrix} 1 \\ -1 \end{bmatrix}$, $S = \begin{bmatrix} s_1 & 0 \\ 0 & s_2 \end{bmatrix}$, where

$s_1^2 = 18$, $s_2^2 = 8$, $U = [u_1 \ u_2]$, where $u_1 = \frac{1}{s_1} A v_1$, $u_2 = \frac{1}{s_2} A v_2$.

b. $V = [v_1 \ v_2 \ v_3]$, where $v_1 = \begin{bmatrix} \frac{2}{\sqrt{5}} \\ \frac{1}{\sqrt{5}} \\ 0 \end{bmatrix}$,

$v_2 = \begin{bmatrix} \frac{-2}{\sqrt{45}} \\ \frac{4}{\sqrt{45}} \\ \frac{5}{\sqrt{45}} \end{bmatrix}$, $v_3 = \frac{1}{3} \begin{bmatrix} -1 \\ 2 \\ -2 \end{bmatrix}$;

$S = \begin{bmatrix} s_1 & 0 & 0 \\ 0 & s_2 & 0 \\ 0 & 0 & s_3 \end{bmatrix}$, where $s_1^2 = 81$, $s_2^2 = 81$,

$s_3^2 = 0$; $U = [u_1 \ u_2 \ u_3]$, where $u_1 = \frac{1}{s_1} A v_1$, $u_2 = \frac{1}{s_2} A v_2$, $u_3 = \frac{1}{s_3} A v_3$.

c. $U = \begin{bmatrix} 0 & 1 & 0 \\ \frac{1}{\sqrt{2}} & 0 & -\frac{1}{\sqrt{2}} \\ \frac{1}{\sqrt{2}} & 0 & \frac{1}{\sqrt{2}} \end{bmatrix}$,

$S = \begin{bmatrix} 3 & 0 & 0 \\ 0 & 2 & 0 \\ 0 & 0 & 1 \end{bmatrix}$,

$V = \begin{bmatrix} 0 & 1 & 0 \\ \frac{1}{\sqrt{2}} & 0 & \frac{1}{\sqrt{2}} \\ \frac{1}{\sqrt{2}} & 0 & \frac{1}{\sqrt{2}} \end{bmatrix}$.

8.3 Exercises

1. a. Hyperbola.   **b.** Hyperbola.

3. a. No.   **b.** Yes.   **c.** No.

8.4 Exercises

1. a. $x = \begin{bmatrix} \frac{1}{2} \\ \frac{1}{2} \end{bmatrix}$.

b. $x = \begin{bmatrix} -\frac{1}{37} \\ \frac{21}{37} \end{bmatrix}$.

c. $x = \begin{bmatrix} -\frac{85}{33} \\ \frac{130}{33} \\ -\frac{40}{33} \end{bmatrix}$.

d. $x = \begin{bmatrix} -\frac{23}{6} \\ \frac{7}{3} \\ \frac{1}{6} \end{bmatrix}$.

2. We have the following system of equations:

$$a + b = 1,$$
$$4b = 1,$$
$$a + b = 1,$$
$$a + 4b = 1.$$

A least-squares solution of $Ax = B$ where

$$A = \begin{bmatrix} 1 & 1 \\ 0 & 4 \\ 1 & 1 \\ 1 & 4 \end{bmatrix}, x = \begin{bmatrix} a \\ b \end{bmatrix}, B = \begin{bmatrix} 1 \\ 1 \\ 1 \\ 1 \end{bmatrix}$$ is

given by $a = \frac{7}{11}, b = \frac{2}{11}$.

3. We have the following system of equations:

$$a + 2b = 1,$$
$$2a + 3b = 1,$$
$$3a + 4b = 1.$$

A least-squares solution of $Ax = B$, where

$$A = \begin{bmatrix} 1 & 2 \\ 2 & 3 \\ 3 & 4 \end{bmatrix}, x = \begin{bmatrix} a \\ b \end{bmatrix}, \text{ and } B = \begin{bmatrix} 1 \\ 1 \\ 1 \end{bmatrix},$$

is given by $a = -1, b = 1$.

4. $\begin{bmatrix} \frac{1}{3} \\ -\frac{1}{3} \\ \frac{1}{3} \end{bmatrix}$.

5. i. $\sqrt{3}$

ii. $\sqrt{\frac{2}{3}}$

6. Every month the approximate revenue is $10,200.

7. The approximate number of absentees in the eighteenth week is 25, and no absentees are predicted after 5 months (Note that the curve of degree 1 is not a good fit. Perhaps a curve of degree 2 would give a better prediction.)

8.5 Exercises

Exercise 1(a) of Section 8.4. The Moore-Penrose Inverse $= \begin{bmatrix} \frac{1}{2} & -\frac{1}{2} & \frac{1}{2} \\ 0 & \frac{1}{2} & 0 \end{bmatrix}$, $X_1 = \frac{1}{2}$, $X_2 = \frac{1}{2}$, Exercise 1(b) of Section 8.4. The Moore-Penrose Inverse $= \frac{1}{185} \begin{bmatrix} 35 & -67 & 24 \\ 5 & 38 & 14 \end{bmatrix}$, $X_1 = \frac{-1}{37}$, $X_2 = \frac{21}{37}$.

Exercise 4 of Section 8.4. The Moore-Penrose Inverse $= \begin{bmatrix} -\frac{1}{6} & \frac{-1}{4} & 0 \\ \frac{-1}{6} & \frac{1}{2} & 0 \\ \frac{1}{6} & 0 & 0 \end{bmatrix}$, $X_1 = \frac{1}{3}$, $X_2 = -\frac{1}{3}$, $X_3 = \frac{1}{3}$.

8.7 Review Questions and Projects

1. a. $\sqrt{3}, \sqrt{2}, \sqrt{6}$.

  b. $\sqrt{2}, \sqrt{2}, 1$.

2. a. $\frac{1}{\sqrt{30}} \begin{bmatrix} 1 \\ 2 \\ 3 \\ 4 \end{bmatrix}$.

  b. $e_1 = \frac{1}{\sqrt{2}} \begin{bmatrix} 0 \\ 1 \\ 0 \\ 1 \end{bmatrix}$, $e_2 = \frac{1}{\sqrt{10}} \begin{bmatrix} -2 \\ -1 \\ 2 \\ 1 \end{bmatrix}$.

  c. $e_1 = \frac{1}{\sqrt{6}} \begin{bmatrix} 1 \\ 2 \\ -2 \end{bmatrix}$, $e_2 = \frac{1}{\sqrt{21}} \begin{bmatrix} 4 \\ -1 \\ 2 \end{bmatrix}$,

  $e_3 = \frac{1}{\sqrt{14}} \begin{bmatrix} -1 \\ 2 \\ 3 \end{bmatrix}$.

3. $\begin{bmatrix} \frac{1}{\sqrt{2}} & -\frac{1}{\sqrt{2}} \\ \frac{1}{\sqrt{2}} & \frac{1}{\sqrt{2}} \end{bmatrix}$, $\begin{bmatrix} \frac{1}{\sqrt{2}} & -\frac{1}{\sqrt{2}} & 0 \\ \frac{1}{\sqrt{2}} & \frac{1}{\sqrt{2}} & 0 \\ 0 & 0 & 1 \end{bmatrix}$.

4. a. $P = \frac{1}{\sqrt{5}} \begin{bmatrix} 2 & 1 \\ 1 & -2 \end{bmatrix}$.

  b. $P = \begin{bmatrix} \frac{1}{\sqrt{2}} & \frac{1}{\sqrt{6}} & \frac{1}{\sqrt{3}} \\ \frac{-1}{\sqrt{2}} & \frac{1}{\sqrt{6}} & \frac{1}{\sqrt{3}} \\ 0 & \frac{-2}{\sqrt{6}} & \frac{1}{\sqrt{3}} \end{bmatrix}$.

5. a. $a \geqslant 5$.

  b. $-2 \leqslant b \leqslant 2$.

6. a. $A = \begin{bmatrix} 1 & \frac{-1}{2} \\ \frac{1}{2} & 1 \end{bmatrix}$ is positive definite.

  b. $A = \begin{bmatrix} 2 & -2 & 1 \\ -2 & 2 & 2 \\ 1 & 2 & 2 \end{bmatrix}$ is neither positive definite nor negative definite.

  c. $A = \begin{bmatrix} -2 & 0 & 1 \\ 0 & -1 & 0 \\ 1 & 0 & -2 \end{bmatrix}$ is negative definite

7. a. $x_1 = \frac{9}{7}$, $x_2 = \frac{6}{7}$.

  b. $a = \frac{-8}{39}$, $b = \frac{10}{13}$.

8. a. Moore-Penrose inverse $= \frac{1}{14} \begin{bmatrix} 5 & 4 & 1 \\ 8 & -2 & -4 \end{bmatrix}$, and the least-squares solution of minimum norm is $x_1 = \frac{9}{7}$, $x_2 = \frac{6}{7}$.

  b. Moore-Penrose inverse $= \frac{1}{39} \begin{bmatrix} -21 & -5 & 11 & 6 & 1 \\ 30 & 9 & -12 & -3 & 6 \end{bmatrix}$, and the least-squares solution of minimum norm is $a = \frac{-8}{39}$, $b = \frac{10}{13}$.

Projects

1. a. $44.34, $44.11, $44.88.

  b. $42.37, $41.07, $39.59.

  c. $53.79, $61.95, $81.11.

CHAPTER 9

9.1 Exercises

1. a. No.   **b.** No.   **c.** No.   **d.** Yes.
e. Yes.

2. $W_1 \cap W_2 = \left\{ \begin{bmatrix} a & 0 \\ 0 & c \end{bmatrix} \mid a, c \in F \right\};$

$W_1 + W_2 = V.$

3. If $x, y \in W$, then $(x+y)^T u = x^T u + y^T u = 0$, and so $x + y \in W$. Similarly, $\alpha x \in W$, where $\alpha \in F$.

4. $W \cap W^\perp = (0).$

9.2 Exercises

1. $W = \{\alpha_0 + \alpha_1 x \mid \alpha_0, \alpha_1 \in R\}$, and so W is generated by $\{1, x\}$. Furthermore, $\{1, x\}$ is a linearly independent set.

2. a. Suppose

$$\alpha x + \beta e^x + \gamma e^{2x} = 0. \qquad (1)$$

Differentiating twice, we get

$$\alpha + \beta e^x + 2\gamma e^{2x} = 0, \qquad (2)$$
$$\beta e^x + 4\gamma e^{2x} = 0. \qquad (3)$$

Solving equations (1)–(3) we get $\alpha = 0 = \beta = \gamma$.

9.3 Exercises

1. a. Here, $f(x_1, x_2) = (2x_1 + 3x_2 + 1, -x_1 + x_2)$; $f(x_1', x_2') = (2x_1' + 3x_2' + 1, -x_1' + x_2')$.
So $f((x_1, x_2) + (x_1', x_2')) = f(x_1 + x_1', x_2 + x_2') = (2x_1 + 2x_1' + 3x_2 + 3x_2' + 1, -x_1 - x_1' + x_2 + x_2')$.
Since $f(x_1, x_2) + f(x_1', x_2') = (2x_1 + 3x_2 + 1 + 2x_1' + 3x_2' + 1, -x_1 + x_2 - x_1' + x_2')$, we obtain $f((x_1, x_2) + (x_1', x_2')) \neq f(x_1, x_2) + f(x_1', x_2')$. This proves that f is not linear.
b. Not linear.
c. Not linear.

d. Here, $f(x_1, x_2) = (x_1 - x_2, 0)$; $f(x_1', x_2') = (x_1' - x_2', 0)$.
So $f((x_1, x_2) + (x_1', x_2')) = f(x_1 + x_1', x_2 + x_2') = (x_1 + x_1' - x_2 - x_2', 0) = (x_1 - x_2 + x_1' - x_2', 0)$.
Furthermore, $f(x_1, x_2) + f(x_1', x_2') = (x_1 - x_2 + x_1' - x_2', 0)$. Thus

$$f((x_1, x_2) + (x_1', x_2')) = f(x_1, x_2) + f(x_1', x_2'). \qquad (1)$$

Also,

$$f(\alpha(x_1, x_2))$$
$$= f((\alpha x_1, \alpha x_2)) = (\alpha x_1 - \alpha x_2, 0)$$
$$= \alpha(x_1 - x_2, 0) = \alpha f(x_1, x_2). \qquad (2)$$

Therefore by equations (1) and (2), it follows that f is linear.

4. a.iv. $f(a, b) = f(a, 0) + f(0, b)$
$= af(1, 0) + bf(0, 1) = (2a - b, 3a + b).$

b.iv. First, we need to express (a, b) as a linear combination of $(1, 2)$ and $(2, -3)$. So let $(a, b) = \alpha(1, 2) + \beta(2, -3)$. This yields

$$\alpha + 2\beta = a \qquad \text{and} \qquad 2\alpha - 3\beta = b.$$

By solving, we obtain $\alpha = \frac{1}{7}(3a + 2b)$, $\beta = \frac{1}{7}(2a - b)$. Therefore $f(a, b) = \alpha f(1, 2) + \beta f(2, -3) = \frac{1}{7}(3a + 2b)(1, -1) + \frac{1}{7}(2a - b)(4, 1) = \frac{1}{7}(11a - 2b, -a - 3b).$

8. a.

$$X = \begin{bmatrix} x \\ y \end{bmatrix} \longrightarrow \begin{bmatrix} 2x \\ 5y \end{bmatrix} = T(X) \qquad (1)$$

$$X' = \begin{bmatrix} x' \\ y' \end{bmatrix} \longrightarrow \begin{bmatrix} 2x' \\ 5y' \end{bmatrix} = T(X'), \qquad (2)$$

$$X + X' = \begin{bmatrix} x + x' \\ y + y' \end{bmatrix} \longrightarrow \begin{bmatrix} 2x + 2x' \\ 5y + 5y' \end{bmatrix}$$
$$= T(X + X'), \qquad (3)$$

$$\alpha X = \begin{bmatrix} \alpha x \\ \alpha y \end{bmatrix} \longrightarrow \begin{bmatrix} 2\alpha x \\ 5\alpha y \end{bmatrix} = T(\alpha X). \qquad (4)$$

Is $T(X + X') = T(X) + T(X')$? Yes, this follows from equations (1)–(3).

Is $T(\alpha X) = \alpha T(X)$? Yes, this follows from equations (1) and (4).

Hence T is a linear mapping.

b.

$$X = (x \ \ y) \longrightarrow (x^2 \ \ y) = T(X), \tag{1}$$

$$X' = (x' \ \ y') \longrightarrow (x'^2 \ \ y') = T(X'), \tag{2}$$

$$X + X' = (x + x' \ \ y + y') \longrightarrow$$
$$((x + x')^2 \ \ y + y') = T(X + X'), \tag{3}$$

$$\alpha X = (\alpha x \ \ \alpha y) \longrightarrow$$
$$(\alpha^2 x^2 \ \ \alpha y) = T(\alpha X). \tag{4}$$

Is $T(X + X') = T(X) + T(X')$? No, because from equation (3), $T(X + X') = (x^2 + 2xx' + x'^2, \ y + y')$, and from equations (1) and (2), $T(X) + T(X') = (x^2 + x'^2, \ y + y')$. Hence T is not a linear mapping. Once any one of the two axioms for a function to be a linear mapping is violated, we may stop and declare that T is not a linear mapping.

c. Yes. **d.** Yes.

9. a.

$$X = (x \ \ y \ \ z) \longrightarrow (0 \ \ 0) = T(X), \tag{1}$$

$$X' = (x' \ \ y' \ \ z') \longrightarrow (0 \ \ 0) = T(X'), \tag{2}$$

$$X + X' = (x + x' \ \ y + y' \ \ z + z') \longrightarrow$$
$$(0 \ \ 0) = T(X + X'), \tag{3}$$

$$\alpha X = (\alpha x \ \ \alpha y \ \ \alpha z) \longrightarrow$$
$$(0 \ \ 0) = T(\alpha X). \tag{4}$$

From equations (1)–(3), $T(X + X') = T(X) + T(X')$.

From equations (1) and (4), $T(\alpha X) = \alpha T(X)$.

Hence T is a linear mapping

b. No.

c.

$$X = (x \ \ y \ \ z) \longrightarrow (x \ \ x + y + z)$$
$$= T(X), \tag{1}$$

$$X' = (x' \ \ y' \ \ z') \longrightarrow$$
$$(x' \ \ x' + y' + z') = T(X'), \tag{2}$$

$$X + X' = (x + x' \ \ y + y' \ \ z + z') \longrightarrow$$
$$(x + x' \ \ x + x' + y + y' + z + z')$$
$$= T(X + X'), \tag{3}$$

$$\alpha X = (\alpha x \ \ \alpha y \ \ \alpha z) \longrightarrow$$
$$(\alpha x \ \ \alpha x + \alpha y + \alpha z) = T(\alpha X). \tag{4}$$

From equations (1)–(4), $T(X + X') = T(X) + T(X')$ and $T(\alpha X) = \alpha T(X)$.

Hence T is a linear mapping.

10. a.

$$x = \begin{bmatrix} a & b \\ c & d \end{bmatrix} \longrightarrow a + d = T(x), \tag{1}$$

$$y = \begin{bmatrix} e & f \\ g & h \end{bmatrix} \longrightarrow e + h = T(y), \tag{2}$$

$$x + y = \begin{bmatrix} a + e & b + f \\ c + g & d + h \end{bmatrix} \longrightarrow$$
$$a + e + d + h = T(x + y), \tag{3}$$

$$\alpha x = \begin{bmatrix} \alpha a & \alpha b \\ \alpha c & \alpha d \end{bmatrix} \longrightarrow$$
$$\alpha a + \alpha d = T(\alpha x). \tag{4}$$

From equations (1)–(3), $T(x + y) = T(x) + T(y)$, and from equations (1) and (4), $T(\alpha x) = \alpha T(x)$. So T is a linear mapping

b. Yes.

c. Yes.

d.

$$x = \begin{bmatrix} a & b \\ c & d \end{bmatrix} \longrightarrow \det \begin{bmatrix} a & b \\ c & d \end{bmatrix}$$

$$= ad - bc = T(x),$$

$$y = \begin{bmatrix} e & f \\ g & h \end{bmatrix} \longrightarrow \det \begin{bmatrix} e & f \\ g & h \end{bmatrix}$$

$$= eh - gf = T(y),$$

$$x + y = \begin{bmatrix} a+e & b+f \\ c+g & d+h \end{bmatrix} \longrightarrow$$

$$\det \begin{bmatrix} a+e & b+f \\ c+g & d+h \end{bmatrix}$$

$$= (a+e)(d+h) - (c+g)(b+f)$$

$$= T(x+y).$$

It is easy to see that $T(x+y) \neq T(x) + T(y)$. Hence T cannot be a linear mapping.

e. No.

9.4 Exercises

1. a. $x = (x_1, x_2, x_3) \in \text{Ker} f$ if and only if $f(x) = 0$, that is, $x_1 = 0 = x_1 + x_2 = x_1 + x_2 + x_3$. This gives $x_1 = 0 = x_2 = x_3$. Hence $\text{Ker} f = \{(0\ 0\ 0)\}$. By convention, the dimension of the zero space is 0.

b. $x = (x_1, x_2, x_3) \in \text{Ker} f$ if and only if $f(x) = 0$, that is, $x_1 - x_2 = 0 = x_2 - x_3 = x_3 - x_1$. This gives $x_1 = x_2 = x_3$. So $\text{Ker} f = \{(x_1, x_1, x_1\} \mid x_1 \in \mathbf{R}\} = \{x_1(1\ 1\ 1) \mid x_1 \subset \mathbf{R}\} = \langle(1\ 1\ 1)\rangle$, yielding $\dim \text{Ker} f = 1$.

c. (See Exercise 8(a) of Section 9.3.)

$$X = \begin{bmatrix} x \\ y \end{bmatrix} \in \text{Ker} T \text{ if and only if } T(X) = 0,$$

that is, $2x = 0 = 5y$. This yields $x = 0 = y$.

Hence $\text{Ker} T = \left\{ \begin{bmatrix} 0 \\ 0 \end{bmatrix} \right\}$. So $\dim \text{Ker} T = 0$.

c. (continued). (See Exercise 8(c) of Section 9.3) $\text{Ker} f = (0)$, $\dim \text{Ker} f = 0$.

c. (continued). (See Exercise 8(d) of Section 9.3.)

$$X = \begin{bmatrix} x_1 \\ x_2 \end{bmatrix} \in \text{Ker} T \text{ if and only if}$$

$T(X) = 0$, that is, $x_1 = 0$. So $\text{Ker} T = \left\{ \begin{bmatrix} 0 \\ x_2 \end{bmatrix} \mid x_2 \in \mathbf{R} \right\} = \left\{ x_2 \begin{bmatrix} 0 \\ 1 \end{bmatrix} \mid x_2 \in \mathbf{R} \right\}$
$= \left\langle \begin{bmatrix} 0 \\ 1 \end{bmatrix} \right\rangle$. Thus $\dim \text{Ker} T = 1$.

d. (See Exercise 9(a) of Section 9.3.) Here, $T(x\ y\ z) = (0\ 0)$. So $\text{Ker} T = \{(x\ y\ z) \mid x, y, z \in \mathbf{R}\} = \mathbf{R}^3$. Hence $\dim \text{Ker} T = 3$.

d. (continued). (See Exercise 9(c) of Section 9.3.)
$X = (x,\ y,\ z) \in \text{Ker} T$ if and only if $T(X) = 0$. That is, $x = 0 = x + y + z$. This gives $x = 0 = y + z$. Choose $z = t$, arbitrary. Then $x = 0$, $y = -t$, $z = t$. This yields $\text{Ker} T = \{(0\ -t\ t) \mid t \in \mathbf{R}\} = \{t(0\ -1\ 1) \mid t \in \mathbf{R}\} = \langle(0\ -1\ 1)\rangle$.
Hence $\dim \text{Ker} T = 1$

e. $\text{Ker} T = \left\{ \begin{bmatrix} a & b \\ c & -a \end{bmatrix} \mid a, b, c \in \mathbf{R} \right\}$, $\dim \text{Ker} T = 3$ (See Exercise 10(a) of Section 9.3);

$\text{Ker} T = \left\{ \begin{bmatrix} a & b \\ -b & d \end{bmatrix} \mid a, b, d \in \mathbf{R} \right\}$, $\dim \text{Ker} T = 3$ (See Exercises 10(b) of Section 9.3).

2. (See Exercise 2 of Section 9.3.) $(a, b, c) \in \text{Im } f$ if and only if $(a, b, c) = f(x_1, x_2, x_3)$ for some x_1, x_2, and x_3. This means by the definition of f that $(a, b, c) = (x_1, x_1 + x_2, x_1 + x_2 + x_3)$. So $a = x_1, b = x_1 + x_2, c = x_1 + x_2 + x_3$. The question is whether we can solve these equations for x_1, x_2, and x_3 with or without any constraints on a, b, c. Clearly, we can solve $a = x_1, b = x_1 + x_2, c = x_1 + x_2 + x_3$ whatever a, b, and c may be. Hence $\text{Im } f = \mathbf{R}^3$.

3. (See Exercise 3 of Section 9.3.) $(a, b, c) \in \text{Im } f$ if and only if $(a, b, c) = f(x_1, x_2, x_3)$ for some x_1, x_2, and x_3. This means by the definition of f that $a = x_1 - x_2, b = x_2 - x_3, c = x_3 - x_1$. The question remains whether we can solve for x_1, x_2, and x_3 for the LS

$$x_1 - x_2 = a,$$
$$x_2 - x_3 = b,$$
$$x_3 - x_1 = c.$$

By our usual method of transforming

$$\begin{bmatrix} 1 & -1 & 0 & a \\ 0 & 1 & -1 & b \\ -1 & 0 & 1 & c \end{bmatrix},$$ we obtain that the sys-

tem is consistent if and only if $a + b + c = 0$. Hence, $(a, b, c) \in \operatorname{Im} f$ if and only if $c = -a - b$, so $(a, b, -a - b) \in \operatorname{Im} f$ for all values of a and b. Writing $(a, b, -a - b) = (a, 0, -a) + (0, b, -b) = a(1, 0, -1) + b(0, 1, -1)$, we find that $\operatorname{Im} f = \langle (1, 0, -1), (0, 1, -1) \rangle$.

4. (Exercise 8(a) of Section 9.3). $\operatorname{Im} T = \mathbf{R}^2$, $\dim \operatorname{Im} T = 2$.

(Exercise 8(c) of Section 9.3). $\operatorname{Im} T = \mathbf{R}^2$, $\dim \operatorname{Im} T = 2$.

(Exercise 8(d) of Section 9.3). $\operatorname{Im} T = \{(x, 6) \mid X \in \mathbf{R}\}$, $\dim \operatorname{Im} T = 1\}$.

6. a. $\{(-1, 1, 0, 1)\}$

b. $\{(1, 0, 1), (0, 1, 0), (0, 0, 1)\}$

7. 0.

9.5 Exercises

1. a. $f(1, 2) = (2, 6) = 2e_1 + 6e_2$ $f(1, 3) = (2, 9) = 2e_1 + 9e_2$.

Thus the matrix of f is $\begin{bmatrix} 2 & 2 \\ 6 & 9 \end{bmatrix}$.

b. $f(e_1) = (2, 0, 0, 0)$; $f(e_2) = (0, 0, 0, 0)$; $f(e_3) = (0, 0, 0, 1)$; $f(e_4) = (0, 0, 0, 1)$.

Thus the matrix of f is $\begin{bmatrix} 2 & 0 & 0 & 0 \\ 0 & 0 & 0 & 0 \\ 0 & 0 & 0 & 0 \\ 0 & 0 & 1 & 1 \end{bmatrix}$.

c. $\begin{bmatrix} \alpha & 0 & \cdots & 0 \\ 0 & \alpha & \cdots & 0 \\ \vdots & \vdots & \ddots & \vdots \\ 0 & 0 & \cdots & \alpha \end{bmatrix}.$

2. a. $A = \begin{bmatrix} 1 & 1 & 0 \\ 0 & 1 & 0 \\ 1 & 0 & 1 \end{bmatrix}.$

b. $T(1, 1, 0) = (2, 1, 1)$
$$= (1, 1, 0) + (1, 0, 1),$$
$T(1, 0, 1) = (1, 1, 2)$
$$= (1, 0, 1) + (0, 1, 1),$$
$T(0, 1, 1) = (1, 2, 1)$
$$= (0, 1, 1) + (1, 1, 0).$$

Thus $B = \begin{bmatrix} 1 & 0 & 1 \\ 1 & 1 & 0 \\ 0 & 1 & 1 \end{bmatrix}.$

c. To find X such that $XB = AX$, we express the new basis $\{(1, 1, 0), (1, 0, 1), (0, 1, 1)\}$ in terms of the standard basis $\{(1, 0, 0), (0, 1, 0), (0, 0, 1)\}$, that is,

$$(1, 1, 0) = (1, 0, 0) + (0, 1, 0),$$
$$(1, 0, 1) = (1, 0, 0) + (0, 0, 1),$$
$$(0, 1, 1) = (0, 1, 0) + (0, 0, 1).$$

Then $X = \begin{bmatrix} 1 & 1 & 0 \\ 1 & 0 & 1 \\ 0 & 1 & 1 \end{bmatrix}$. One can check that $XB = AX$.

3. a. $\begin{bmatrix} 1 & 1 & 0 & 0 \\ 2 & 0 & -1 & 0 \\ 0 & 0 & 1 & -1 \\ 0 & 0 & 0 & 1 \end{bmatrix}$

b. $\begin{bmatrix} \frac{2}{3} & \frac{1}{6} & -\frac{1}{3} & \frac{1}{6} \\ 1 & 2 & 1 & 0 \\ 0 & -\frac{1}{2} & -1 & -\frac{1}{2} \\ -\frac{2}{3} & \frac{1}{3} & -\frac{2}{3} & \frac{4}{3} \end{bmatrix}$

4. $\varphi \begin{bmatrix} 1 & 0 \\ 0 & 0 \end{bmatrix} = \begin{bmatrix} 1 & 2 \\ 3 & -1 \end{bmatrix} \begin{bmatrix} 1 & 0 \\ 0 & 0 \end{bmatrix} = \begin{bmatrix} 1 & 0 \\ 3 & 0 \end{bmatrix}$

$= \begin{bmatrix} 1 & 0 \\ 0 & 0 \end{bmatrix} + 0 \begin{bmatrix} 0 & 1 \\ 0 & 0 \end{bmatrix}$

$+ 3 \begin{bmatrix} 0 & 0 \\ 1 & 0 \end{bmatrix} + 0 \begin{bmatrix} 0 & 0 \\ 0 & 1 \end{bmatrix}.$

So the first column of the desired matrix is $\begin{bmatrix} 1 \\ 0 \\ 3 \\ 0 \end{bmatrix}$. Similarly, the second, third, and fourth

columns are $\begin{bmatrix} 0 \\ 1 \\ 0 \\ 3 \end{bmatrix}$, $\begin{bmatrix} 2 \\ 0 \\ -1 \\ 0 \end{bmatrix}$, and $\begin{bmatrix} 0 \\ 2 \\ 0 \\ -1 \end{bmatrix}$,

respectively.

5. a. $\begin{bmatrix} 0 & -1 \\ 1 & 0 \end{bmatrix}$. **b.** $\frac{1}{3} \begin{bmatrix} -1 & -2 \\ 5 & 1 \end{bmatrix}$.

6. $D(f_1) = D(1) = 0 = 0f_1 + 0f_2 + 0f_3 + 0f_4$,

$D(f_2) = D(x) = 1 = 1f_1 + 0f_2 + 0f_3 + 0f_4$,

$D(f_3) = D(x^2) = 2x = 0f_1 + 2f_2 + 0f_3 + 0f_4$,

$D(f_4) = D(x^3) = 3x^2 = 0f_1 + 0f_2 + 3f_3 + 0f_4$.

Thus the matrix is $\begin{bmatrix} 0 & 1 & 0 & 0 \\ 0 & 0 & 2 & 0 \\ 0 & 0 & 0 & 3 \\ 0 & 0 & 0 & 0 \end{bmatrix}$.

8. a. $\begin{bmatrix} -1 & 0 \\ 0 & 3 \end{bmatrix}$.

b. $\begin{bmatrix} 1 & 1 \\ 0 & 1 \end{bmatrix}$.

c. $\begin{bmatrix} 0 & 1 & 0 \\ 0 & 0 & 1 \\ 0 & 0 & 0 \end{bmatrix}$.

d. $\begin{bmatrix} 0 & 1 & 0 & 0 & 0 \\ 0 & 0 & 0 & 0 & 0 \\ 0 & 0 & 1 & 0 & 0 \\ 0 & 0 & 0 & 2 & 1 \\ 0 & 0 & 0 & 0 & 2 \end{bmatrix}$.

e. $\begin{bmatrix} 0 & -1 \\ 1 & 0 \end{bmatrix}$.

9.6 Chapter Review Questions and Projects

1. a. Yes **b.** Yes **c.** No **d.** Yes
e. Yes **f.** Yes, if $a = 0$.

2. a. Linearly independent.
 b. Linearly independent.
 c. Linearly independent.
 d. Linearly independent.

3. a. $\begin{bmatrix} 1 & 0 \\ 0 & 0 \end{bmatrix}$, $\begin{bmatrix} 1 & 0 \\ 0 & 0 \end{bmatrix}$, $\begin{bmatrix} 0 & 0 \\ 1 & 0 \end{bmatrix}$, and

$\begin{bmatrix} 0 & 0 \\ 0 & 1 \end{bmatrix}$.

 b. The set E_{ij}, $1 \leqslant i, j \leqslant 3$, of 3×3 matrices where E_{ij} is a matrix all of whose entries are zero except the (i, j) entry which is 1.

 c. Suppose E_{ij}, $1 \leqslant i, j \leqslant 3$, denote the 3×3 matrix all of whose entries are zero except the (i, j) entry which is 1. Then E_{11}, $E_{12}, E_{13}, E_{22}, E_{23}, E_{33}$ form a basis of the 3×3 upper triangular matrices, and E_{11}, $E_{21}, E_{22}, E_{31}, E_{32}, E_{33}$ form a basis of the 3×3 lower triangular matrices.

5. a. Not a linear transformation.
 b. Linear transformation.
 c. Linear transformation.

6. $(S + T)(x_1, x_2) = (-x_1 + 2x_2, 2x_1 + x_2)$, $(ST)(x_1, x_2) = (-x_1 - 3x_2, 3x_1 - x_2)$. Both $S + T$ and ST are linear transformations.

7. a. $T(x, y, z) = \left(z - y, -\frac{x}{2} - \frac{z}{2}\right)$.
 b. $T(x, y, z) = (2x - 3y + z, 3x - 5y + 2z)$.

8. (Exercise 7(a)) Ker $T = \left\{\left(-\frac{t}{2}, t, t\right) \mid t \in \mathbf{R}\right\}$.
(Exercise 7(b)) Ker $T = \{(t, t, t) \mid t$ is arbitrary $\}$.

9. a. $\begin{bmatrix} 1 & 0 \\ 0 & -1 \end{bmatrix}$. **b.** $\begin{bmatrix} 1 & 2 \\ 1 & -1 \end{bmatrix}$ **c** $\begin{bmatrix} -1 & 0 \\ 3 & 0 \end{bmatrix}$.

10. i. $\begin{bmatrix} 0 & 1 & 0 & 0 \\ 0 & 0 & 2 & 0 \\ 0 & 0 & 0 & 3 \\ 0 & 0 & 0 & 0 \end{bmatrix}$.

ii.
$$\begin{bmatrix} 0 & 1 & -2 & -3 \\ 0 & 0 & 2 & 0 \\ 0 & 0 & 0 & 3 \\ 0 & 0 & 0 & 0 \end{bmatrix}.$$

Projects

2. a. Matrices of S:
$$\begin{bmatrix} 2 & 3 & 0 & 0 & 0 & 0 \\ 0 & 3 & 4 & 0 & 0 & 0 \\ 0 & 0 & 4 & 5 & 0 & 0 \\ 0 & 0 & 0 & 5 & 6 & 0 \\ 0 & 0 & 0 & 0 & 6 & 7 \\ 8 & 0 & 0 & 0 & 0 & 7 \end{bmatrix},$$

$$\begin{bmatrix} 4 & 0 & 0 & 5 & 5 & 0 \\ -5 & -5 & -8 & -9 & -5 & 0 \\ 6 & 10 & 10 & 7 & 0 & 0 \\ 8 & 8 & 8 & 7 & 0 & 0 \\ -8 & -8 & -8 & -2 & 5 & 6 \\ 0 & 0 & 0 & 7 & 0 & 6 \end{bmatrix}$$

Matrices of T:
$$\begin{bmatrix} 1 & -1 & 0 & 0 & 0 & 0 \\ 0 & 1 & -1 & 0 & 0 & 0 \\ 0 & 0 & 1 & -1 & 0 & 0 \\ 0 & 0 & 0 & 1 & -1 & 0 \\ 0 & 0 & 0 & 0 & 1 & -1 \\ -1 & 0 & 0 & 0 & 0 & 1 \end{bmatrix},$$

$$\begin{bmatrix} 1 & 0 & 0 & -1 & -1 & 0 \\ 0 & 2 & 1 & 1 & 1 & 0 \\ -1 & -2 & 0 & -1 & 0 & 0 \\ -1 & -1 & -1 & 1 & 0 & 0 \\ 1 & 1 & 1 & 0 & 1 & -1 \\ 0 & 0 & 0 & -1 & 0 & 1 \end{bmatrix}$$

b. $(S + T)(x_1, \ldots, x_6) = (3x_1 + 2x_2, 4x_2 + 3x_3, 5x_3 + 4x_4, 6x_4 + 5x_5, 7x_5 + 6x_6, 8x_6 + 7x_1)$

$(ST)(x_1, \ldots, x_6) = (2x_1 + x_2 - 3x_3, 3x_2 + x_3 - 4x_4, 4x_3 + x_4 - 5x_5, 5x_4 + x_5 - 6x_6, 6x_5 + x_6 - 7x_1, 7x_6 + x_1 - 8x_2)$.

c. S is invertible and onto.

T is not invertible, nor onto.

d. $T^n(x_1, x_2, \ldots, x_6) = (y_1, y_2, \ldots, y_6)$, where $y_1 = \binom{n}{0} x_1 - \binom{n}{1} x_2 + \binom{n}{2} x_3 - \cdots + (-1)^n \binom{n}{n} x_n$, $y_2 = \binom{n}{0} x_2 - \binom{n}{1} x_3 + \cdots + (-1)^{n-1} \binom{n}{n-1} x_n + (-1)^n \binom{n}{n} x_1$, and so on where $x_{6m} = x_6$ for all $m > 0$ and $x_{6m+r} = x_r$ for $1 \leqslant r < 6$.

e. $T^6 - 6T^5 + 15T^4 - 20T^3 + 15T^2 - 6T = 0$.

Drill Solutions Using Matlab

CHAPTER 1

Solution to Drill 1.1

» **format rat**

Step 1 Enter the three equations in the list form.

» **eq1 = [1 2 -2 1], eq2 = [3 -1 2 7], eq3 = [2 -3 -4 5]**
```
eq1=
   1     2    -2     1
eq2=
   3    -1     2     7
eq3=
   2    -3    -4     5
```

Step 2 Remove the variable x from eq2 and eq3 as follows. Multiply eq1 by -3 and add to eq2 and multiply eq1 by -2; add to eq3.

» **eq1, eq2 = (-3)*eq1 + eq2, eq3 = (-2) * eq1 + eq3**
```
eq1 =
   1     2    -2     1
eq2 =
   0    -7     8     4
eq3 =
   0    -7     0     3
```

The reduced system is

$$x + 2y - 2z = 1, \tag{1}'$$
$$-7y + 8z = 4, \tag{2}'$$
$$-7y = 3. \tag{3}'$$

Step 3 Use eq3 to find the value of y. First, enter the equations of the reduced system.

» **eq1 = 'x + 2*y - 2*z = 1', eq2 = '- 7*y + 8*z = 4', eq3 = '- 7*y = 3'**
```
eq1 =
   x+2*y-2*z=1
eq2 =
   -7*y + 8*z = 4
eq3 =
   -7*y = 3
```

Solve equation $(3)'$ for y.

» **y = solve(eq3, 'y')**
```
y =
   -3/7
```

Step 4 Substitute the value of y in equation $(2)'$ and solve for z.

» **eq2=subs(eq2, 'y', y); z = solve(eq2, 'z')**

z =
   1/8

Step 5 Substitute the values of z and y in equation $(1)'$ and solve for x.

» **eq1= subs(eq1, 'z', z);**

» **eq1 = subs(eq1, 'y', y);**

» **x = solve(eq1, 'x')**

x =
   59/28

This yields the final result **x = 59/28, y = -3/7, z = 1/8.**

Solution to Drill 1.2

» **format rat**

Step 1 Enter the three equations in the list form.

» **eq1=[2 -5 3 -4 2 4], eq2=[3 -7 2 -5 4 9], eq3=[5 -10 -5 -4 7 22]**

eq1 =
   2   -5   3   -4   2   4
eq2 =
   3   -7   2   -5   4   9
eq3 =
   5   -10   -5   -4   7   22

Step 2 Divide eq1 by 2.

» **format rat**   **% display in rational mode**
» **eq1=(1/2)*eq1, eq2, eq3**

eq1 =
   1   -5/2   3/2   -2   1   2
eq2 =
   3   -7   2   -5   4   9
eq3 =
   5   -10   -5   -4   7   22

Step 3 Remove the coefficients corresponding to x from eq2 and eq3 using eq1.

» **eq1,eq2=(-3)*eq1+eq2, eq3=(-5)*eq1+eq3**   **% multiply eq1 by -3 add to eq2**
»   **% multiply eq1 by -5 add to eq3**

eq1 =
   1   -5/2   3/2   -2   1   2
eq2 =
   0   1/2   -5/2   1   1   3
eq3 =
   0   5/2   -25/2   6   2   12

Step 4 Divide eq2 by (1/2), i.e, multiply equation 2 by 2.

» **eq1, eq2=2*eq2, eq3**
eq1 =

| 1 | -5/2 | 3/2 | -2 | 1 | 2 |

eq2 =

| 0 | 1 | -5 | 2 | 2 | 6 |

eq3 =

| 0 | 5/2 | -25/2 | 6 | 2 | 12 |

Step 5 Multiply eq2 by $(-5/2)$ and add to eq3.

» **eq1, eq2, eq3=(-5/2)*eq2+eq3**
eq1 =

| 1 | -5/2 | 3/2 | -2 | 1 | 2 |

eq2 =

| 0 | 1 | -5 | 2 | 2 | 6 |

eq3 =

| 0 | 0 | 0 | 1 | -3 | -3 |

Step 6 The resulting system is in row echelon form. First enter the reduced system.

» **eq1='x+(-5/2)*y+(3/2)*z-2*u+v = 2'; eq2='y-5*z+2*u+2*v=6'; eq3='u-3*v=-3';**

Choose v to be a free variable and solve for u.

» **v='t'; u = solve(subs(eq3,'v',v),'u'),**　　% **substitute the value of v in eq3**
u =　　　　　　　　　　　　　　　　　　　　% **and solve for u**
3*t-3

Substitute the values of u, v in eq2 and choose a value for free variable z and solve eq2 for y.

» **eq2=subs(eq2, 'u', u); eq2=subs(eq2, 'v', v)**　　% **substitute u and v in eq2**
eq2 =
y-5*z+6*t-6+2*(t)=6
» **z='n'; eq2 = subs(eq2, 'z', z);**　　　　　　　% **substitute z in eq2**
» **y=solve(eq2,'y')**　　　　　　　　　　　　% **solve eq2 for y**
y =
5*n-8*t+12

Substitute the values of u, v, z, y in eq1 and solve for x.

» **eq1=subs(eq1,'u',u); eq1=subs(eq1,'v',v); eq1=subs(eq1,'z',z);**
eq1=subs(eq1,'y',y);
» **x = solve(eq1,'x')**　　　　　　　　　　% **substitute u, v, z, and y in eq1 and**
　　　　　　　　　　　　　　　　　　　% **solve for x**

x =
11*n-15*t+26

Solution to Drill 1.3

» **format rat**

Step 1 Enter four equations (only coefficients) in list form.

» **eq1 = [1 2 1 2 0]; eq2 = [-1 3 4 -1 0]; eq3 = [4 3 4 3 0]; eq4 = [3 3 2 1 0];**

Drill Solutions Using Matlab

Step 2 Remove the variable x_1 from eq2, eq3, and eq4 as follows. Multiply eq1 by 1 and add to eq2. Multiply eq1 by -4 and add to eq3. Multiply eq1 by -3 and add to eq4.

» eq1, eq2 = (1)*eq1 + eq2, eq3 = (-4)*eq1 + eq3, eq4 = (-3)*eq1 + eq4

```
eq1 =
   1     2     1     2     0
eq2 =
   0     5     5     1     0
eq3 =
   0    -5     0    -5     0
eq4 =
   0    -3    -1    -5     0
```

Step 3 Divide eq2 by 5 and remove x_2 from eq3 and eq4 using eq2.

» format rat;   % display in rational format
» eq1, eq2 = (1/5)*eq2, eq3 = 5*eq2 + eq3, eq4 = 3*eq2 + eq4

```
eq1 =
   1     2     1     2       0
eq2 =
   0     1     1     1/5     0
eq3 =
   0     0     5    -4       0
eq4 =
   0     0     2    -22/5    0
```

Step 4 Divide equation 3 by 5 and remove x_3 from eq4 as follows.

» eq1, eq2, eq3= (1/5)*eq3, eq4 = (-2)*eq3 + eq4

```
eq1 =
   1     2     1     2        0
eq2 =
   0     1     1     1/5      0
eq3 =
   0     0     1    -4/5      0
eq4 =
   0     0     0    -14/5     0
```

Step 5 Divide eq4 by $-14/5$.

» eq1, eq2, eq3, eq4 = (-5/14)*eq4

```
eq1 =
   1     2     1     2       0
eq2 =
   0     1     1     1/5     0
eq3 =
   0     0     1    -4/5     0
eq4 =
   0     0     0     1       0
```

The last equation implies $x_4 = 0$. Substitute $x_4 = 0$ in eq3 and get $x_3 = 0$. Similarly, eq1 and eq2 imply $x_1 = 0$ and $x_2 = 0$.

Note: if you are not interested in showing steps, you may use direct Matlab commands to obtain the solution. In the latter case we will proceed as below.

» eq1= 'x1 + 2*x2 + x3 + 2*x4 = 0';
» eq2= '-x1 + 3*x2 + 4*x3 - x4 = 0';
» eq3= '4*x1 + 3*x2 + 4*x3 + 3*x4 = 0';
» eq4= '3*x1 + 3*x2 + 2*x3 + x4 = 0';
» [x1, x2, x3, x4] = solve(eq1, eq2, eq3, eq4)

This command is valid only if the number of equations is greater than or equal to the number of unknowns.

Solution to Drill 1.4

Step 1 Enter four equations (only coefficients) in list form.

» eq1=[1 2 1 3 2 0], eq2=[2 5 2 1 2 0], eq3=[1 3 1 -2 1 0], eq4=[3 2 3 -1 1 0]

```
eq1 =
    1    2    1    3    2    0
eq2 =
    2    5    2    1    2    0
eq3 =
    1    3    1   -2    1    0
eq4 =
    3    2    3   -1    1    0
```

Step 2 Remove variable x_1 from eq2, eq3, and eq4 as follows. Multiply eq1 by -2 and add to eq2. Multiply eq1 by -1 and add to eq3. Multiply eq1 by -3 and add to eq4.

» eq1, eq2 = (-2)*eq1 +eq2, eq3 = (-1)*eq1 + eq3,eq4 = (-3)*eq1 + eq4

```
eq1 =
    1    2    1    3    2    0
eq2 =
    0    1    0   -5   -2    0
eq3 =
    0    1    0   -5   -1    0
eq4 =
    0   -4    0  -10   -5    0
```

Step 3 Remove x_2 from eq3 and eq4 using eq2 as follows.

» **format rat**
» eq1, eq2, eq3 = -1*eq2 + eq3, eq4 = 4*eq2 + eq4

```
eq1 =
    1    2    1    3    2    0
eq2 =
    0    1    0   -5   -2    0
eq3 =
    0    0    0    0    1    0
eq4 =
    0    0    0  -30  -13    0
```

Step 4 Note that the coefficient corresponding to x_4 is zero. Therefore, interchange eq3 and eq4.

» **eq1,eq2, temp = eq3; eq3=eq4, eq4= temp**

```
eq1 =
   1    2    1    3    2    0
eq2 =
   0    1    0   -5   -2    0
eq3 =
   0    0    0  -30  -13    0
eq4 =
   0    0    0    0    1    0
```

Step 5 Divide eq3 by -30. The system is indeed already in row echelon form and this step is optional.

» **eq1, eq2, eq3 = (-1/30)*eq3, eq4**

```
eq1 =
   1    2    1    3    2      0
eq2 =
   0    1    0   -5   -2      0
eq3 =
   0    0    0    1   13/30   0
eq4 =
   0    0    0    0    1      0
```

Step 6 Trivially, eq4 implies $x_5 = 0$, and hence from eq3 we have $x_4 = 0$. Also, eq2 implies $x_2 = 0$. The variable x_3 is arbitrary, say a. Then from eq1 we get $x_1 = -a$. Therefore, the solution set is $x_1 = -a, x_2 = 0, x_3 = a, x_4 = 0, x_5 = 0$.

Note: If we were to use direct Matlab commands as given at the end of Drill 1.3, we must introduce one virtual equation, say eq5 = '0 = 0', so that the number of equations is $\geq$ the number of unknowns.

Solution to Drill 1.5

» **format rat**

Step 1 Enter the augmented matrix.

» **AUG = [1 2 1 3 0; 2 4 3 1 0; 3 6 6 2 0]**

```
AUG=
   1    2    1    3    0
   2    4    3    1    0
   3    6    6    2    0
```

Step 2 Using AUG(1,1), convert the entries in the first column below the AUG(1,1) to zero.

» **AUG(2,:) = (-2)*AUG(1,:) + AUG(2,:)**

```
AUG =
   1    2    1    3    0
   0    0    1   -5    0
   3    6    6    2    0
```

» **AUG(3,:) = (-3)*AUG(1,:) + AUG(3,:)**

AUG =

1	2	1	3	0
0	0	1	-5	0
0	0	3	-7	0

Step 3 Since all entries below the nonzero entry in column 2 are zero, let us reduce the third column. Using the (2,3) entry, perform the following operations. Multiply row 2 by -3 and add to row 3.

» **AUG(3,:) = (-3)*AUG(2,:) + AUG(3,:)**

AUG =

1	2	1	3	0
0	0	1	-5	0
0	0	0	8	0

Step 4 Divide row 3 by 8.

» **format rat, AUG(3,:)=(1/8)*AUG(3,:)**

AUG =

1	2	1	3	0
0	0	1	-5	0
0	0	0	1	0

Step 5 The resulting matrix is in row echelon form. Now we reduce the matrix to its reduced row echelon form. Multiply row 2 by -1 and add it to row 1. Also multiply row 3 by 5 and add it to row 2 and multiply row 3 by -8 and add it to row 1.

» **AUG(1,:)=(-1)*AUG(2,:) + AUG(1,:); AUG(2,:)=(5)*AUG(3,:) + AUG(2,:);**
» **AUG(1,:)=(-8)*AUG(3,:) + AUG(1,:)**

AUG =

1	2	0	0	0
0	0	1	0	0
0	0	0	1	0

The matrix is in reduced row echelon form.

Solution to Drill 1.6

» **format rat**

Step 1 Enter the augmented matrix.

» **AUG = [2 4 3 2 2; 3 6 5 2 2; 2 5 2 -3 3; 4 5 14 14 11]**

AUG =

2	4	3	2	2
3	6	5	2	2
2	5	2	-3	3
4	5	14	14	11

Step 2 Multiply row 1 by 1/2. Using AUG(1,1) convert the entries in first column below AUG(1,1) to zero.

» **AUG(1,:) = (1/2)*AUG(1,:); AUG(2,:) = (-3)*AUG(1,:) + AUG(2,:);**

AUG(3,:) = (-2)*AUG(1,:) + AUG(3,:); AUG(4,:) = (-4)*AUG(1,:) + AUG(4,:)

AUG =

1	2	3/2	1	1
0	0	1/2	-1	-1
0	1	-1	-5	1
0	-3	8	10	7

Step 3 Interchange rows 2 and 3 to make AUG(2,2) nonzero.

» AUG = AUG([1,3,2,4],:)

AUG =

1	2	3/2	1	1
0	1	-1	-5	1
0	0	1/2	-1	-1
0	-3	8	10	7

Step 4 Multiply row 2 by 3 and add to row 4.

» AUG(4,:) = 3*AUG(2,:) + AUG(4,:)

AUG =

1	2	3/2	1	1
0	1	-1	-5	1
0	0	1/2	-1	-1
0	0	5	-5	10

Step 5 Divide row 3 by 1/2 (or multiply by 2). Then multiply row 3 by −5 and add to row 4.

» AUG(3,:) = (2/1)*AUG(3,:), AUG(4,:) = (-5)*AUG(3,:) + AUG(4,:)

AUG =

1	2	3/2	1	1
0	1	-1	-5	1
0	0	1	-2	-2
0	0	5	-5	10

AUG =

1	2	3/2	1	1
0	1	-1	-5	1
0	0	1	-2	-2
0	0	0	5	20

Step 6 Divide row 4 by 5.

» AUG(4,:) = (1/5)*AUG(4,:)

AUG =

1	2	3/2	1	1
0	1	-1	-5	1
0	0	1	-2	-2
0	0	0	1	4

Step 7 Using AUG(4,4), make all entries in the 4th column zero (except AUG(4,4))

» AUG(3,:)=2*AUG(4,:) +AUG(3,:); AUG(2,:)=5*AUG(4,:) + AUG(2,:);
AUG(1,:)=(-1)*AUG(4,:) + AUG(1,:)

```
AUG =
   1    2    3/2    0    -3
   0    1    -1     0    21
   0    0    1      0    6
   0    0    0      1    4
```

Step 8 Using AUG(3,3), make all entries in the 3rd column zero (except AUG(3,3)).
» AUG(2,:)=1*AUG(3,:) + AUG(2,:); AUG(1,:)=(-3/2)*AUG(3,:) + AUG(1,:)

```
AUG =
   1    2    0    0    -12
   0    1    0    0    27
   0    0    1    0    6
   0    0    0    1    4
```

Step 9 Using AUG(2,2), make all entries in the 2nd column zero (except AUG(2,2))
» AUG(1,:)=(-2)*AUG(2,:) + AUG(1,:)

```
AUG =
   1    0    0    0    -66
   0    1    0    0    27
   0    0    1    0    6
   0    0    0    1    4
```

The above matrix is in reduced row echelon form and gives $x_1 = -66$, $x_2 = 27$, $x_3 = 6$, and $x_4 = 4$.

Note: Matlab command rref(AUG), can directly yield the reduced row echelon form of AUG.

Solution to Drill 1.7

» **format rat**

Step 1 Enter the augmented matrix corresponding to the system of equations.
» AUG=[4 -1 -1 0 0 1; -1 4 -1 -1 0 2; -1 -1 4 -1 -1 3; 0 -1 -1 4 -1 4; 0 0 -1 -1 4 5]

```
AUG =
   4    -1    -1    0     0    1
  -1     4    -1   -1     0    2
  -1    -1     4   -1    -1    3
   0    -1    -1    4    -1    4
   0     0    -1   -1     4    5
```

Step 2 Divide row 1 by 4 (or AUG(1,1)). Using AUG(1,1) convert the entries in the first column below AUG(1,1) to zero.
» AUG(1,:) – AUG(1,:)/AUG(1,1), AUG(2,:) = (1)*AUG(1,:) + AUG(2,:),
AUG(3,:) = (1)*AUG(1,:) + AUG(3,:)

```
AUG =
   1    -1/4    -1/4    0     0    1/4
  -1      4     -1     -1     0    2
  -1     -1      4     -1    -1    3
   0     -1     -1      4    -1    4
   0      0     -1     -1     4    5
```

AUG =

1	-1/4	-1/4	0	0	1/4
0	15/4	-5/4	-1	0	9/4
-1	-1	4	-1	-1	3
0	-1	-1	4	-1	4
0	0	-1	-1	4	5

AUG =

1	-1/4	-1/4	0	0	1/4
0	15/4	-5/4	-1	0	9/4
0	-5/4	15/4	-1	-1	13/4
0	-1	-1	4	-1	4
0	0	-1	-1	4	5

Step 3 Divide row 2 by 15/4. i.e., multiply row 2 by 4/15.

» AUG(2,:) = (4/15)*AUG(2,:)

AUG =

1	-1/4	-1/4	0	0	1/4
0	1	-1/3	-4/15	0	3/5
0	-5/4	15/4	-1	-1	13/4
0	-1	-1	4	-1	4
0	0	-1	-1	4	5

Step 4 Multiply row 2 by (5/4) and add to row 3. Multiply row 2 by 1 and add to row 4.

» AUG(3,:) = (5/4)*AUG(2,:) + AUG(3,:), AUG(4,:) = (1)*AUG(2,:) + AUG(4,:)

AUG =

1	-1/4	-1/4	0	0	1/4
0	1	-1/3	-4/15	0	3/5
0	0	10/3	-4/3	-1	4
0	-1	-1	4	-1	4
0	0	-1	-1	4	5

AUG =

1	-1/4	-1/4	0	0	1/4
0	1	-1/3	-4/15	0	3/5
0	0	10/3	-4/3	-1	4
0	0	-4/3	56/15	-1	23/5
0	0	-1	-1	4	5

Step 5 Divide row 3 by 10/3. Multiply row 3 by (4/3) and add to row 4. Multiply row 3 by 1 and add to row 5.

» AUG(3,:) =(3/10)*AUG(3,:), AUG(4,:) = (4/3)*AUG(3,:) + AUG(4,:),
AUG(5,:) = (1)*AUG(3,:) + AUG(5,:)

AUG =

1	-1/4	-1/4	0	0	1/4
0	1	-1/3	-4/15	0	3/5
0	0	1	-2/5	-3/10	6/5
0	0	-4/3	56/15	-1	23/5
0	0	-1	-1	4	5

AUG =

1	-1/4	-1/4	0	0	1/4
0	1	-1/3	-4/15	0	3/5

0	0	1	-2/5	-3/10	6/5
0	0	0	16/5	-7/5	31/5
0	0	-1	-1	4	5

AUG =

1	-1/4	-1/4	0	0	1/4
0	1	-1/3	-4/15	0	3/5
0	0	1	-2/5	-3/10	6/5
0	0	0	16/5	-7/5	31/5
0	0	0	-7/5	37/10	31/5

Step 6a Divide row 4 by 16/5. Multiply row 4 by (7/5) and add to row 5.

» AUG(4,:) = (5/16)*AUG(4,:), AUG(5,:) =(7/5)*AUG(4,:)+AUG(5,:)

AUG =

1	-1/4	-1/4	0	0	1/4
0	1	-1/3	-4/15	0	3/5
0	0	1	-2/5	-3/10	6/5
0	0	0	1	-7/16	31/16
0	0	0	-7/5	37/10	31/5

AUG =

1	-1/4	-1/4	0	0	1/4
0	1	-1/3	-4/15	0	3/5
0	0	1	-2/5	-3/10	6/5
0	0	0	1	-7/16	31/16
0	0	0	0	247/80	713/80

Step 6b Divide row 5 by 247/80.

» AUG(5,:) = (80/247)*AUG(5,:)

AUG =

1	-1/4	-1/4	0	0	1/4
0	1	-1/3	-4/15	0	3/5
0	0	1	-2/5	-3/10	6/5
0	0	0	1	-7/16	31/16
0	0	0	0	1	713/247

Step 7 The augmented matrix is in row echelon form. Therefore, we may apply back substitution algorithm to find the solution. The reduced augmented matrix is equivalent to

$$x + (-1/4)*y + (-1/4)*z = 1/4$$
$$y + (-1/3)*z + (-4/15)*m = 3/5$$
$$z + (-2/5)*m + (-3/10)*n = 6/5$$
$$m + (-7/16)*n = 31/16$$
$$n = 713/247$$

Enter the four equations.

» eq1 = 'x + (-1/4)*y + (-1/4)*z = 1/4';
» eq2 = 'y + (-1/3)*z + (-4/15)*m = 3/5';
» eq3 = 'z + (-2/5)*m + (-3/10)*n = 6/5';

```
» eq4 = 'm + (-7/16)*n = 31/16';
» eq5 = 'n = 713/247';
```

Step 8   Solve eq5 for n.

```
» n = solve(eq5,'n')

n =
     713/247
```

Step 9   Substitute the value of n in eq4 and solve eq4 for m.

```
» eq4 = subs(eq4,'n', n);
» m = solve(eq4,'m')

m =
     158/494
```

Step 10   Substitute the values of m and n in eq3 and solve eq3 for z.

```
» eq3 = subs(eq3, 'm', m);
» eq3 = subs(eq3, 'n', n);
» z = solve(eq3,'z')

z =
     87/26
```

Step 11   Substitute the values of m, n, and z in eq2 and solve eq2 for y.

```
» eq2 = subs(eq2, 'm', m);
» eq2 = subs(eq2, 'n', n);
» eq2 = subs(eq2, 'z', z);
» y = solve(eq2,'y')

y =
     1269/494
```

Step 12   Substitute the values of m, n, z, and y in eq1 and solve eq1 for x.

```
» eq1 = subs(eq1, 'm', m);
» eq1 = subs(eq1, 'n', n);
» eq1 = subs(eq1, 'z', z);
» eq1 = subs(eq1, 'y', y);
» x = solve(eq1,'x')

x =
     427/247
```

The solution to the system is $x = 427/247$, $y = 1269/494$, $z = 87/26$, $m = 1581/494$, $n = 713/247$. Part (b) can be solved using the same method.

Note: As stated at the end of Drill 1.6, we could use the Matlab command rref(AUG) to reduce AUG into reduced row echelon form, and then solve the system. On the other hand, if one is not required to show any steps, one may invoke direct Matlab command:

```
» [m, n, x, y, z] = solve(eq1, eq2, eq3, eq4, eq5)
```

Solution to Drill 1.8

» **format rat**

Step 1 Write down the system of linear equations. Suppose x, y, z, and t are the amounts from foods A, B, C, and D. Then we may write the following equation for the calcium content: $20x + 10y + 10z + 15t = 70$.

Similarly equations for iron and vitamins A and D are

$$5x + 5y + 10z + 15t = 35,$$
$$5x + 15y + 5z + 10t = 35,$$
$$8x + 10y + 10z + 20t = 50.$$

Step 2 Enter the augmented matrix corresponding to the above system of equations.

» AUG=[20 10 10 15 70; 5 5 10 15 35; 5 15 5 10 35; 8 10 10 20 50]

AUG =

20	10	10	15	70
5	5	10	15	35
5	15	5	10	35
8	10	10	20	50

Step 3 Find the reduced row echelon form of AUG.

» rref(AUG)

AUG =

1	0	0	0	60/29
0	1	0	0	23/29
0	0	1	0	18/29
0	0	0	1	28/29

We deduce from the reduced row echelon form that $x = 60/29$, $y = 23/29$, $z = 18/29$, $t = 28/29$.

CHAPTER 2

Solution to Drill 2.1

Step 1 Enter matrices A and B.

» A=[2000 500;1300 100]

A –

2000	500
1300	100

» B=[2500 300;1400 200]

B =

2500	300
1400	200

Step 2 Write down the difference expression $B - A$.

» **B-A**

ans =
```
   500     -200
   100      100
```

Step 3 If from 1990 to 1991 the population changes in the opposite direction by 75% of the population change from 1989 to 1990, find the population matrix for 1991.

» **B - 0.75*(B-A)**

ans =
```
  2125      450
  1325      125
```

Solution to Drill 2.2

» **A=[7000   6000   4000; 8000   5000   3000]**   % **original matrix**

A =
```
  7000      6000      4000
  8000      5000      3000
```

» **B = [0.5   0.3   0.4; 0.5   0.3   0.3]**   % **matrix of reject percentages**

B =
```
  0.5000      0.3000      0.4000
  0.5000      0.3000      0.3000
```

» **C = [0.005*7000   0.003*6000   0.004*4000;**
        0.005*8000 0.003*5000 0.003*3000]   % **matrix of rejects**

C =
```
  35      18      16
  40      15       9
```

» **A - C**   % **shippable items**

ans =
```
  6965      5982      3984
  7960      4985      2991
```

Solution to Drill 2.3

a. » **A = [0.89 2.7 0.95; -0.19 -0.7 0.57; 0.18 0.7 0.31; -0.11 -0.5 0.49; -0.53 -2.3 0.8;**
        0.42 2.11 0.51; 0.37 2.69 0.31; -1.02 -10.1 1.57; 1.66 18.26 1.68; 0.03 0.52 1.16];

» **B = [-0.55 -2.5 0.84; -0.97 -3.6 1.18; 0.42 2.0 0.4; 0.44 2.49 0.58;**
        0.44 3.31 0.56; -0.88 -6.67 0.73;-0.06 -0.64 0.36;4.81 75.84 6.48;
        0.42 11.45 1.07; 0.66 27.98 1.58];

» **C = (A+B)/2**

C =
```
    0.1700      0.1000      0.8950
   -0.5800     -2.1500      0.8750
```

0.3000	1.3500	0.3550
0.1650	0.9950	0.5350
-0.0450	0.5050	0.6800
-0.2300	-2.2800	0.6200
0.1550	1.0250	0.3350
1.8950	32.8700	4.0250
1.0400	14.8550	1.3750
0.3450	14.2500	1.3700

Each entry in *C* represents the corresponding average change from August to September in a given year.

b. » u = (A(1,3) + A(2,3) + A(3,3) + A(4,3) + A(5,3) + A(6,3) + A(7,3) + A(8,3)

  + A(9,3) + A(10,3))/10

u =

   0.8350

» v = (B(1,3) + B(2,3) + B(3,3) + B(4,3) + B(5,3) + B(6,3) + B(7,3)

  + B(8,3) + B(9,3)+ B(10,3))/10

v =

   1.3780

Solution to Drill 2.4

» A = [1 3 0; 2 1 1; -1 -2 0];

» B = [3 5 7; 9 11 1; 0 0 1];

» C = A*B

C =
30	38	10
15	21	16
-21	-27	-9

» D = B*A

D =
6	0	5
30	36	11
-1	-2	0

Solution to Drill 2.5

Step 1 Pick any three matrices of order 3 (with random entries).

  » A = rand(3), B = rand(3), C = rand(3)

  A =
0.9501	0.4860	0.4565
0.2311	0.8913	0.0185
0.6068	0.7621	0.8214

  B =
0.4447	0.9218	0.4057
0.6154	0.7382	0.9355
0.7919	0.1763	0.9169

C =
0.4103	0.3529	0.1389
0.8936	0.8132	0.2028
0.0579	0.0099	0.1987

Step 2 Find $(AB)C$.

» lhs = (A*B)*C

lhs =
1.6924	1.4640	0.6672
1.1092	0.9553	0.4575
1.8012	1.5372	0.7901

Step 3 Find $A(BC)$.

» rhs = A*(B*C)

rhs =
1.6924	1.4640	0.6672
1.1092	0.9553	0.4575
1.8012	1.5372	0.7901

Step 4 Find the difference $(A * B) * C - A * (B * C)$.

» lhs - rhs

ans =
1.0e-015 *
0	0	-0.1110
-0.2220	0	-0.0555
0.2220	0	0

Disregarding the small computational error, we conclude that $(AB)C = (AB)C$. You may repeat the process with different matrices.

Solution to Drill 2.6

» A = [2 6; 3 9]; B = [3 -1; -1 2]; C = [3 2; 1 0];
» A * (B+C) - (A*B + A*C)

ans =
0	0
0	0

This proves that $A(B + C) = AB + AC$.

» (B+C) * A - (B*A + C*A)

ans =
0	0
0	0

This proves that $(B + C)A = BA + CA$. Choosing A, B, C as rand(6), one can verify these distributive laws of multiplication over addition.

Solution to Drill 2.7

Step 1 Enter the two matrices A and B.

» A = [1 1.5 0.75; 0.75 1 1.5]

A =
   1.0000   1.5000   0.7500
   0.7500   1.0000   1.5000

» B = [0.15 0.2 0.2; 0.12 0.2 0.3]
B =
   0.1500   0.2000   0.2000
   0.1200   0.2000   0.3000

Step 2 Obtain the total number of cars sold in January and February.
» January = A(1,1) + A(1,2) + A(1,3) + B(1,1) + B(1,2) + B(1,3);
» February = A(2,1) + A(2,2) +A(2,3) + B(2,1) + B(2,2) + B(2,3);
» M = [January; February]

M =
  3.8
  3.7

Step 3 Enter 3×1 price matrices for dealer I and dealer II.
» PriceA = [15000; 16000; 15000]

PriceA =
  15000
  16000
  15000

» PriceB = [35000; 45000; 35000]

PriceB =
  35000
  45000
  35000

Step 4 Multiply the matrices A and PriceA.
» C = A*PriceA

C =
  50250
  49750

Step 5 Multiply the matrices B and PriceB.
» D = B*PriceB

D =
  21250
  23700

Step 6 Find the sum of the entries of matrices C and D.
» sum(C), sum(D)

ans =
  100000
ans =
  44950

Solution to Drill 2.8

The relations are given by

$$A[t+1] = 0.6 * A[t] + 0.7 * B[t],$$
$$B[t+1] = 0.1 * A[t] + 1.2 * B[t].$$

Step 1 Enter the transition matrix.

» A = [0.6 0.7; 0.1 1.2]

A =

   0.6000   0.7000
   0.1000   1.2000

Step 2 Enter the present population vector.

» P0 = [50000; 100000]

P0 =

   50000
   100000

Step 3 Find the population in the cities after one year.

» P1 = A*P0

P1 =

   100000
   125000

Step 4 Now find the population in the cities after 2, 3, 4, and 5 years using

» P2 = A*P1, P3 = A*P2 , P4 =A*P3, P5 = A*P4

P2 =

   147500
   160000

P3 =

   200500
   206750

P4 =

   265025
   268150

P5 =

   1.0e+005 *

   3.4672
   3.4828

Step 5 In general, the population after n years is given by the product of the nth power of A and P_0. Therefore, the populations after 4 and 5 years can be calculated using

» P4 = A^4*P0, P5 = A^5*P0

P4 =

   265025
   268150

P5 =
  1.0e+005 *
  3.4672
  3.4828

Solution to Drill 2.9

Step 1 The Vanguard Airlines routes are represented by the matrix, where cities are represented by Minnesota-1, Chicago-2, Kansas City-3, Denver-4, Dallas-5, Atlanta-6, Myrtle Beach-7, Cincinnati-8, Pittsburgh-9, Buffalo-10. Enter each row of the adjacency matrix A.

```
» row1=[0 1 1 0 0 0 0 0 0 0];
» row2=[1 0 1 0 0 0 0 1 1 1];
» row3=[1 1 0 1 1 1 0 0 0 0];
» row4=[0 0 1 0 0 0 0 0 0 0];
» row5=[0 0 1 0 0 0 0 0 0 0];
» row6=[0 0 1 0 0 0 1 0 0 0];
» row7=[0 0 0 0 0 1 0 0 1 0];
» row8=[0 1 0 0 0 0 0 0 0 0];
» row9=[0 1 0 0 0 0 1 0 0 0];
» row10=[0 1 0 0 0 0 0 0 0 0];
» A=[row1;row2;row3;row4;row5;row6;row7;row8;row9;row10]
A =
   0   1   1   0   0   0   0   0   0   0
   1   0   1   0   0   0   0   1   1   1
   1   1   0   1   1   1   0   0   0   0
   0   0   1   0   0   0   0   0   0   0
   0   0   1   0   0   0   0   0   0   0
   0   0   1   0   0   0   1   0   0   0
   0   0   0   0   0   1   0   0   1   0
   0   1   0   0   0   0   0   0   0   0
   0   1   0   0   0   0   1   0   0   0
   0   1   0   0   0   0   0   0   0   0
```

Step 2 Find the powers of A. The square of A is

```
» A2=A*A
A2 =
   2   1   1   1   1   1   0   1   1   1
   1   5   1   1   1   1   1   0   0   0
   1   1   5   0   0   0   1   1   1   1
   1   1   0   1   1   1   0   0   0   0
   1   1   0   1   1   1   0   0   0   0
   1   1   0   1   1   2   0   0   1   0
   0   1   1   0   0   0   2   0   0   0
   1   0   1   0   0   0   0   1   1   1
   1   0   1   0   0   1   0   1   2   1
   1   0   1   0   0   0   0   1   1   1
```

Step 3 The nonnegative entries in the square of the matrix A represent the cities that are connected by two flights. The value of any entry represents (for example, the (2,2) entry of the square of A is 5) the number of different flight paths that connect the two cities in two flights (one connection). In this case, there are five different ways you can fly from Chicago to some city and fly back to Chicago.

Step 4 Find other powers of A until the powers have no zero entries.

» A3=A^3

A3 =

2	6	6	1	1	1	2	1	1	1
6	2	9	1	1	2	1	5	6	5
6	9	2	5	5	6	1	1	2	1
1	1	5	0	0	0	1	1	1	1
1	1	5	0	0	0	1	1	1	1
1	2	6	0	0	0	3	1	1	1
2	1	1	1	1	3	0	1	3	1
1	5	1	1	1	1	1	0	0	0
1	6	2	1	1	1	3	0	0	0
1	5	1	1	1	1	1	0	0	0

» A4=A^4

A4 =

12	11	11	6	6	8	2	6	8	6
11	31	12	9	9	10	8	2	3	2
11	12	31	2	2	3	8	9	10	9
6	9	2	5	5	6	1	1	2	1
6	9	2	5	5	6	1	1	2	1
8	10	3	6	6	9	1	2	5	2
2	8	8	1	1	1	6	1	1	1
6	2	9	1	1	2	1	5	6	5
8	3	10	2	2	5	1	6	9	6
6	2	9	1	1	2	1	5	6	5

(a) Since (2,3) entries of first three powers of A are, 1,1,9 respectively, it follows that there is one path each of lengths 1 and 2, and 9 paths of length 3.

Note: The 4th power of A is the first power of A with no zero entries. This means that all cities of the flight map can be visited in four or fewer flights.

Solution to Drill 2.10

Step 1 Construct two matrices A and B of order 3 with random entries.

» A = rand(3), B = rand(3)

A =

0.6038	0.0153	0.9318
0.2722	0.7468	0.4660
0.1988	0.4451	0.4186

B =

0.8462	0.6721	0.6813
0.5252	0.8381	0.3795
0.2026	0.0196	0.8318

Step 2 Find their sum and product.

» **AplusB = A + B**

AplusB =

1.4500	0.6874	1.6131
0.7973	1.5849	0.8455
0.4015	0.4647	1.2504

» **AtimesB = A*B**

AtimesB =

0.7078	0.4369	1.1922
0.7169	0.8180	0.8564
0.4868	0.5149	0.6526

Step 3 Find the difference of $(A + B)^T - (A^T + B^T)$. Note that in Matlab the transpose of a matrix can be found by A'.

» **AplusB'- (A'+ B')**

ans =

0	0	0
0	0	0
0	0	0

This proves that $(A+B)^T = A^T + B^T$. Similarly, verify the other property by invoking the Matlab command.

(A*B)'- B'*A'.

CHAPTER 3

Solution to Drill 3.1

» **format rat**

Step 1 Enter the matrix A.

» **A=[-1 2 3 4; 5 0 -1 -1; 8 -6 -10 -13]**

A =

-1	2	3	4
5	0	-1	-1
8	-6	-10	-13

The idea is to find scalars c_1 and c_2 such that $(8 \ -6 \ -10 \ -13) - c_1(-1 \ 2 \ 3 \ 4) + c_2(5 \ 0 \ -1 \ -1)$. This is equivalent to solving the system of equations

$$8 = -c_1 + 5c_2$$
$$-6 = 2c_1$$
$$-10 = 3c_1 - c_2$$
$$-13 = 4c_1 - c_2$$

Step 2 We now solve the linear system by reducing the augmented matrix of the linear system to reduced row echelon form.

» **rref([-1 5 8; 2 0 -6; 3 -1 -10; 4 -1 -13])**
ans =

1	0	-3
0	1	1
0	0	0
0	0	0

This gives $c_1 = -3$, $c_2 = 1$.

Therefore, the system is consistent, and the last row can be written as a linear combination of the other two as follows.

$$(8 -6 -10 -13) = (-3) (-1\ 2\ 3\ 4) + (1) (5\ 0\ -1\ -1)$$

For parts (b) and (c), set up the equations that represent the vector equations (as in part (a)) and solve the corresponding linear systems.

Solution to Drill 3.2

» **format rat**

Step 1 We need to find whether there exist a, b, c such that

$$(1, 1, 1) = a(1, 3, 4) + b(4, 0, 1) + c(3, 1, 2).$$

This is equivalent to the set of equations

$$a + 4b + 3c = 1,$$
$$3a + c = 1,$$
$$4a + b + 2c = 1.$$

Step 2 Obtain the reduced row echelon form of the augmented matrix.

» **rref([1 4 3 1; 3 0 1 1; 4 1 2 1])**
ans =

1	0	1/3	0
0	1	2/3	0
0	0	0	1

Since the third row gives $0 = 1$, it follows that the above linear system is not consistent. Therefore the vector $(1, 1, 1)$ is not in the span of $\{(1, 3, 4), (4, 0, 1), (3, 1, 2)\}$.

Solution to Drill 3.3

» **format rat**

Step 1 Enter the matrix A.

a. » A = [1 1 3 1; 2 1 5 4; 1 2 4 -1];

Step 2 Construct a row vector by adding to row 1, 2 times row 2 and -3 times row 3. Also, construct another row vector by adding to row 1, one times row 2 and one times row 3.

» A(1, :) + 2 * A(2, :) - 3 * A(3, :), A(1, :) + A(2, :) + A(3, :)

ans =

   2   -3   1   12

ans =

   4   4   12   4

Step 3 Construct a column vector by adding to column 1, -1 times column 2. Construct another column vector by adding to column 1, -2 times column 2 and 3 times column 3.

b. » A(:, 1) - A(:, 2), A(:, 1) - 2 * A(:, 2) + 3 * A(:, 3)

ans =

   0

   1

  -1

ans =

   8

  15

  9

c. Let $A^{(4)} = aA^{(1)} + bA^{(2)} + cA^{(3)}$. This yields the linear system

$$a + b + 3c = 1$$
$$2a + b + 5c = 4$$
$$a + 2b + 4c = -1.$$

To solve this linear system we obtain the reduced row echelon form of the augmented matrix.

» rref(A)

ans =

   1   0   2   3

   0   1   1   -2

   0   0   0   0

This gives

$$a + 2c = 3, b + c = -2.$$

So there are infinitely many solutions. One may, for example, choose $c = 1; b = -3; a = 1$. This shows that the fourth column is in the subspace spanned by the first three columns of A.

d. Let $[\ 1\ 1\ 0\ 1] = a[1\ 3\ 1\ 1] + b[2\ 1\ 5\ 4] + c[1\ 2\ 4\ -1]$

This gives the linear system

$$a + 2b + c = -1$$
$$3a + b + 2c = 1$$
$$a + 5b + 4c = 0$$
$$a + 4b - c = 1.$$

To solve this linear system we obtain the reduced row echelon form of the augmented matrix.

```
» AUG = [1 2 1 -1; 3 1 2 1; 1 5 4 0; 1 4 -1 1];
» rref(AUG)

ans =
   1   0   0   0
   0   1   0   0
   0   0   1   0
   0   0   0   1
```

Since the fourth row of the augmented matrix gives $0 = 1$, it follows that the linear system is inconsistent, and thus the given vector is not in the subspace spanned by the rows.

e. We proceed as in part (c) and write the given column vector as a linear combination of the columns of the matrix, yielding the system of equations

$$a + b + 3c + d = 1$$
$$2a + b + 5c + 4d = 0$$
$$a + 2b + 4c - d = -1.$$

```
» b = [1; 0; -1];
» AUG = [A b];
» rref(AUG)

ans =
   1   0   2   3    0
   0   1   1  -2    0
   0   0   0   0    1
```

Since the third row of the augmented matrix gives $0 = 1$, it follows that the linear system is inconsistent, and thus the given vector is not in the subspace spanned by the columns.

Since the zero vector belongs to each subspace, it belongs to the subspace spanned by the columns.

f. The row vector [1 0 0 0] is in the subspace spanned the rows of A^T is equivalent to saying that the column vector $[1\ 0\ 0\ 0]^T$ is in the subspace spanned by the columns of A which is not possible because the columns of A are 3×1 column vectors.

Thus the vector $[1\ 0\ 0\ 0]^T$ is not in the subspace spanned by the columns of A.

Solution to Drill 3.4

```
» format rat
```

Step 1 Any vector in the subspace W spanned by the vectors can be written as a linear combination of the vectors $(1, 4, -1, 3)$, $(2, 1, -3, 1)$, and $(0, 2, 1, -5)$. First check whether the given vectors are linearly independent. By definition of linear independence we may write, $a(1, 4, -1, 3) +$

$b(2, 1, -3, 1) + c(0, 2, 1, -5) = (0, 0, 0, 0)$, leading to the system of equations

$$a + 2b = 0,$$
$$4a + b + 2c = 0,$$
$$-a - 3b + c = 0,$$
$$3a + b - 5c = 0.$$

We solve the above homogeneous linear system by reducing the coefficient matrix to reduced row echelon form.

» **rref([1 2 0; 4 1 2; -1 -3 1; 3 1 -5])**

ans =

1	0	0
0	1	0
0	0	1
0	0	0

This gives $a = 0, b = 0, c = 0$. Therefore, the three vectors are linearly independent and hence form a basis of the subspace spanned by the given vectors.

Solution to Drill 3.5

In order to find a basis of S we need to find a maximal subset of linearly independent vectors of the set $\{v_1, v_2, v_3, v_4\}$. We proceed to find whether the vectors v_1, v_2, v_3, v_4 are linearly independent or not. So let

(1) $\qquad a[3 \;\; -2 \;\; 2 \;\; -1] + b[2 \;\; -6 \;\; 4 \;\; 0] + c[4 \;\; 8 \;\; -4 \;\; -3]$
$\qquad\qquad + d[1 \;\; 10 \;\; -6 \;\; -2] = 0.$

This gives the following homogeneous linear system of equations:

$$3a + 2b + 4c + d = 0$$
$$-2a - 6b + 8c + 10d = 0$$
$$2a + 4b - 4c - 6d = 0$$
$$-a - 3c - 2d = 0.$$

We now obtain the reduced row echelon form of the coefficient matrix.

» **format rat;**

» **rref([3 2 4 1; -2 -6 8 10; 2 4 -4 -6; -1 0 -3 -2])**

ans =

1	0	0	-1
0	1	0	0
0	0	1	1
0	0	0	0

Reduced row echelon form yields the following system of equations:

$$a - d = 0$$
$$b = 0$$
$$c + d = 0.$$

Solving by backward substitution, we begin with the last equation $c + d = 0$. Since there are two unknowns we can choose one of the variables, say d, as arbitrary. So let $d = 1$. Then $c = -1, b = 0, a = 1$. From (1) above this gives

$$[3 \ -2\ 2\ -1] + 0[2\ -6\ 4\ 0] - [4\ 8\ -4\ -3] + [1\ 10\ -6\ -2] = 0.$$

This implies that the vectors are linearly dependent and any one of the vectors v_1, v_3, or v_4 can be expressed as a linear combination of others. So we now omit one of them and check for linear independence of the remaining vectors, say v_1, v_2, v_3. So we proceed as above with the understanding that v_4 and d are absent in the above work. Then the new system of equations immediately yields $a = 0, b = 0, c = 0$. Hence v_1, v_2, v_3 are linearly independent and they form a basis of S.

To find a basis of T we need to find a maximal subset of linearly independent vectors of the set $\{v_4, v_5, v_6\}$. We proceed to find whether the vectors v_4, v_5, v_6 are linearly independent or not. So let $a[1\ 10\ -6\ -2] + b[1\ -1\ 8\ 5] + c[6\ -2\ 4\ 8] = 0$. This gives the following homogeneous linear system:

$$a + b + 6c = 0$$
$$10a - b - 2c = 0$$
$$-6a + 8b + 4c = 0$$
$$-2a + 5b + 8c = 0.$$

We now obtain the reduced row echelon form of the coefficient matrix.

» **rref([1 1 6; 10 -1 -2; -6 8 4; -2 5 8])**

ans =

1	0	0
0	1	0
0	0	1
0	0	0

This gives $a = 0, b = 0, c = 0$. Thus the vectors v_4, v_5, v_6 are linearly independent and form a basis of T.

Finally, to find a basis of $S + T$ we need to find a maximal subset of linearly independent vectors of the set $v_1, v_2, v_3, v_4, v_5, v_6$. As before, we first find whether the vectors $v_1, v_2, v_3, v_4, v_5, v_6$ are linearly independent or not. So let

$$(1) \quad \begin{aligned} &a[3\ -2\ 2\ -1] + b[2\ -6\ 4\ 0] + c[4\ 8\ -4\ -3] \\ &+ d[1\ 10\ -6\ -2] + e[1\ -1\ 8\ 5] + f[6\ -2\ 4\ 8] = 0. \end{aligned}$$

This will give the homogeneous linear system of four equations in six unknowns. This implies that the linear system has a nontrivial solution. Therefore the given vectors are linearly dependent. However we need the values of a, b, c, d, e, and f in the dependency relation (1) to determine which vector can be dropped. The relation (1) gives the following linear system

$$3a + 2b + 4c + d + e + 6f = 0$$
$$-2a - 6b + 8c + 10d - e - 2f = 0$$
$$2a + 4b - 4c - 6d + 8e + 4f = 0$$
$$-a - 3c - 2d + 5e + 8f = 0.$$

We now obtain the reduced row echelon form of the coefficient matrix.

» **rref([3 2 4 1 1 6; -2 -6 8 10 -1 -2; 2 4 -4 -6 8 4; -1 0 -3 -2 5 8])**

ans =

1	0	0	-1	0	-664/7
0	1	0	0	0	535/7
0	0	1	1	0	236/7
0	0	0	0	1	20/7

This gives

$$a - d - \frac{664}{7}f = 0, b + \frac{535}{7}f = 0, c + d + \frac{236}{7}f = 0, e + \frac{20}{7}f = 0.$$

By choosing f and d arbitrarily, say $f = 1$ and $d = 1$ we obtain that the values of a, b, c, d, e, and f are nonzero. Hence any one of the vectors can be dropped.

Drop v_6 and proceed as above with the understanding that v_6 and f are absent in the relation (1). In this case we will obtain

$$a - d = 0, b = 0, c + d = 0, e = 0.$$

Beginning from the last equation we get $e = 0$, c or d can be chosen arbitrarily, say $c = 1$. Then $d = -1, b = 0, a = -1$. This shows that any one of the vectors v_1, v_3, or v_4 (that have nonzero coefficients in the relation (1)) can be expressed as a linear combination of the others. So we drop v_1 and proceed as in relation (1) with the understanding that v_1 and v_6 are absent. In this case we reduce the coefficient matrix of the linear system in b, c, d, and e to reduced row echelon form.

» **rref([2 4 1 1; -6 8 10 -1; 4 -4 -6 8; 0 -3 -2 5])**

ans =

1	0	0	0
0	1	0	0
0	0	1	0
0	0	0	1

This shows that $b = 0, c = 0, d = 0, e = 0$. Hence the vectors v_2, v_3, v_4, v_5 are linearly independent, forming a basis of $S + T$.

Solution to Drill 3.6

a. » **format rat**

Let $aA^{(1)} + bA^{(2)} + cA^{(3)} + dA^{(4)} + eA^{(5)} = 0$, where $A^{(i)}$ denotes the ith column of A. This yields the linear system

$$-a + 2b + 3c + d + 11e = 0$$
$$2b + 3c + 4d + 8e = 0$$
$$a - 2b + 3c - d + e = 0$$
$$a - 2b + 9c - d + 13e = 0$$
$$-a + 4b + 6c + 5d + 19e = 0.$$

To solve this linear system we obtain the reduced row echelon form of the coefficient matrix.

» **rref([-1 2 3 1 11; 0 2 3 4 8; 1 -2 3 -1 1; 1 -2 9 -1 13; -1 4 6 5 19])**

ans =

1	0	0	3	-3
0	1	0	2	1
0	0	1	0	2
0	0	0	0	0
0	0	0	0	0

This gives $a + 3d - 3e = 0$, $b + 2d + e = 0$, and $c + 2e = 0$. This shows that we can obtain nonzero values of each of the unknowns a, b, c, d, e. Thus any one of the column vectors can be dropped. We may drop the last column and the unknown e in the relation $aA^{(1)} + bA^{(2)} + cA^{(3)} + dA^{(4)} + eA^{(5)} = 0$.

The reduced row echelon form of the coefficient matrix of new linear system is the same as the first four columns of the reduced row echelon form of the coefficient matrix of the original system, yielding

(1) $a + 3d = 0, b + 2d = 0, \text{ and } c = 0.$

This gives that d can be chosen arbitrarily, say $d = 1$. Then $b = -2, a = -3$. Therefore any one of the first, second, or the fourth column can be dropped. We drop the fourth and fifth columns and unknowns d and e from the original relation leading to the system of equations $a = 0$, $b = 0$, and $c = 0$ (from (1)). This shows that the first three columns form a maximal linearly independent set of columns.

b. We observe that the subspace generated by the rows of A is the same as the subspace generated by the rows of the reduced row echelon form (formally proved in Section 4.4, Theorem 1). We reduce the given matrix to reduced row echelon form as in the part (a). The nonzero rows, namely $[1 \ 0 \ 0 \ 3 \ -3]$, $[0 \ 1 \ 0 \ 2 \ 1]$, $[0 \ 0 \ 1 \ 0 \ 2]$, are 'clearly' linearly independent (verify) and form a basis of the row space of A.

Solution to Drill 3.7

a. » **format rat**

Step 1 Enter the matrix A and find its reduced row echelon form.

» **A=[1 2 3; 4 5 6; 7 8 9]**

A =

1	2	3
4	5	6
7	8	9

» **rref(A)**

ans =

1	0	-1
0	1	2
0	0	0

As stated earlier in Drill 3.6 that the row space of any matrix is the same as the row space of its reduced row echelon form (proved formally in Section 4.4, Theorem 1), the two nonzero rows in the reduced form a basis of the row space. Hence the dimension of the row space of A is 2.

Step 2 To find the space spanned by the columns of A, we find the reduced row echelon form of A^T.

» **rref(A')**

ans =

1	0	-1
0	1	2
0	0	0

The transpose of each of the first two rows of the reduced form of A^T together form a basis for the subspace spanned by the columns of A. Therefore the dimension of this subspace is 2.

For the second matrix, proceed as before and observe that the dimension of the subspace spanned by the rows of A is equal to the subspace spanned by the columns of A.

b. To create a random matrix of size $m \times n$, use the Matlab command rand(m,n). Then proceed as in Parts (a) and (b).

Solution to Drill 3.8

Any vector on S is $[x \ \ y \ \ 2y - x]$ which can be expressed as $x[1 \ 0 \ -1] + y[0 \ 1 \ 2]$, showing that the plane is generated by $[1 \ 0 \ -1]$ and $[0 \ 1 \ 2]$. To check for these two vectors to be linearly independent or not, we solve for a and b the following equation:

$$a[1 \ 0 \ -1] + b[0 \ 1 \ 2] = 0.$$

This yields, $a = 0$, $b = 0$. So the vectors are linearly independent and thus form a basis of S. Similarly, any vector on the plane T is $[x \ \ y \ \ x - 2y]$ and is a linear combination of vectors $[1 \ 0 \ 1]$ and $[0 \ 1 \ -2]$ which forms a basis of T. If the point $[u \ v \ w]$ is in the intersection of the planes then we must have $u \ \ 2v + w = 0$, and $-u + 2v + w = 0$. By adding these equations we obtain $w = 0$, and so $u = 2v$, yielding that any point on the intersection of planes is given by $u[2 \ 1 \ 0]$. To find a basis of $S + T$ we have to find a basis of the subspace spanned by the collection of vectors generating S and generating T. Therefore, we need a maximal linearly independent subset of the set containing the vectors $[1 \ 0 \ -1]$, $[0 \ 1 \ 2]$, $[1 \ 0 \ 1]$ and $[0 \ 1 \ -2]$. Let

(1) $\qquad a[1 \ 0 \ -1] + b[0 \ 1 \ 2] + c[1 \ 0 \ 1] + d[0 \ 1 \ -2] = 0.$

So we obtain the linear system of equations given as below.

$$a + c = 0$$
$$b + d = 0$$
$$-a + 2b + c - 2d = 0.$$

We now reduce the coefficient matrix into reduced row echelon form.
» **rref([1 0 1 0; 0 1 0 1; -1 2 1 -2])**

```
ans =
    1    0    0    2
    0    1    0    1
    0    0    1   -2
```

This gives $a + 2d = 0, b + d = 0, c - 2d = 0$. Choose $d = 1$. Then $a = -2, b = -1, c = 2$. Thus any one of the four vectors can be dropped. Dropping the fourth vector and the unknown d, the reduced row echelon form of the coefficient matrix of the new system is simply the first three columns of the reduced row echelon form of the original system. So we obtain $a = 0, b = 0, c = 0$. Therefore, the vectors $[1 \ 0 \ -1]$, $[0 \ 1 \ 2]$, and $[1 \ 0 \ 1]$ are linearly independent, and so they form a basis of $S + T$. We now show how to plot the graphs of the planes S and T.

Step 1 Create 20 equally spaced points on the x axis from -2 to 2 with $\Delta x = 0.2$.
» **x = -2:0.2:2;**

Step 2 Create a mesh grid of evenly spaced points on the y-axis as created on the x axis.

» **y = x;**
» **[X,Y]=meshgrid(x,y);**

Step 3 Compute Z at the mesh points and plot the graph of S.
» **Z = -X+2*Y;**
» **mesh(X,Y,Z)**

Step 4 Now plot the graph of T on the same axes.
» **Z1 = X-2*Y;**
» **hold on**
» **mesh(X,Y,Z1)**

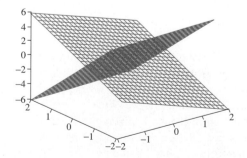

CHAPTER 4

Solution to Drill 4.1

a. Step 1 Enter the matrix A and find its reduced row echelon form.

```
» A=[2 3 4 5; 0 1 5 6; 0 0 7 8; 0 0 5 3]
A =
    2    3    4    5
    0    1    5    6
    0    0    7    8
    0    0    5    3

» rref(A)
ans =
    1    0    0    0
    0    1    0    0
    0    0    1    0
    0    0    0    1
```

Step 2 Since there are four independent rows in the reduced form, all four rows in the original matrix are linearly independent. Therefore, row space of A is spanned by the four independent rows of A.

To find a basis for the column space of A, note that the column space of A is the same as the row space of A^T. Therefore, we find the reduced row echelon form of A^T.

```
» rref(A')
ans =
    1    0    0    0
    0    1    0    0
    0    0    1    0
    0    0    0    1
```

Since there are four independent rows in the reduced form of A^T, the row space of A^T (= the column space of A) is spanned by four independent rows of A^T (= columns of A). Therefore, the column space of A is spanned by the four columns of A.

b. Step 1 Similarly, the reduced row echelon form of B and B^T will determine the row and column spaces of B.

```
» B=[1 2 3 4; 2 4 6 8; 3 5 7 9; 4 6 8 10]
B =
    1    2    3    4
    2    4    6    8
    3    5    7    9
    4    6    8   10

» rref(B)
ans =
    1    0   -1   -2
    0    1    2    3
```

$$\begin{array}{cccc} 0 & 0 & 0 & 0 \\ 0 & 0 & 0 & 0 \end{array}$$

The first two rows of the reduced form of B form a basis for the row space of A.

Step 2 Find the reduced row echelon form of B^T.

» **rref(B')**

ans =

$$\begin{array}{cccc} 1 & 2 & 0 & -2 \\ 0 & 0 & 1 & 2 \\ 0 & 0 & 0 & 0 \\ 0 & 0 & 0 & 0 \end{array}$$

Taking the transpose of each of the first two rows of the reduced form of B^T, we obtain a basis for the column space of B.

c. Step 1 Augment the matrices A and B to form C and find the row space and column space of C as above.

» **C=[A B]**

C =

$$\begin{array}{cccccccc} 2 & 3 & 4 & 5 & 1 & 2 & 3 & 4 \\ 0 & 1 & 5 & 6 & 2 & 4 & 6 & 8 \\ 0 & 0 & 7 & 8 & 3 & 5 & 7 & 9 \\ 0 & 0 & 5 & 3 & 4 & 6 & 8 & 10 \end{array}$$

Step 2 Find the reduced row echelon form of C and C^T.

» **rref(C)**

ans =

Columns 1 through 4

$$\begin{array}{cccc} 1 & 0 & 0 & 0 \\ 0 & 1 & 0 & 0 \\ 0 & 0 & 1 & 0 \\ 0 & 0 & 0 & 1 \end{array}$$

Columns 5 through 8

-11/38	-24/19	-85/38	-61/19
1/19	13/19	25/19	37/19
23/19	33/19	43/19	53/19
-13/19	-17/19	-21/19	-25/19

» **rref(C')**

ans =

$$\begin{array}{cccc} 1 & 0 & 0 & 0 \\ 0 & 1 & 0 & 0 \\ 0 & 0 & 1 & 0 \\ 0 & 0 & 0 & 1 \\ 0 & 0 & 0 & 0 \\ 0 & 0 & 0 & 0 \\ 0 & 0 & 0 & 0 \\ 0 & 0 & 0 & 0 \end{array}$$

Since there are four nonzero rows of reduced row echelon form of C (as well as of C^T), rank $C = 4$. Thus all the four rows of C form a basis of the row space of C. A basis of the column space of C consists of the transpose of the nonzero rows of reduced row echelon form of C^T.

Solution to Drill 4.2

» **format rat**

Step 1 Enter the corresponding augmented matrix and reduce the matrix to its reduced row echelon form.

» **A=[1 2 0 3 0; 2 5 2 1 0; 1 3 1 -1 0]**

```
A =
    1    2    0    3    0
    2    5    2    1    0
    1    3    1   -1    0
```

» **rref(A)**

```
ans =
    1    0    0    9    0
    0    1    0   -3    0
    0    0    1   -1    0
```

Step 2 Enter the corresponding reduced system and solve it using the back substitution method. Since we have four variables and three independent equations, we must choose one free variable (say $x_4 = t$).

» **eq1='x1 + 9*x4 = 0', eq2='x2 - 3*x4 = 0', eq3='x3 - x4 = 0'**

```
eq1 =
   x1 + 9*x4 = 0
eq2 =
   x2 - 3*x4 = 0
eq3 =
   x3 - x4 = 0
```

Step 3 Substitute $x_4 = t$ in eq3 and solve eq3 for x_3.

» **x4='t'; eq3=subs(eq3,'x4',x4); x3=solve(eq3,'x3')**
» **x3 =**
```
   t
```

Step 4 Substitute $x_4 = t$ in eq2 and solve eq2 for x_2.

» **eq2=subs(eq2,'x4',x4); x2=solve(eq2,'x2')**
```
x2 =
   3*t
```

Step 5 Substitute $x_4 = t$ in eq1 and solve eq1 for x_1.

» **eq1=subs(eq1,'x4',x4); x1=solve(eq1,'x1')**
```
x1 =
   -9*t
```

The solution space of the homogeneous system is given by $S = \{(-9t, 3t, t, t) \mid t \in \mathbf{R}\}$. By writing the general vector $(-9t, 3t, t, t) = t(-9, 3, 1, 1)$, it is clear that the set consisting of the vector $\{(-9, 3, 1, 1)\}$ is linearly independent that spans the solution space and is therefore a basis for the solution space.

Solution to Drill 4.3

» **format rat**

Step 1 Enter the corresponding augmented matrix and reduce the matrix to its reduced row echelon form.

» **A=[1 2 0 0; 2 -1 3 0]**

A =
```
 1    2    0    0
 2   -1    3    0
```

» **format rat**
» **rref(A)**

ans =
```
 1    0    6/5    0
 0    1   -3/5    0
```

Step 2 Enter the corresponding reduced system and solve it using the back substitution method. Since we have three variables and two independent equations, we must choose one free variable (say $x_3 = t$).

» **eq1='x1 + (6/5)*x3 = 0', eq2='x2-(3/5)*x3 = 0'**

eq1 =
  x1 + (6/5)*x3 = 0

eq2 =
  x2 -(3/5)*x3 = 0

Step 3 Substitute $x_3 = t$ in eq2 and eq1 and solve for x_1 and x_2.

» **x3='t'; eq2=subs(eq2,'x3',x3);**
» **x2=solve(eq2,'x2')**

x2 =
  3/5*t

» **eq1=subs(eq1,'x3',x3);**
» **x1=solve(eq1,'x1')**

x1 =
  -6/5*t

The solution set is given by $\{((-6/5)t, (3/5)t, t) \mid t \in \mathbf{R}\}$

Step 4 To plot the graph with z-coordinate in $[-2, 2]$ and $\Delta = 1/10$, use the following commands.

» **t = -2:1/10:2;**   % **define a set of points in [-2 2] with △=0.1**
» **plot3((-6/5)*t,(3/5)*t,t);**   % **plot the graph in 3-dimension space**

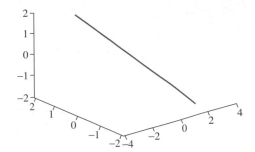

Solution to Drill 4.4

» **format rat**

Step 1 The null space of a matrix A is the set $W = \{x \text{ in } F^n \mid Ax = 0\}$. Therefore, finding the null space of the matrix A is equivalent to finding the solution of the homogeneous system $Ax = 0$. Enter the coefficient matrix A and reduce the matrix to its row echelon form.

» **A=[1 1 1 1 1; 1 2 1 2 1; 1 3 3 1 1; 1 4 4 1 1]**

A =

1	1	1	1.	1
1	2	1	2	1
1	3	3	1	1
1	4	4	1	1

» **rref(A)**

ans =

1	0	0	1	1
0	1	0	1	0
0	0	1	-1	0
0	0	0	0	0

Step 2 The reduced form of the matrix is equivalent to the following system.

» **eq1='x1+x4+x5=0', eq2='x2+x4=0',eq3='x3-x4=0'**

eq1 =
   x1+x4+x5=0

eq2 =
   x2+x4=0

eq3 =
   x3-x4=0

Step 3 Since there are five variables and three equations in the reduced form, we must choose two free variables. Choose $x_4 = t$ and solve eq3.

» **x4='t';**
» **eq3=subs(eq3,'x4',x4);**
» **x3=solve(eq3,'x3')**

x3 =
   t

Substitute $x_3 = t$ in eq2 and solve for x_2.

```
» eq2=subs(eq2,'x4',x4);
» x2=solve(eq2,'x2')

x2 =
 -t
```

Step 4 To solve eq1 for the other unknown variables x_1 and x_5 choose one of them to be the second free variable. Let $x_5 = a$.

```
» x5='a';
» eq1=subs(eq1,{'x4', 'x5'},{x4,x5});
» x1=solve(eq1,'x1')

x1 =
 -t-a
```

Therefore, the null space of A is given by the set $\{(-t - a, -t, t, t, a) \mid a, t \in R\}$. To find a basis for the null space of A, write the general vector $(-t - a, -t, t, t, a)$ as follows: $(-t - a \ -t \ t \ t \ a) = t(-1 \ -1 \ 1 \ 1 \ 0) + a(-1 \ 0 \ 0 \ 0 \ 1)$. Therefore a basis for the null space is the set consisting of the two independent vectors $\{(-1 \ -1 \ 1 \ 1 \ 0), (-1 \ 0 \ 0 \ 0 \ 1)\}$.

Solution to Drill 4.5

a. Step 1 Enter the matrix A and find its reduced row echelon form.

```
» A=[1 2 1 0; 2 5 3 -1; 2 2 0 2; 0 1 1 -1]
A =
 1    2    1    0
 2    5    3   -1
 2    2    0    2
 0    1    1   -1
» rref(A)
ans =
 1    0   -1    2
 0    1    1   -1
 0    0    0    0
 0    0    0    0
```

Step 2 Based on the reduced form of A, the matrix A has two linearly independent rows and therefore the row rank of A = rank of A = 2. To find the null space of A, solve the homogeneous system, $Ax = 0$, and we obtain two independent equations in four variables. Therefore, we choose two free variables and the dimension of the null space is 2 (see Remark 4.2.4). Therefore, rank A + nullity of A = number of columns in A = 4.

b. Step 1 Enter 5×7 random matrix with integral entries, say from 1–9.

```
» A=round(10*rand(5,7))
A =
 10    8    6    4    1    2    0
  2    5    8    9    4    2    7
```

6	0	9	9	8	6	4
5	8	7	4	0	3	9
9	4	2	9	1	2	5

Step 2 Compute the rank of A, and reduce A into reduced row echelon form in order to solve the linear system $Ax = 0$.

```
» format rat, rank(A), rref(A)
```

ans =

   5

ans =

Columns 1 through 6

1	0	0	0	0	134/289
0	1	0	0	0	-179/123
0	0	1	0	0	355/218
0	0	0	1	0	37/160
0	0	0	0	1	-267/158

Column 7

   26/227

   -767/246

   839/218

   643/400

   -2095/366

Step 3 To find nullity we solve $Ax = 0$. The reduced row echelon form of A gives rise to a new system that we solve by backward substitution. So we begin with the last equation.

```
» lasteq='x5+(-267/1580)*x6+(-2095/366)*x7=0'
lasteq =
   x5+(-267/1580)*x6+(-2095/366)*x7=0
```

Since this equation has three variables, we can choose any two of them arbitrarily. Then we can determine the values of other variables from the rest of the equations. By Remark 4.2.2 it follows that the nullity is 2. Hence rank + nullity is 7, the number of columns. This verifies rank-nullity theorem.

Solution to Drill 4.6

a. Step 1 Create two random matrices using the random function and find their product.

```
» A=round(10*rand(4,6))
A =
```

3	2	0	7	4	9
5	4	8	5	2	7
8	0	7	4	8	5
7	4	3	2	2	7

```
» B=round(10*rand(6,3))
```

B =

3	5	2
7	4	9
9	6	5
9	8	6
8	7	3
3	3	5

» AB=A*B

AB =

145	134	123
197	164	157
202	185	124
131	120	118

Step 2 Find the reduced row echelon form of A, B, and AB.

» rref(A)

ans =

1	0	0	0	187/249	356/743
0	1	0	0	-793/743	538/743
0	0	1	0	-24/743	-123/370
0	0	0	1	117/211	145/166

» rref(B)

ans =

1	0	0
0	1	0
0	0	1
0	0	0
0	0	0
0	0	0

» rref(AB)

ans =

1	0	0
0	1	0
0	0	1
0	0	0

Step 3 From the reduced row echelon forms, it follows that the ranks of A, B, and AB are 4, 3, 3 (number of nonzero rows in the reduced form). Therefore, $\text{rank}(AB) \leqslant \min(\text{rank}(A), \text{rank}(B))$.

b. Step 1 Enter two random matrices A and B of the same size.

» A=round (10*rand (4,5)), B=round (10*rand (4,5))

A =

10	9	8	9	9
2	8	4	7	9
6	5	6	2	4
5	0	8	4	9

B =

1	1	3	4	8
4	2	2	9	5
8	2	0	5	2
0	6	7	4	7

Step 2 Compute $\text{rank}(A + B) - \text{rank}(A) - \text{rank}(B)$.

» **rank (A+B)-rank (A)-rank (B)**

ans =

-4

Thus $\text{rank}(A + B) < \text{rank}(A) + \text{rank}(B)$. In general, equality is also possible.

c. Step 1 Enter a random matrix by giving the command

» **A=round(10*rand(4,6))**

A =

10	9	8	9	9	1
2	8	4	7	9	4
6	5	6	2	4	8
5	0	8	4	9	0

Step 2 Compute $\text{rank}(AA') - \text{rank}(A)$ and $\text{rank}(A'A) - \text{rank}(A)$ by giving the following commands.

» **rank(A*A')-rank(A),rank(A'*A)-rank(A)**

ans =

0

ans =

0

Thus $\text{rank}(AA') = \text{rank}(A) = \text{rank}(A'A)$

Solution to Drill 4.7

a. We first verify Theorem 1 for the operation $(1/2)R_1$.

» **format rat**

Step 1 Enter the matrix and the corresponding identity matrix

» **A=[2 3 7 6; 5 4 2 0; 1 -1 1 -1]**

A =

2	3	7	6
5	4	2	0
1	-1	1	-1

» **ID=eye(3)**

ID =

1	0	0
0	1	0
0	0	1

Step 2 Divide row 1 of A by 2 and perform the same operation on ID.

» **A1=A;**

» A1(1,:)=(1/2)*A1(1,:)

A1 =

1	3/2	7/2	3
5	4	2	0
1	-1	1	-1

» E1=ID;
» E1(1,:)=(1/2)*E1(1,:)

E1 =

1/2	0	0
0	1	0
0	0	1

Step 3 Show that $A1$ is equal to $E1A$.

» E1*A

ans =

1	3/2	7/2	3
5	4	2	0
1	-1	1	-1

This verifies Theorem 1 for the elementary row operation $(1/2)R_2$.

b. » format rat

Step 1 Next, we perform $(-3)C_2 + C_3$ on A. Call the new matrix $A1$.

» A1=A;
» A1(:,3)=(-3)*A1(:,2) + A1(:,3)

A1 =

2	3	-2	6
5	4	-10	0
1	-1	4	-1

Step 2 Apply the same operation to the 4×4 identity matrix. Call the new matrix $E1$.

» ID=eye(4)

ID =

1	0	0	0
0	1	0	0
0	0	1	0
0	0	0	1

» E1=ID;
» E1(:,3)=(-3)*E1(:,2) + E1(:,3)

E1 =

1	0	0	0
0	1	-3	0
0	0	1	0
0	0	0	1

Step 3 Verify that $A1$ is indeed equal to $AE1$ (postmultiplying A by the elementary matrix $E1$).

» A*E1-A1

ans =
$$
\begin{array}{cccc}
0 & 0 & 0 & 0 \\
0 & 0 & 0 & 0 \\
0 & 0 & 0 & 0
\end{array}
$$

Thus $AE1 = A1$.

Solution to Drill 4.8

Step 1 Enter the matrix A and reduce the matrix to its row echelon form.

» A=[1 1 1 1; 2 3 1 2; 1 -1 3 2]

A =
$$
\begin{array}{cccc}
1 & 1 & 1 & 1 \\
2 & 3 & 1 & 2 \\
1 & -1 & 3 & 2
\end{array}
$$

» A(2,:)=-2*A(1,:) + A(2,:)

A =
$$
\begin{array}{cccc}
1 & 1 & 1 & 1 \\
0 & 1 & -1 & 0 \\
1 & -1 & 3 & 2
\end{array}
$$

» A(3,:)=-1*A(1,:) + A(3,:)

A =
$$
\begin{array}{cccc}
1 & 1 & 1 & 1 \\
0 & 1 & -1 & 0 \\
0 & -2 & 2 & 1
\end{array}
$$

» A(3,:)=2*A(2,:) + A(3,:)

A =
$$
\begin{array}{cccc}
1 & 1 & 1 & 1 \\
0 & 1 & -1 & 0 \\
0 & 0 & 0 & 1
\end{array}
$$

Step 2 For each elementary row operation performed on A, we construct the corresponding elementary matrix.

» E1 = eye(3)

E1 =
$$
\begin{array}{ccc}
1 & 0 & 0 \\
0 & 1 & 0 \\
0 & 0 & 1
\end{array}
$$

» E1(2,:) = -2* E1(1,:) + E1(2,:)

E1 =
$$
\begin{array}{ccc}
1 & 0 & 0 \\
-2 & 1 & 0 \\
0 & 0 & 1
\end{array}
$$

» E2 = eye(3)

E2 =
$$
\begin{array}{ccc}
1 & 0 & 0 \\
0 & 1 & 0 \\
0 & 0 & 1
\end{array}
$$

» E2(3,:) = -1* E2(1,:) + E2(3,:)

E2 =

1	0	0
0	1	0
-1	0	1

» E3 = eye(3)

E3 =

1	0	0
0	1	0
0	0	1

» E3(3,:) = 2*E3(2,:) + E3(3,:)

E3 =

1	0	0
0	1	0
0	2	1

Step 3 $P = E3\ E2\ E1$ is a product of elementary matrices such that PA is in row echelon form. The original matrix A must be entered again. Otherwise, in our calculations of the product PA, the matrix A obtained at the end of Step 1 will be used in Matlab computations.

» P = E3 * E2 * E1

P =

1	0	0
-2	1	0
-5	2	1

» A=[1 1 1 1; 2 3 1 2; 1 -1 3 2]

A =

1	1	1	1
2	3	1	2
1	-1	3	2

» P*A

ans =

1	1	1	1
0	1	-1	0
0	0	0	1

Solution to Drill 4.9

Step 1 Enter the matrix A and reduce the matrix to its row echelon form.

» A=[1 1 2 6; 3 4 -1 5; -1 1 1 2]

A =

1	1	2	6
3	4	-1	5
-1	1	1	2

» A(2,:)=-3*A(1,:) + A(2,:)

A =

1	1	2	6
0	1	-7	-13
-1	1	1	2

```
» A(3,:)=1*A(1,:) + A(3,:)
A =
   1   1    2    6
   0   1   -7  -13
   0   2    3    8
» A(3,:)=-2*A(2,:) + A(3,:)
A =
   1   1    2    6
   0   1   -7  -13
   0   0   17   34
» A(3,:)=(1/17)*A(3,:)
A =
   1   1    2    6
   0   1   -7  -13
   0   0    1    2
```

Step 2 Apply the same operations (in that order) to the 3×3 identity matrix. For each operation create the elementary matrix E_i.

```
» ID=eye(3);
» E1=ID;
» E1(2,:)=-3*E1(1,:) + E1(2,:)
E1 =
    1   0   0
   -3   1   0
    0   0   1
» E2=ID;
» E2(3,:)=1*E2(1,:) + E2(3,:)
E2 =
    1   0   0
    0   1   0
    1   0   1
» E3=ID;
» E3(3,:)=-2*E3(2,:) + E3(3,:)
E3 =
    1   0   0
    0   1   0
    0  -2   1
» E4=ID;
» E4(3,:)=(1/17)*E4(3,:)
E4 =
    1   0    0
    0   1    0
    0   0  1/17
```

Step 3 Show that the product of the elementary matrices $P = E4\ E3\ E2\ E1$ will indeed convert A to its row echelon form. We must reenter the original matrix A. Otherwise, in our calculations of the product PA, the matrix A obtained at the end of Step 1 will be used in Matlab computations.

```
» A=[1 1 2 6; 3 4 -1 5; -1 1 1 2]
```

```
A =
   1    1    2    6
   3    4   -1    5
  -1    1    1    2
» E4*E3*E2*E1*A
ans =
   1    1    2    6
   0    1   -7  -13
   *    *    1    2
```

Note that the *'s in the final answer are small quantities created by computational error.

CHAPTER 5

Solution to Drill 5.1

> **» format rat**

Step 1 Enter the matrix A and augment A with the identity matrix.

> **» A=[1 0 1; 0 -1 1; 2 3 4]**

```
A =
   1    0    1
   0   -1    1
   2    3    4
» AUG=[A eye(3)]
AUG =
   1    0    1    1    0    0
   0   -1    1    0    1    0
   2    3    4    0    0    1
```

Step 2 Apply Gauss elimination and reduce the matrix to its reduced row echelon form.

> **» AUG(3,:)=(-2)*AUG(1,:)+AUG(3,:)**   % multiply row 1 by -2 and add to row 3

```
AUG =
   1    0    1    1    0    0
   0   -1    1    0    1    0
   0    3    2   -2    0    1
```

> **» AUG(2,:)=(-1)*AUG(2,:)**   % multiply row 2 by -1

```
AUG =
   1    0    1    1    0    0
   0    1   -1    0   -1    0
   0    3    2   -2    0    1
```

> **» AUG(3,:)=(-3)*AUG(2,:)+AUG(3,:)**   % multiply row 2 by -3 and add to row 3

```
AUG =
   1    0    1    1    0    0
   0    1   -1    0   -1    0
   0    0    5   -2    3    1
```

> **» format rat**
> **» AUG(3,:)=(1/5)*AUG(3,:)**   % divide row 3 by 5

```
AUG =
   1    0    1    1    0    0
   0    1   -1    0   -1    0
   0    0    1  -2/5  3/5  1/5
```

» AUG(2,:)=(1)*AUG(3,:)+AUG(2,:)   % multiply row 3 by 1 and add to row 2

```
AUG =
   1    0    1    1    0    0
   0    1    0  -2/5 -2/5  1/5
   0    0    1  -2/5  3/5  1/5
```

» AUG(1,:)=(-1)*AUG(3,:)+AUG(1,:)   % multiply row 3 by -1 and add to row 1

```
AUG =
   1    0    0   7/5 -3/5 -1/5
   0    1    0  -2/5 -2/5  1/5
   0    0    1  -2/5  3/5  1/5
```

Step 3 The augmented matrix $[A \ I]$ was reduced to a matrix of the form $[I \ B]$, and we claim that B is the inverse of A. To verify our claim, let us compute AB and BA.

» B=AUG(:,[4,5,6])   % extract matrix B from the reduced
                                      % matrix AUG

```
B =
   7/5   -3/5  -1/5
  -2/5   -2/5   1/5
  -2/5    3/5   1/5
```

» B*A
```
ans =
   1    0    *
   0    1    *
   0    0    1
```

» A*B
```
ans =
   1    0    0
   0    1    0
   *    *    1
```

Clearly, the result is the identity matrix in both cases (disregard the small computational error shown as *). Since the inverse is unique, we conclude that $B = A^{-1}$.

Note: We can also use the Matlab command rref to find the reduced row echelon form of AUG as follows.

» AUG=[A eye(3)];
» rref(AUG)

```
ans =
   1    0    0   7/5 -3/5 -1/5
   0    1    0  -2/5 -2/5  1/5
   0    0    1  -2/5  3/5  1/5
```

Solution to Drill 5.2

» **format rat**

Step 1 Enter the matrix A and augment the matrix with the identity matrix.

» **A=[1 1 2 1; 2 -1 1 2; -1 2 1 -2; 1 -1 1 -1]**

```
A =
    1     1     2     1
    2    -1     1     2
   -1     2     1    -2
    1    -1     1    -1
```

» **AUG=[A eye(4)]**

```
AUG =
Columns 1 through 6
    1     1     2     1     1     0
    2    -1     1     2     0     1
   -1     2     1    -2     0     0
    1    -1     1    -1     0     0
Columns 7 through 8
    0     0
    0     0
    1     0
    0     1
```

Step 2 Find the reduced row echelon form of AUG.

» **AUG=rref(AUG)**

```
AUG =
Columns 1 through 6
    1     0     0     0    -3     4
    0     1     0     0  -5/3   7/3
    0     0     1     0   7/3  -8/3
    0     0     0     1     1    -1
Columns 7 through 8
    3    -1
    2    -1
   -2     1
   -1     0
```

Step 3 Extract the submatrix B containing the last four columns of the reduced form of AUG.

» **B=AUG(:,[5,6,7,8])**

```
B =
   -3      4      3     -1
 -5/3    7/3      2     -1
  7/3   -8/3     -2      1
    1     -1     -1      0
```

» **B*A**

```
ans =
    1     0     0     0
    0     1     0     0
```

```
*   0   1   *
0   0   0   1
```

» A*B
ans =
```
1   *   0   0
*   1   0   0
0   *   1   0
*   0   0   1
```

Since the inverse is unique, we conclude that $B = A^{-1}$

Solution to Drill 5.3

a. **Step 1** Enter the matrix A.

» A = sym('[a 1 1 1; 1 b 2 -1; 2 2 3 1; -1, 1 1 1]')

A =
[a, 1, 1, 1]
[1, b, 2, -1]
[2, 2, 3, 1]
[-1, 1, 1, 1]

Step 2 Interchange rows 1 and 4.

» A = A([4, 2, 3, 1], :)

A =
[-1, 1, 1, 1]
[1, b, 2, -1]
[2, 2, 3, 1]
[a, 1, 1, 1]

Step 3 Using $A(1, 1)$, convert the entries $A(2, 1)$, $A(3, 1)$, and $A(4, 1)$ to zero.

» A(2, :) = A(2, :) + A(1, :); A(3, :) = A(3, :) + 2*A(1, :); A(4, :) = A(4, :) + symmul('a', A(1, :))

A =
[-1, 1,   1, 1]
[0, b+1, 3, 0]
[0, 4,   5, 3]
[0, 1+a, 1+a, 1+a]

Step 4 Divide fourth row by $1 + a$, in order to make $A(4, 2)$ equal to 1. Note that since we are dividing by $1 + a$, $1 + a$ must be different from zero, that is, a must be different from -1.

» A(4, :) = symmul('1/(1+a)', A(4, :))

A =
[-1, 1,   1, 1]
[0, b+1, 3, 0]
[0, 4,   5, 3]
[0, 1,   1, 1]

Step 5 Interchange rows 2 and 4 to make $A(2, 2)$ equal to 1.

```
» A = A([1, 4, 3, 2], :)
A =
[-1,  1,  1,  1]
[ 0,  1,  1,  1]
[ 0,  4,  5,  3]
[ 0,  b+1,  3,  0]
```

Step 6 Using $A(2, 2)$, convert the entries $A(3, 2)$ and $A(4, 2)$ to zero.

```
» A(3,:) = A(3,:) + (-4)*A(2,:); A(4,:) = A(4, :) + symmul('-(b+1)', A(2,:))
A =
[-1,  1,  1,  1]
[ 0,  1,  1,  1]
[ 0,  0,  1,  -1]
[ 0,  0,  2-b,  -b-1]
```

Step 7 Using $A(3, 3)$, convert $A(4, 3)$ to zero.

```
» A(4,:) = A(4, :) + symmul('-(2-b)', A(3,:))
A =
[-1,  1,  1,  1]
[ 0,  1,  1,  1]
[ 0,  0,  1,  -1]
[ 0,  0,  0,  -2*b+1]
```

We conclude that for A to be invertible, $-2b + 1$ must be different from zero. That is, b must be different from $-1/2$.

Note: An alternative solution using determinants is given in Chapter 6 (Drill 6.9).

b. The solution of linear system is quite complex in terms of a and b, although one can compute it by giving the command *inv(A) * [1 2 3 4]'*. We solve the system in a special case by choosing $a = 0 = b$.

Step 1 Enter the matrix A.

```
» A = sym('[a, 1, 1, 1; 1, b, 2, -1; 2, 2, 3, 1; -1, 1, 1, 1]')
A =
[ a,  1,  1,  1]
[ 1,  b,  2,  -1]
[ 2,  2,  3,  1]
[-1,  1,  1,  1]
```

Step 2 Choose $a = 0 = b$ and substitute these values in A.

```
» a = 0; b = 0;
» A = subs(A, 'a', a); A = subs(A, 'b', b)
A =
   0    1    1    1
   1    0    2   -1
   2    2    3    1
  -1    1    1    1
```

Step 2 Solve the linear system using the matlab command *inv(A) * [1 2 3 4]'*.

» inv(A) * [1 2 3 4]'

ans =
-3
12
-2
-9

Solution to Drill 5.4

Step 1 Enter the matrix A.

» A = [1 2 4; 3 0 1; 1 -4 -7]
A =
1   2   4
3   0   1
1   -4   -7

Step 2 Reduce A to row echelon form.

» A(2, :) = (-3)*A(1, :) + A(2, :)
A =
1   2   4
0   -6   -11
1   -4   -7

» A(3, :) = (-1)*A(1, :) + A(3, :)
A =
1   2   4
0   -6   -11
0   -6   -11

» format rat
» A(2, :) = (-1/6)*A(2, :)

A =
1   2   4
0   1   11/6
0   -6   -11

» A(3, :) = 6*A(2, :) + A(3, :)
A =
1   2   4
0   1   11/6
0   0   0

Step 3 Extract the submatrix G, consisting of the nonzero rows of A.

» G = A([1 2], :)
G =
1   2   4
0   1   11/6

Step 4 To obtain the matrix F, perform on the 3×3 identity matrix the inverse of the row operations performed on A but in the reverse order.

» I = eye(3)

I =
```
1    0    0
0    1    0
0    0    1
```
» I(3, :) = (-6)*I(2, :) + I(3, :)
I =
```
1    0    0
0    1    0
0   -6    1
```
» I(2, :) = (-6)*I(2, :)
I =
```
1    0    0
0   -6    0
0   -6    1
```
» I(3, :) = I(1, :) + I(3, :)
I =
```
1    0    0
0   -6    0
1   -6    1
```
» I(2, :) = 3*I(1, :) + I(2, :)
I =
```
1    0    0
3   -6    0
1   -6    1
```

Step 5 Since there are only two rows in G ($=$ number of nonzero rows in the row echelon form of A), form the 3×2 matrix F whose columns are the first two columns of I obtained in Step 3.

» F=I(:,[1 2])
F =
```
1    0
3   -6
1   -6
```

Step 6 Verify $A = FG$.

» A = [1 2 4; 3 0 1; 1 -4 -7]
A =
```
1    2    4
3    0    1
1   -4   -7
```
» F*G
ans =
```
1    2    4
3    0    1
1   -4   -7
```

Solution to Drill 5.5

Step 1 Enter the matrix A.

» A=[2 1 -1 0; 1 0 2 3; -1 1 0 4; 3 0 1 0; 1 2 3 4]

A =

2	1	-1	0
1	0	2	3
-1	1	0	4
3	0	1	0
1	2	3	4

Step 2 Reduce *A* to row echelon form.

» A(1,:) = (1/2)*A(1,:)

A =

1	1/2	-1/2	0
1	0	2	3
-1	1	0	4
3	0	1	0
1	2	3	4

» A(2,:)=(-1)*A(1,:)+A(2,:);
» A(3,:)=1*A(1,:)+A(3,:);
» A(4,:)=(-3)*A(1,:)+A(4,:);
» A(5,:)=(-1)*A(1,:)+A(5,:)

A =

1	1/2	-1/2	0
0	-1/2	5/2	3
0	3/2	-1/2	4
0	-3/2	5/2	0
0	3/2	7/2	4

» A(2,:)=(-2)*A(2,:)

A =

1	1/2	-1/2	0
0	1	-5	-6
0	3/2	-1/2	4
0	-3/2	5/2	0
0	3/2	7/2	4

» A(3,:)=(-3/2)*A(2,:)+A(3,:);
» A(4,:)=(3/2)*A(2,:)+A(4,:);
» A(5,:)=(-3/2)*A(2,:)+A(5,:)

A =

1	1/2	-1/2	0
0	1	-5	-6
0	0	7	13
0	0	-5	-9
0	0	11	13

» A(3,:)=(1/7)*A(3,:)

A =

1	1/2	-1/2	0
0	1	-5	-6
0	0	1	13/7
0	0	-5	-9
0	0	11	13

```
» A(4,:)=5*A(3,:)+A(4,:);
» A(5,:)=(-11)*A(3,:)+A(5,:)
A =
   1    1/2   -1/2     0
   0     1     -5     -6
   0     0      1    13/7
   0     0      0    2/7
   0     0      0   -52/7
» A(4,:)=(7/2)*A(4,:)
A =
   1    1/2   -1/2     0
   0     1     -5     -6
   0     0      1    13/7
   0     0      0     1
   0     0      0   -52/7
» A(5,:)=(52/7)*A(4,:)+A(5,:)
A =
   1    1/2   -1/2     0
   0     1     -5     -6
   0     0      1    13/7
   0     0      0     1
   0     0      0     *
```

Step 3 Extract the submatrix G by choosing the nonzero rows of A.

```
» G = A([1:4],:)
G =
   1    1/2   -1/2     0
   0     1     -5     -6
   0     0      1    13/7
   0     0      0     1
```

Step 4 To obtain the matrix F, perform on the 4×4 identity matrix the inverse of the row operations performed on A but in reverse order. Note, however, that since rank$A = 4$, A is a full-column-rank matrix. Thus we may choose $F = A$ and $G = I$.

Solution to Drill 5.6

Step 1 Enter the matrix A.

```
» A=[1 2 3 0; 2 9 6 10; 3 7 10 7]
A =
   1    2    3    0
   2    9    6   10
   3    7   10    7
```

Step 2 Augment the matrix A with the 3×3 identity matrix.

```
» AUG=[A eye(3)]
AUG =
Columns 1 through 6
   1    2    3    0    1    0
   2    9    6   10    0    1
   3    7   10    7    0    0
```

Column 7
  0
  0
  1

Step 3 Reduce AUG into its row echelon form (not row reduced echelon form).

» **AUG(2, :)=(-2)*AUG(1, :) + AUG(2, :)**

AUG =

Columns 1 through 6

1	2	3	0	1	0
0	5	0	10	-2	1
3	7	10	7	0	0

Column 7
  0
  0
  1

» **AUG(3,:)=(-3)*AUG(1,:)+AUG(3,:)**

AUG =

Columns 1 through 6

1	2	3	0	1	0
0	5	0	10	-2	1
0	1	1	7	-3	0

Column 7
  0
  0
  1

» **format rat; AUG(2,:)=(1/5)*AUG(2,:)**

AUG =

Columns 1 through 6

1	2	3	0	1	0
0	1	0	2	-2/5	1/5
0	1	1	7	-3	0

Column 7
  0
  0
  1

» **AUG(3,:)=(-1)*AUG(2,:)+AUG(3,:)**

AUG =

Columns 1 through 6

1	2	3	0	1	0
0	1	0	2	-2/5	1/5
0	0	1	5	-13/5	-1/5

Column 7
  0
  0
  1

Step 4   Extract submatrices U and V consisting of columns 1 through 4 (corresponding to the columns of A) and columns 5 through 7 (corresponding to I_3).

» U=AUG(:,1:4)

U =

1	2	3	0
0	1	0	2
0	0	1	5

» V=AUG(:,5:7)

V =

1	0	0
-2/5	1/5	0
-13/5	-1/5	1

Step 5   Compute inverse of V, which is the desired lower triangular matrix L.

» L=inv(V)

L =

1	0	0
2	5	*
3	1	1

Step 6   Verify that $A = LU$.

» L*U

ans =

1	2	3	0
2	9	6	10
3	7	10	7

» A

A =

1	2	3	0
2	9	6	10
3	7	10	7

Note: Alternatively, we could have produced U by reducing A to row echelon form in the same manner as we have done with the augmented $B = [A \ I_3]$. To obtain L, we would then perform the inverse of the elementary operations on I_3 but in the reverse order.

Solution to Drill 5.7

We may either proceed as in Drill 5.6, or follow the steps given in a note at the end of the solution of the Drill. For the sake of illustrating an approach other than the one shown in Drill 5.6, we will follow the method in the Note. Indeed Example 5.4.2 follows the same steps. Note that the matrix A is not a square matrix. We will obtain a 3×3 lower triangular matrix L, and a 3×4 upper triangular matrix U. (U is upper triangular in the sense that for all $i > j$, the (i, j) entry of U is zero.)

Step 1   Enter the matrix A.

» A = [1 1 1 7; 6 12 18 0; 3 9 6 10]

A =
```
1    1    1    7
6   12   18    0
3    9    6   10
```

Step 2 Perform elementary row operations to reduce *A* into row echelon form.

» A(2,:) = A(2,:) + (-6)*A(1,:); A(3,:) = A(3,:) + (-3)*A(1,:)

A =
```
1    1    1     7
0    6   12   -42
0    6    3   -11
```

» A(3,:) = A(3,:) + (-1)*A(2,:)

A =
```
1    1    1     7
0    6   12   -42
0    0   -9    31
```

» U = A

U =
```
1    1    1     7
0    6   12   -42
0    0   -9    31
```

Step 3 Perform, in reverse order, the inverse of the above elementary operations on the identity matrix to get *L*.

» L = eye(3)

L =
```
1    0    0
0    1    0
0    0    1
```

» L(3,:) = L(3,:) + L(2,:)

L =
```
1    0    0
0    1    0
0    1    1
```

» L(2,:) = L(2,:) + 6*L(1,:); L(3,:) = L(3,:) + 3*L(1,:)

L =
```
1    0    0
6    1    0
3    1    1
```

Solution to Drill 5.8

Step 1 Enter the matrix *A*

» A=[1 2 3; 2 9 6; 3 2 10]

A =
```
1    2    3
2    9    6
3    2   10
```

Step 2 Find the LU-decomposition of *A* using the Matlab lu command.

» [L,U]=lu(A)

$$L =$$

1/3	4/23	1
2/3	1	0
1	0	0

$$U =$$

3	2	10
0	23/3	-2/3
0	0	-5/23

Step 3 The system $Ax = b$ is equivalent to $LUx = b$. Let $Ux = a$. Then $La = b$. Solve $La = b$ for a using forward substitution.

```
» eq1='(1/3)*a1+(4/23)*a2+a3=0';
» eq2='(2/3)*a1+a2=10';
» eq3='a1=7';
» a1=solve(eq3,'a1')
```

a1 =
  7

```
» eq2=subs(eq2,'a1',a1)
```

eq2 =
  14/3+a2 = 10

```
» a2=solve(eq2,'a2')
```

a2 =
  16/3

```
» eq1=subs(eq1,{'a1','a2'},{a1,a2});
» a3=solve(eq1,'a3')
```

a3 =
  -75/23

Step 4 Now solve the system $Ux = a$ for x.

```
» eq1='3*x1 + 2*x2 + 10*x3 = 7'; eq2='(23/3)*x2+(-2/3)*x3=16/3';
» eq3='(-5/23)*x3=-75/23';
» x3=solve(eq3,'x3')
```

x3 =
  15

```
» eq2=subs(eq2,'x3',x3);
» x2=solve(eq2,'x2')
```

x2 =
  2

```
» eq1=subs(eq1,'x3',x3);
» eq1=subs(eq1,'x2',x2);
» x1=solve(eq1,'x1')
```

x1 =
  -49

Therefore, the solution to the system is given by $x = [-49, 2, 15]$.

CHAPTER 6

Solution to Drill 6.1

Step 1 The determinant of the matrix A can be found using the formula

$$\det A = a_{11}(-1)^{(1+1)} \cdot \det(B_{11}) + a_{12}(-1)^{(1+2)} \cdot \det(B_{12})$$
$$+ a_{13}(-1)^{(1+3)} \cdot \det(B_{13}) + a_{14}(-1)^{(1+4)} \cdot \det(B_{14}) \qquad (1)$$

where B_{ij} is the submatrix obtained from A by deleting the ith row and jth column of A. But since a_{12}, a_{13}, a_{14} are all zero, we only need to find $a_{11}(-1)^{1+1} \det(B_{11})$. Find the submatrix B_{11} by deleting row 1 and column 1 of A. That is, construct the submatrix of A consisting of rows $\{2, 3, 4\}$ and columns $\{2, 3, 4\}$.

```
» A = [ 1 0 0 0; 2 1 2 -1; 0 0 4 5; 0 0 0 6]
A =
    1    0    0    0
    2    1    2   -1
    0    0    4    5
    0    0    0    6
» B11=A([2:4],[2:4])
B11 =
    1    2   -1
    0    4    5
    0    0    6
```

Step 2 To find the determinant of B_{11}, expand B_{11} by the third row (since the third row has the greatest number of zeros).

$$\det B_{11} = b_{31}(-1)^{(3+1)} \det(C_{31}) + b_{32}(-1)^{(3+2)} \det(C_{32})$$
$$+ b_{33}(-1)^{(3+3)} \det(C_{33}), \qquad (2)$$

where C_{ij} is a submatrix obtained from B_{11} by deleting row i and column j of B_{11}. Since $b_{31} = b_{32} = 0$, we only need to find the $\det(C_{33})$.

```
» C33=B11([1:2],[1:2])
C33 =
    1    2
    0    4
```

Step 3 Since $\det(C_{33}) = (1)(4)-(2)(0) = 4$, we have from equation (2), $\det(B_{11})$ $= (6)(+1)(4) = 24$ and from equation (1), $\det(A) = (1)(+1)(24) = 24$. We can also verify the determinant using the Matlab command det.

```
» det(A)
ans =
   24
```

Solution to Drill 6.2

Step 1 Enter the matrix A and set P to 1, where P is the determinant of the identity matrix.

```
» A=[2 0 -1 3; 4 0 1 -1; -3 1 0 1; 1 4 1 1], P = 1
A =
   2    0    -1    3
   4    0    1    -1
  -3    1    0    1
   1    4    1    1
P =
   1
```

Step 2 Reduce A into row echelon form, and calculate its effect on the value of the determinant by multiplying P with the appropriate number.

```
» format rat;                    % display in rational mode
» A(1,:)=(1/2)*A(1,:), P = P*2   % divide row 1 by 2 and calculate the effect on P

A =
   1     0    -1/2    3/2
   4     0    1       -1
  -3     1    0       1
   1     4    1       1

P =
   2

» A(2,:)=(-4)*A(1,:)+A(2,:);     % multiply row 1 by -4 add it to row 2
» A(3,:)=3*A(1,:)+A(3,:);        % multiply row 1 by 3 add it to row 3
» A(4,:)=(-1)*A(1,:)+A(4,:);     % multiply row 1 by -1 add it to row 4
» A, P = P*1;                    % no change in the value of the determinant

A =
   1     0    -1/2    3/2
   0     0    3       -7
   0     1    -3/2    11/2
   0     4    3/2     -1/2

» A=A([1,3,2,4],:),              % interchange rows 2 and 3
  P =(-1)*P                      % determinant changes sign

A =
   1     0    -1/2    3/2
   0     1    -3/2    11/2
   0     0    3       -7
   0     4    3/2     -1/2

P =
  -2

» A(4,:)=(-4)*A(2,:)+A(4,:),     % multiply row 2 by -4 and add to row 4
  P = P*1;                       % no change in the value of the determinant

A =
   1     0    -1/2    3/2
   0     1    -3/2    11/2
   0     0    3       -7
   0     0    15/2    -45/2

» A(3,:)=(1/3)*A(3,:);           % divide row 3 of A by 3
» P = 3*P;                       % determinant is multiplied by 3
```

» A(4,:)=(-15/2)*A(3,:)+A(4,:),   % multiply row 3 by -15/2 and add to row 4
  P = 1*P   % no change in the value of the determinant

A =
1   0   -1/2   3/2
0   1   -3/2   11/2
0   0   1   -7/3
0   0   0   -5

P =
 -6

» A(4,:)=(-1/5)*A(4,:),   % multiply row 4 by -1/5
  P = -5*P   % determinant is multiplied by -5

A =
1   0   -1/2   3/2
0   1   -3/2   11/2
0   0   1   -7/3
0   0   0   1

P =
 30

Therefore the determinant of the original matrix A is 30.

Step 3 Verify that 30 is indeed the determinant of A by directly computing the determinant of A using the Matlab command $\det(A)$.

» A=[2 0 -1 3; 4 0 1 -1; -3 1 0 1; 1 4 1 1]

A =
 2   0   -1   3
 4   0   1   -1
-3   1   0   1
 1   4   1   1
» det(A)

ans =
 30

Solution to Drill 6.3

a. Step 1 Enter any 5×5 random matrix.

» A=round(10*rand(5,5))

A =
10   2   6   5   9
 8   5   0   8   4
 6   8   9   7   2
 4   9   9   4   9
 1   4   8   0   1

Step 2 Interchange any two rows (say the first and third rows) of A and call the new matrix B. Find the determinants of A and B.

» B=A([3,2,1,4,5],:)

B =

6	8	9	7	2
8	5	0	8	4
10	2	6	5	9
4	9	9	4	9
1	4	8	0	1

» **det(A)**

ans =

5972

» **det(B)**

ans =

-5972

This verifies part (a).

b. Step 1 Multiply second row of A by a scalar, say -3 and call the new matrix B. Find the determinants of A and B.

» **B=A;**   % **make a copy of A**

» **B(2,:)=(-3)*B(2,:)**   % **multiply row 2 of B by 3**

B =

10	2	6	5	9
-24	-15	0	-24	-12
6	8	9	7	2
4	9	9	4	9
1	4	8	0	1

» **det(B)**

ans =

-17916

» **det(A)**

ans =

5972

Clearly, $\det(B) = -(3)\det(A)$.

c. Step 1 Enter any upper triangular matrix, say.

» **B= [10 2 6 5 9; 0 5 0 8 4; 0 0 9 7 2; 0 0 0 4 9; 0 0 0 0 1]**

B =

10	2	6	5	9
0	5	0	8	4
0	0	9	7	2
0	0	0	4	9
0	0	0	0	1

» **det(B)**   % **the determinant of the matrix B**

ans =

1800

» **B(1,1)*B(2,2)*B(3,3)*B(4,4)*B(5,5)**   % **is the same as the product of the**
                                              % **diagonal elements**

ans =

1800

d. **Step 1** Consider the matrix as in part (a).

» **det(A)**   % **the determinant of A**

ans =
    5972

» **det(A')**   % **is the same as the determinant of A^T**

ans =
    5972

Solution to Drill 6.4

Step 1 Enter the matrix A and set P to 1, where P is the determinant of the identity matrix.

» **A=[4 5 0 1 0; 0 0 0 0 1; 4 1 8 2 0; 1 0 0 1 0; 4 8 0 1 0], P = 1**

A =

4	5	0	1	0
0	0	0	0	1
4	1	8	2	0
1	0	0	1	0
4	8	0	1	0

P =
    1

Step 2 Reduce A into row echelon form, and calculate its effect on the value of the determinant by multiplying P with the appropriate number.

» **A(1,:) = (1/4)*A(1,:);   P = P*4;** % **divide row 1 of A by 4**
» **A(3,:)=(-4)*A(1,:)+A(3,:); P = P*1;** % **multiply row 1 of A by -4 and add to row 3**
» **A(4,:)=(-1)*A(1,:)+A(4,:); P = P*1;** % **multiply row 1 by -1 and add to row 4**
» **A(5,:)=(-4)*A(1,:)+A(5,:), P = P*1** % **multiply row 1 by -4 and add to row 5**

A =

1	5/4	0	1/4	0
0	0	0	0	1
0	-4	8	1	0
0	-3/4	0	3/4	0
0	3	0	0	0

P =
    4

» **A=A([1,3,2,4,5],:), P = -P**   % **interchange second and third rows**

A =

1	5/4	0	1/4	0
0	-4	8	1	0
0	0	0	0	1
0	-5/4	0	3/4	0
0	3	0	0	0

P =
    -4

```
» A(2,:) = (-1/4)*A(2,:);        P = -4*P;     % divide row 2 by -4
» A(4,:)=(5/4)*A(2,:)+A(4,:); P = 1*P;       % multiply row 2 by 5/4 and add to row 1
» A(5,:)=(-3)*A(2,:)+A(5,:),  P = 1*P        % multiply row 2 by -3 and add to row 5
```

A =
1	5/4	0	1/4	0
0	1	-2	-1/4	0
0	0	0	0	1
0	0	-5/2	7/16	0
0	0	6	3/4	0

P =
   16

```
» A=A([1,2,4,3,5],:), P = -P                    % interchange row 3 and row 4
```
A =
1	5/4	0	1/4	0
0	1	-2	-1/4	0
0	0	-5/2	7/16	0
0	0	0	0	1
0	0	6	3/4	0

P =
  -16

```
» A(3,:) = (-2/5)*A(3,:);        P = (-5/2)*P;  % divide row 3 by -5/2
» A(5,:)=(-6)*A(3,:)+A(5,:), P = 1*P            % multiply row 3 by -6 and add to row 5
```
A =
1	5/4	0	1/4	0
0	1	-2	-1/4	0
0	0	1	-7/40	0
0	0	0	0	1
0	0	0	9/5	0

P =
   40

```
» A=A([1:3,5,4],:), P = -P                     % interchange row 4 and row 5
```

A =
1	5/4	0	1/4	0
0	1	-2	-1/4	0
0	0	1	-7/40	0
0	0	0	9/5	0
0	0	0	0	1

P =
  -40

```
» A(4,:)=(5/9)*A(4,:), P = (9/5)*P             % divide row 4 by 9/5
```

A =
1	5/4	0	1/4	0
0	1	-2	-1/4	0
0	0	1	-7/40	0
0	0	0	1	0
0	0	0	0	1

P =
     -72

Step 3 Verify that -72 is indeed the determinant of A by directly computing the determinant of A using the Matlab command $\det(A)$.

» A=[4 5 0 1 0; 0 0 0 0 1; 4 1 8 2 0; 1 0 0 1 0; 4 8 0 1 0]

A =
     4   5   0   1   0
     0   0   0   0   1
     4   1   8   2   0
     1   0   0   1   0
     4   8   0   1   0

» det(A)

ans =
     -72

Solution to Drill 6.5

Step 1 Enter the matrix A.

» A=[0 -3 4; 0 5 0; 1 -2 0]

A =
     0   -3   4
     0   5   0
     1   -2   0

Step 2 Find the matrix $B = A - xI$, where I is the 3×3 identity matrix.

» I=eye(3);   % define the 3x3 identity matrix I
» B=A-sym('x')*I   % find the matrix B = A-xI

B =
[-x, -3, 4]
[0, 5-x, 0]
[1, -2, -x]

Step 3 Find the determinant of the symbolic matrix B.

» determ(B)

ans =
     5*x^2-x^3-20+4*x

Step 4 Solve the equation $\text{determ}(B) = 0$ for x.

» solve('5*x^2-x^3-20+4*x=0','x')

ans =
[5]
[2]
|-2|

Solution to Drill 6.6

Step 1 Enter two arbitrary random matrices A and B.

» A=rand(3,3)

A =
10	2	6
5	9	8
5	0	8

» B=rand(3,3)

B =
4	6	8
9	7	2
4	9	9

Step 2 Evaluate $\det(AB) - \det(A) * \det(B)$.

» det(A*B) - det(A)*det(B)

ans =

0

Step 3 Note that the value of $\det(AB) - \det(A) * \det(B)$ is zero.

Solution to Drill 6.7

a. Step 1 Let A be a 4×4 random matrix.

» A=round(10*rand(4,4))

A =
10	2	6	5
9	8	5	0
8	4	6	8
9	7	2	4

Step 2 In the first inequality, the expression on right side is equivalent to $\prod \left(\sqrt{(c_i' c_i)} \right)$, c_i' is the transpose of c_i and the product $(\prod)$ is taken over all columns c_i. Extract all columns of A.

» c1=A(:,1);
» c2=A(:,2);
» c3=A(:,3);
» c4=A(:,4);

Step 3 Find the product of all terms of the form $\sqrt{(c_i' c_i)}$ for each column c_i.

» RHS=sqrt(c1'*c1)*sqrt(c2'*c2) *sqrt(c3'*c3)*sqrt(c4'*c4)

RHS =

2.1443e+004

Step 4 Find the determinant of A and compare.

» det(A)

ans =

1366

Clearly the first inequality holds.

b. The inequality is not true, in general. For example, for $A = I$, $\det A = 1$ but the right hand side is 0.

Solution to Drill 6.8

Step 1 Enter the matrix A and find the cofactor A_{ij} by removing ith row and jth column from the matrix A.

» A=[1 -1 1 2; 1 0 1 3; 0 0 2 4; 1 1 -1 1]

```
A =
   1    -1    1    2
   1     0    1    3
   0     0    2    4
   1     1   -1    1
```

» A11 = (-1)^(1+1)*det(A([2:4],[2:4]))

```
A11 =
  -2
```

» A12=(-1)^(1+2)*det(A ([2:4],[1,3:4]))

```
A12 =
  -4
```

» A13=(-1)^(1+3)*det(A ([2:4],[1:2,4]))

```
A13 =
  -4
```

» A14=(-1)^(1+4)*det(A ([2:4],[1:3]))

```
A14 =
   2
```

» A21= (-1)^(2+1)*det(A([1,3:4],[2:4]))

```
A21 =
   6
```

» A22= (-1)^(2+2)*det(A([1,3:4],[1,3:4]))

```
A22 =
   6
```

» A23= (-1)^(2+3)*det(A([1,3:4],[1:2,4]))

```
A23 =
   8
```

» A24= (-1)^(2+4)*det(A([1,3:4],[1:3]))

```
A24 =
  -4
```

» A31= (-1)^(3+1)*det(A([1:2,4],[2:4]))

```
A31 =
  -3
```

» A32= (-1)^(3+2)*det(A([1:2,4],[1,3:4]))

```
A32 =
  -2
```

» A33= (-1)^(3+3)*det(A([1:2,4],[1:2,4]))

```
A33 =
  -3
```

» **A34= (-1)^(3+4)*det(A([1:2,4],[1:3]))**
A34 =
  2

» **A41= (-1)^(4+1)*det(A([1:3],[2:4]))**
A41 =
  -2

» **A42= (-1)^(4+2)*det(A([1:3],[1,3:4]))**
A42 =
  -2

» **A43= (-1)^(4+3)*det(A([1:3],[1:2,4]))**
A43 =
  -4

» **A44= (-1)^(4+4)*det(A([1:3],[1:3]))**
A44 =
  2

Step 2 Form the cofactor matrix B using the A_{ij}'s.

» **B=[A11 A12 A13 A14; A21 A22 A23 A24;A31 A32 A33 A34; A41 A42 A43 A44]**
B =
-2	-4	-4	2
6	6	8	-4
-3	-2	-3	2
-2	-2	-4	2

Step 3 The adjoint matrix is the transpose of the cofactor matrix.

» **adj A=B'**

adj A =
-2	6	-3	-2
-4	6	-2	-2
-4	8	-3	-4
2	-4	2	2

Step 4 Inverse of A is the adjoint matrix divided by the determinant of A.

» **invA=(1/det(A))*adjA**
invA =
-1	3	-3/2	-1
-2	3	-1	-1
-2	4	-3/2	-2
1	-2	1	1

Step 5 Verify the results by finding A^{-1} directly using Matlab.

» **inv(A)**
ans =
-1	3	-3/2	-1
-2	3	-1	-1
-2	4	-3/2	-2
1	-2	1	1

Solution to Drill 6.9

Step 1 Enter the matrix A in symbolic form.

» **A=sym('[a, 1, 1, 1; 1, b, 2, -1; 2, 2, 3, 1; -1,1,1,1]')**
A =
[a, 1, 1, 1]
[1, b, 2, -1]
[2, 2, 3, 1]
[-1, 1, 1, 1]

Step 2 Find the determinant of the matrix A.

» **determ(A)**
ans =
  2*a*b-a + 2*b-1

Step 3 Verify that $a = -1$ or $b = 1/2$ will make the determinant zero.

Solution to Drill 6.10

Step 1 Enter the coefficient matrix A and the right vector b.

» **A=[2 3 1 -1; 1 2 5 3; -1 0 3 1; 1 -2 1 0]**
A =
   2   3   1   -1
   1   2   5   3
  -1   0   3   1
   1   -2   1   0
» **b=[2 5 1 -2]'**

b =
   2
   5
   1
  -2

Step 2 Let A_1 be the matrix obtained from the matrix A by replacing its first column with the vector b.

» **A1=[b A(:,[2:4])]**
A1 =
   2   3   1   -1
   5   2   5   3
   1   0   3   1
  -2   -2   1   0

Step 3 By Cramer's rule, the first component of the solution is given by

» **x1=det(A1)/det(A)**
x1 =
   0

Step 4 Similarly, find the other components of the solution.

» **A2=[A(:,1) b A(:,[3:4])]**   % **replace second column of A by vector b**

```
A2 =
    2    2    1   -1
    1    5    5    3
   -1    1    3    1
    1   -2    1    0
» x2=det(A2)/det(A)                    % find the second component of the solution
x2 =
    1
» A3=[A(:,[1:2]) b A(:,4)]             % replace third column of A by vector b
A3 =
    2    3    2   -1
    1    2    5    3
   -1    0    1    1
    1   -2   -2    0
» x3=det(A3)/det(A)                    % find the third component of the solution
x3 =
    0
» A4=[A(:,[1:3]) b]                    % replace fourth column of A by vector b
A4 =
    2    3    1    2
    1    2    5    5
   -1    0    3    1
    1   -2    1   -2
» x4=det(A4)/det(A)                    % find the fourth component of the solution
x4 =
    1
```

Therefore the solution to the system is $(0, 1, 0, 1)$.

CHAPTER 7

Solution to Drill 7.1

Step 1 Enter the matrix A and form the matrix $B = A - xI$, where I is the 3×3 identity matrix.

```
» A=[1 -3 3; 0 -1 2; 0 -3 4]
A =
    1   -3    3
    0   -1    2
    0   -3    4
» B=A-sym('x')*eye(3)
B =
[1-x, -3, 3]
[0, -1-x, 2]
[0, -3, 4-x]
```

Step 2 Solve the equation $\det(A - xI) = 0$ and find the eigenvalues of A.

```
» P = determ(B)
```

P =
   (1-x)*(2-3*x+x^2)
» **solve(P, 'x')**
ans =
   [1]
   [2]

Step 3 To find the eigenvector corresponding to each eigenvalue x, we shall solve the homogeneous system $(A - xI)v = 0$ for v.

Case $x = 1$: Consider $(A - 1I)v = 0$. Reduce the matrix $A - I$ to its reduced row echelon form.

» **B=A-eye(3)**
B =
   0   -3   3
   0   -2   2
   0   -3   3
» **B = rref(B)**
B =
   0   1   -1
   0   0   0
   0   0   0

The reduced system, $Bv = 0$ is equivalent to $v_2 - v_3 = 0$, and v_1 is arbitrary. Let $v_1 = 1$ and $v_2 = 1$. Then $v_3 = 1$. Therefore, an eigenvector corresponding to eigenvalue $\lambda = 1$ is $(1, 1, 1)$.

Case $x = 2$: Consider $(A - 2I)v = 0$. Reduce the matrix $A - 2I$ to its reduced row echelon form.

» **B=A-2*eye(3)**
B =
   -1   -3   3
    0   -3   2
    0   -3   2
» **B = rref(B)**
B =
   1   0   -1
   0   1   -2/3
   0   0   0

The reduced system $Bv = 0$ is equivalent to $v_1 - v_3 = 0$ and $v_2 - (2/3)v_3 = 0$. Choose an arbitrary value for $v_3 = 1$. Then we get $v_2 = (2/3)$ and $v_1 = 1$. Therefore, an eigenvector corresponding to eigenvalue $x = 2$ is $(1, 2/3, 1)$.

Solution to Drill 7.2

Step 1 Enter the matrix A and form the matrix $B = A - xI$, where I is the 3×3 identity matrix.

» **A=[1 -2 -2 -2; -2 1 -2 -2; -2 -2 1 -2; -2 -2 -2 1]**

```
A =
   1    -2    -2    -2
  -2     1    -2    -2
  -2    -2     1    -2
  -2    -2    -2     1
» B=A-sym('x')*eye(4)

B =
[1-x,    -2,    -2,    -2]
[ -2,   1-x,    -2,    -2]
[-2,    -2,    1-x,    -2]
[-2,    -2,    -2,    1-x]
```

Step 2 Solve the equation $\det(A - xI) = 0$ and find the eigenvalues of A.

```
» P = determ(B)
P =
  -135+108*x-18*x^2-4*x^3+x^4

» solve(P,'x')
ans =
  [ -5]
  [3]
  [3]
  [3]
```

Step 3 To find the eigenvector corresponding to each eigenvalue x, we shall solve the homogeneous system $(A - xI)v = 0$ for v.

Case $x = -5$: Consider $(A + 5I)v = 0$. Reduce the matrix $A + 5I$ to its reduced row echelon form.

```
» B=A+5*eye(4)
B =
   6    -2    -2    -2
  -2     6    -2    -2
  -2    -2     6    -2
  -2    -2    -2     6
» B = rref(B)
B =
   1     0     0    -1
   0     1     0    -1
   0     0     1    -1
   0     0     0     0
```

The reduced system $Bv = 0$ is equivalent to

$$v_1 - v_4 = 0$$
$$v_2 - v_4 = 0$$
$$v_3 - v_4 = 0$$

By choosing $v_4 = 1$, we obtain $v_1 = v_2 = v_3 = 1$. Therefore, an eigenvector corresponding to eigenvalue $x = -5$ is $(1, 1, 1, 1)$.

Case $x = 3$: Consider $(A - 3I)v = 0$. Reduce the matrix $A - 3I$ to its row echelon form.

» **B=A-3*eye(4)**

B =

-2	-2	-2	-2
-2	-2	-2	-2
-2	-2	-2	-2
-2	-2	-2	-2

» **B = rref(B)**

B =

1	1	1	1
0	0	0	0
0	0	0	0
0	0	0	0

The reduced system $Bv = 0$ is equivalent to $v_1 + v_2 + v_3 + v_4 = 0$. Since there is only one equation in four unknowns, three of the unknowns, say, v_2, v_3, v_4 can be chosen arbitrarily. By choosing (v_2, v_3, v_4) as $(1, 0, 0)$, $(0, 1, 0)$, and $(0, 0, 1)$ in succession, we obtain three linearly independent eigenvectors $(-1, 1, 0, 0)$, $(-1, 0, 1, 0)$, and $(-1, 0, 0, 1)$ corresponding to eigenvalue $x = 3$.

Solution to Drill 7.3

Step 1 Enter the matrix A and form the matrix $B = A - xI$, where I is the 3×3 identity matrix.

» **A=[1 1 2; -1 2 1; 0 1 3]**

A =

1	1	2
-1	2	1
0	1	3

» **B=A-sym('x')*eye(3)**

B =

[1-x, 1, 2]
[-1, 2-x, 1]
[0, 1, 3-x]

Step 2 Solve the equation $\det(A - xI) = 0$ and find the eigenvalues of A.

» **P = determ(B)**

P =

  6-11*x+6*x^2-x^3

» **solve(P, 'x')**

ans =

 [1]
 [2]
 [3]

Step 3 To find the eigenvector corresponding to each eigenvalue x, we shall solve the homogeneous system $(A - xI)v = 0$ for v.

Case $x = 1$: Consider $(A - I)v = 0$. Reduce the matrix $A - I$ to its reduced row echelon form.

```
» B=A-eye(3)
B =
   0    1    2
  -1    1    1
   0    1    2
» rref(B)
ans =
   1    0    1
   0    1    2
   0    0    0
```

The reduced system $Bv = 0$ is equivalent to

$$v_1 + v_3 = 0,$$
$$v_2 + 2v_3 = 0.$$

By choosing $v_3 = 1$, we obtain $v_2 = -2$ and $v_1 = -1$. Therefore, an eigenvector corresponding to eigenvalue $x = 1$ is $(-1, -2, 1)$.

Case $x = 2$: Consider $(A - 2I)v = 0$. Reduce the matrix $A - 2I$ to its row echelon form.

```
» B=A-2*eye(3)
B =
  -1    1    2
  -1    0    1
   0    1    1
» rref(B)
ans =
   1    0   -1
   0    1    1
   0    0    0
```

The reduced system $Bv = 0$ is equivalent to $v_1 - v_3 = 0$ and $v_2 + v_3 = 0$. By choosing $v_3 = 1$, we obtain $v_2 = -1$ and $v_1 = 1$. Therefore, an eigenvector for eigenvalue 2 is $(1, -1, 1)$. Similarly, we can find the eigenvector for eigenvalue 3.

Solution to Drill 7.4

Step 1 Enter the matrix A and find its eigenvalues and eigenvectors by using Matlab command $[U, D] = \text{eig}(A)$.

```
» A=[1 1 1 1; 0 0 1 1; 0 0 1 0; 0 0 1 2]
A =
   1    1    1    1
   0    0    1    1
```

$$\begin{array}{cccc} 0 & 0 & 1 & 0 \\ 0 & 0 & 1 & 2 \end{array}$$

» [U, D]=eig(A)

U =

$$\begin{array}{cccc} 1.0000 & -0.7071 & 0.8018 & 0 \\ 0 & 0.7071 & 0.2673 & 0 \\ 0 & 0 & 0 & 0.7071 \\ 0 & 0 & 0.5345 & -0.7071 \end{array}$$

D =

$$\begin{array}{cccc} 1 & 0 & 0 & 0 \\ 0 & 0 & 0 & 0 \\ 0 & 0 & 2 & 0 \\ 0 & 0 & 0 & 1 \end{array}$$

Step 2 Find the reduced row echelon form of U^T, and determine (by inspection) the number of linearly independent rows of U^T.

» rref(U')

ans =

$$\begin{array}{cccc} 1 & 0 & 0 & 0 \\ 0 & 1 & 0 & 0 \\ 0 & 0 & 1 & 0 \\ 0 & 0 & 0 & 1 \end{array}$$

It follows that all the rows of U^T, and hence the columns of U, are linearly independent.

Solution to Drill 7.5

Step 1 Find the eigenvalues of A, A^2, A^3, A^4, and A^{-1}.

» A=[1 1 2; -1 2 1; 0 1 3]

A =

$$\begin{array}{ccc} 1 & 1 & 2 \\ -1 & 2 & 1 \\ 0 & 1 & 3 \end{array}$$

» eig(A)

ans =

1
2
3

» eig(A^2)

ans =

4
9
1

» eig(A^3)

ans =

8
27
1

» **eig(A^4)**

ans =

   81
   16
   1

» **format rat**
» **eig(inv(A))**

ans =

   1
   1/2
   1/3

Conclusion: The eigenvalues of A^n = (eigenvalues of A)^n and the eigenvalues of A^{-1} = 1/(eigenvalues of A).

Solution to Drill 7.6

Step 1 Find the coefficients of the characteristic polynomial using the Matlab command poly(A).

» **format rat**
» **A=[1 -1 2; 0 3 2; 2 1 2]**

A =

   1   -1   2
   0   3   2
   2   1   2

» **poly(A)**

ans =

   1   -6   5   12

The coefficients of the characteristic polynomial are given in highest (exponent) to lowest (exponent) order. Therefore, the characteristic polynomial is $p(x) = x^3 - 6x^2 + 5x + 12$. Using Matlab, show that the matrix A satisfies $p(x) = 0$.

» **A^3-6*A^2+5*A+12*eye(3)**

ans =

   0   0   0
   0   0   0
   0   0   0

Step 2 By Cayley-Hamilton theorem, $A^3 - 6A^2 + 5A + 12I = 0$. So $A(A^2 - 6A + 5I) = -12I$. Since the inverse of A is unique, $A^{-1} = (A^2 - 6A + 5I)/(-12)$. We shall verify this using Matlab.

» **inv(A)**

ans =

   -1/3   -1/3   2/3
   -1/3   1/6   1/6
    1/2   1/4   -1/4

» **(-1/12)*(A^2-6*A+5*eye(3))**

ans =
-1/3	-1/3	2/3
-1/3	1/6	1/6
1/2	1/4	-1/4

Solution to Drill 7.7

Step 1 Find the coefficients of the characteristic polynomial using the Matlab command poly(A).

» A=[1 0 0; 5 0 1; 3 0 1]

A =

1	0	0
5	0	1
3	0	1

» poly(A)

ans =

1	-2	1	0

Here the coefficients are given in highest (exponent) to lowest (exponent) order. Therefore, the characteristic equation is $p(x) = x^3 - 2x^2 + x = 0$. Using Matlab, show that the matrix A satisfies $p(x) = 0$. That is, $p(A) = 0$.

» A^3-2*A^2+A

ans =

0	0	0
0	0	0
0	0	0

Since the constant term of the characteristic polynomial of A is zero, it follows that $\det A = 0$. Hence by Property 6.2.9, A^{-1} does not exist.

Solution to Drill 7.8

Step 1 The sum of the eigenvalues is given by the sum of the diagonal entries (known as the trace) of the matrix.

» A=[2 0 1; 1 0 -1; 1 5 3]

A =

2	0	1
1	0	-1
1	5	3

» A(1,1)+A(2,2)+A(3,3)

ans =

5

Step 2 We can verify this property by finding the eigenvalues of the matrix using Matlab.

» eig(A)

```
ans =
  1096/327
  539/654      + 1132/581i
  539/654      - 1132/581i
» ans(1)+ans(2)+ans(3)
ans =
  5
```

Step 3 Similarly, we can verify that the product of the eigenvalues of A is indeed given by the determinant of A.

```
» det(A)
ans =
  15
```

```
» eig(A)
ans =
  1096/327
  539/654      + 1132/581i
  539/654      - 1132/581i
» ans(1)*ans(2)*ans(3)
ans =
  15 + 1/1125899906842624i
```

Note that the extremely small imaginary part of the above complex number is a result of computational error.

Solution to Drill 7.9

Step 1 Find the eigenvalues of A.

```
» A = [1 1 2; -1 2 1; 0 1 3]
A =
   1    1    2
  -1    2    1
   0    1    3
» eig(A)
ans =
   1
   2
   3
```

Step 2 Now we find the eigenvectors for each eigenvalue x. Consider $x = 1$. Solve the system $(A - I_3)v = 0$ for v.

```
» rref(A-1*eye(3))
ans =
   1    0    1
   0    1    2
   0    0    0
```

It follows that $v_1 + v_3 = 0$ and $v_2 + 2v_3 = 0$.

By choosing $v_3 = 1$, we get an eigenvector $[-1 \ -2 \ 1]$ for $x = 1$. Call vector1 $= [-1 \ -2 \ 1]'$.

Case $x = 2$:

» **rref(A-2*eye(3))**

ans =

1	0	-1
0	1	1
0	0	0

It follows that $v_1 - v_3 = 0$ and $v_2 + v_3 = 0$.

By choosing $v_3 = 1$, we get an eigenvector $[1 \ -1 \ 1]$ for $x = 2$. Call vector2 $= [1 \ -1 \ 1]'$.

Case $x = 3$:

» **rref(A-3*eye(3))**

ans =

1	0	-1
0	1	0
0	0	0

This gives $v_1 = v_3, v_2 = 0$. By choosing $v_3 = 1$, we get an eigenvector $[1 \ 0 \ 1]$ for $x = 3$. Call vector3 $= [1 \ 0 \ 1]'$.

Step 3 Let P be the matrix whose columns are the three eigenvectors vector1, vector2, and vector3 obtained above.

» **vector1 = [-1 -2 1]'; vector2 = [1 -1 1]'; vector3 = [1 0 1]';**

» **P = [vector1 vector2 vector3]**

P =

-1	1	1
-2	-1	0
1	1	1

» **inv(P) * A * P**

ans =

1	*	*
*	2	*
*	*	3

Solution to Drill 7.10

Step 1 Find the eigenvalues of A.

» **A=[4 -1 -2; 2 1 -2; 1 -1 1]**

A =

4	-1	-2
2	1	-2
1	-1	1

» **eig(A)**

ans =

1

3

2

Step 2 Now we find the eigenvectors for each eigenvalue x. Consider $x = 1$. Solve the system $(A - I_3)v = 0$ for v.

» **rref(A-1*eye(3))**

ans =

1	0	-1
0	1	-1
0	0	0

It follows that $v_1 - v_3 = 0$ and $v_2 - v_3 = 0$.

By choosing $v_3 = 1$, we get an eigenvector $(1, 1, 1)$ for $x = 1$.

Case $x = 2$:

» **rref(A-2*eye(3))**

ans =

1	0	-1
0	1	0
0	0	0

This gives $v_1 - v_3 = 0$ and $v_2 = 0$.

By choosing $v_3 = 1$, we get an eigenvector $(1, 0, 1)$ for $x = 2$.

Case $x = 3$:

» **rref(A-3*eye(3))**

ans =

1	-1	0
0	0	1
0	0	0

It follows that $v_1 = v_2$ and $v_3 = 0$.

By choosing $v_2 = 1$, we get an eigenvector $(1\ \ 1\ \ 0)$ for $x = 3$.

Step 3 Let P be the matrix whose columns are the three eigenvectors.

» **v1=[1 1 1]';**
» **v2=[1 0 1]';**
» **v3=[1 1 0]';**
» **P=[v1 v2 v3]**

P =

1	1	1
1	0	1
1	1	0

Step 4 Show that $P^{-1}AP$ is a diagonal matrix whose diagonal entries are the eigenvalues of A.

» **inv(P)*A*P**

ans =

1	0	0
0	2	0
0	0	3

Note that the eigenvectors of the matrix can be found directly using Matlab command,

» **[U,D] = eig(A)**

The columns of U are the eigenvectors and D is a diagonal matrix consisting of the corresponding eigenvalues.

Solution to Drill 7.11

Step 1 The product of the eigenvalues is given by the determinant of the matrix.

» A = sym('[4, 3, 5, 7; 2, x, 0, 1; 2, 2, 3, 8; 1, 3, 6, 2]')

A =
[4, 3, 5, 7]
[2, x, 0, 1]
[2, 2, 3, 8]
[1, 3, 6, 2]

» determ(A)
ans =
-85*x+15

Step 2 Since the product of the eigenvalues is given to be 355, we solve the following equation to find x.

» x=solve('-85*x+15=355','x')

x =
 -4

Solution to Drill 7.12 (Gershgorin Theorem)

Step 1 Choose a 5×5 random matrix B as below.

» B=round(10*rand(5))

B =
10	8	6	4	1
2	5	8	9	4
6	0	9	9	8
5	8	7	4	0
9	4	2	9	1

Step 2 Form the symmetric matrix $A = B + B^T$.

» A=B+B'

A =
20	10	12	9	10
10	10	8	17	8
12	8	18	16	10
9	17	16	8	9
10	8	10	9	2

Step 3 Find the eigenvalues of A and five intervals given by the formula, where the center of each interval is given by a_{ii} and radius is given by the absolute sum of the entries $a_{ij}, i \neq j$.

» eig(A)

```
ans =
  -10.0792
  -4.3400
  6.3022
  9.5625
  56.5545
» L1 = sum(A(1,:))-A(1,1), L2 = sum(A(2,:))-A(2,2), L3 = sum(A(3,:))-A(3,3)
  L4 = sum(A(4,:))-A(4,4), L5 = sum(A(5,:))-A(5,5)

L1 =
  41

L2 =
  43

L3 =
  46

L4 =
  51

L5 =
  37
```

	Center	Radius	Interval boundaries
Interval 1	A(1,1)=20	sum(A(1,:)) -A(1,1)=41	(-21,61)
Interval 2	A(2,2)=10	sum(A(2,:)) -A(2,2)=43	(-33,53)
Interval 3	A(3,3)=18	sum(A(3,:))-A(3,3)=46	(-28,64)
Interval 4	A(4,4)=8	sum(A(4,:)) -A(4,4)=51	(-43,59)
Interval 5	A(5,5)=2	sum(A(5,:)) -A(5,5)=37	(-35,39)

Observe that the union of these intervals $(-43, 64)$ encloses all eigenvalues.

CHAPTER 8

Solution to Drill 8.1

Step 1 Enter the four vectors.

» **v1=[1 0 1 2]; v2=[1 1 -1 0]; v3=[1 -2 -1 0]; v4=sym('[a, b, c, d]');**

Step 2 Since the vectors are mutually orthogonal, the inner products $\langle v_1, v_4 \rangle$, $\langle v_2, v_4 \rangle$, $\langle v_3, v_4 \rangle$ are all zero.

» **v1*transpose(v4)**
ans =
  a+c+2*d

» **v2*transpose(v4)**
ans =
  a+b-c

» **v3*transpose(v4)**

ans =
   a-2*b-c

Step 3 Solve the homogeneous system of equations for a, b, and c.

» eq1='a+c+2*d=0';
» eq2='a+b-c=0';
» eq3= 'a-2*b-c =0';
» eq4='1=1';   % since we have 3 equations in 4 variables, we introduce
                      % a virtual equation
» [a,b,c,d]=solve(eq1,eq2,eq3,eq4)

a =
   -d

b =
    0

c =
   -d

d =
    d

Therefore, for any value in the set $\{(-d, 0, -d, d)\}$ where d is arbitrary, the vectors are mutually orthogonal.

Solution to Drill 8.2

Step 1 Enter the three vectors in symbolic form.

» v1=sym('[2, b, 1]'); v2=sym('[a, 1, -1]'); v3=sym('[1, 3, c]');

Step 2 Since the vectors are mutually orthogonal, the inner products $\langle v_1, v_2 \rangle$, $\langle v_2, v_3 \rangle$, $\langle v_3, v_1 \rangle$ are all zero.

» v1*transpose(v2)
ans =
   2*a+b-1

» v2*transpose(v3)
ans =
   a+3-c

» v3*transpose(v1)
ans =
   2+3*b+c

Step 3 Solve the resulting homogeneous system for a, b, and c.

» eq1='2*a+b-1=0';
» eq2='a+3-c=0';
» eq3='2+3*b+c=0';
» [a,b,c]=solve(eq1,eq2,eq3)

a =
   8/5

$$b =$$
$$-11/5$$
$$c =$$
$$23/5$$

Solution to Drill 8.3

Step 1 First check to see whether the three vectors are linearly independent. Solve the homogeneous system derived from $a(1\ 2\ 1) + b(1\ 0\ 1) + c(3\ 1\ 0) = (0\ 0\ 0)$.

```
» eq1='a+b+3*c=0';
» eq2='2*a+c=0';
» eq3='a+b=0';
» [a,b,c]=solve(eq1,eq2,eq3)
a =
  0
b =
  0
c =
  0
```

Step 2 Therefore, the three vectors are linearly independent. We now apply the Gram-Schmidt method to obtain an orthonormal basis. Pick the first vector of the orthogonal basis to be $v_1 = (1\ 2\ 1)$ and normalize v_1.

```
» v1=[1 2 1]
v1 =
  1    2    1
» u1=v1/norm(v1)
u1 =
  0.4082    0.8165    0.4082
```

Step 3 Obtain the second orthonormal vector u_2 as follows,

$$w_2 = v_2 - \langle v_2, u_1 \rangle u_1 \text{ and } u_2 = w_2/\text{norm}(w_2).$$

```
» v2=[1 0 1];
» w2=v2-(v2*u1')*u1
w2 =
  0.6667    -0.6667    0.6667
» u2=w2/norm(w2)        % normalize the vector
u2 =
  0.5774    -0.5774    0.5774
```

Step 4 Obtain the third orthonormal vector u_3 as follows:

$$w_3 = v_3 - \langle v_3, u_1 \rangle u_1 - \langle v_3, u_2 \rangle u_2 \text{ and } u_3 = w_3/\text{norm}(w_3).$$

```
» v3=[3 1 0];
» w3=v3-(v3*u1')*u1-(v3*u2')*u2
w3 =
  1.5000    0.0000    -1.5000
```

» **u3=w3/norm(w3)**   % **normalize the vector**

u3 =
  0.7071   0.0000   -0.7071

Three vectors $\{u_1, u_2, u_3\}$ given by

$u_1 = [0.4082\ \ 0.8165\ \ 0.4082]$,
$u_2 = [0.5774\ -0.5774\ \ 0.5774]$, and
$u_3 = [0.7071\ \ 0.0000\ -0.7071]$ form an orthonormal basis.

Solution to Drill 8.4

Step 1 First check to see whether the three vectors are linearly independent. Solve the homogeneous system derived from $a(0, 0, 1, 0) + b(1, 0, 1, 1) + c(1, 1, 2, 1) = (0, 0, 0, 0)$.

» **eq1='b+c=0';**
» **eq2='c=0';**
» **eq3='a+b+2*c=0';**
» **eq4='b+c=0';**
» **[a,b,c]=solve(eq1,eq2,eq3,eq4)**

Warning: 4 equations in 3 variables.
» **In C:\MATLAB5/toolbox/symbolic/solve.m at line 110**

a =
  0

b =
  0

c =
  0

Step 2 Therefore, the three vectors are linearly independent. We now apply the Gram-Schmidt method to obtain an orthonormal basis.

» **v1=[0 0 1 0]; v2=[1 0 1 1]; v3=[1 1 2 1];**
» **u1=v1/norm(v1)**   % **find the first normalized vector**
u1 =
  0   0   1   0

» **w2=v2-(v2*u1')*u1**   % **find the second orthogonal vector**
w2 =
  1   0   0   1

» **u2=w2/norm(w2)**   % **normalize it**
u2 =
  0.7071   0   0   0.7071

» **w3=v3-(v3*u1')*u1-(v3*u2')*u2**   % **find the third orthogonal vector**
w3 =
  0.0000   1.0000   0   0.0000

» **u3=w3/norm(w3)**   **% normalize it**
u3 =
  0.0000   1.0000   0   0.0000

The set $\{u_1, u_2, u_3\}$ forms an orthonormal basis for the subspace generated by the vectors

$\{(0\ 0\ 1\ 0), (1\ 0\ 1\ 1),$ and $(1\ 1\ 2\ 1)\}.$

Solution to Drill 8.5

Step 1 Enter the matrix A and find its eigenvalues.

» **A = [5 2 2; 2 5 2; 2 2 5]**
A =
   5   2   2
   2   5   2
   2   2   5
» **eig(A)**

ans =
   3
   3
   9

Step 2 Find the eigenvector(s) for eigenvalue 3. That is, solve the system $(A - 3I)v = 0$ for v.

» **rref(A - 3*eye(3))**

ans =
   1   1   1
   0   0   0
   0   0   0

This gives $x + y + z = 0$. Thus $x = -y - z$. Choosing $y = 1, z = 0$ and $y = 0, z = 1$, successively we obtain two eigenvectors $(-1, 0, 1)$ and $(-1, 1, 0)$ corresponding to the eigenvalue 3.

Step 3 Find the eigenvector(s) for eigenvalue 9. That is, solve the system $(A - 9I)v = 0$ for v.

» **rref(A - 9*eye(3))**

ans =
   1   0   -1
   0   1   -1
   0   0   0

This gives $x - z = 0, y - z = 0$. Thus $x = -y - z$. Choosing $z = 1$ we get $x = 1, y = 1$. Thus $(1, 1, 1)$ is an eigenvector corresponding to the eigenvalue 9.

Step 4 The set of eigenvectors $\{(-1\ 0\ 1), (-1\ 1\ 0), (1\ 1\ 1)\}$ can be tested to be linearly independent, and one may invoke Gram-Schmidt process to obtain an orthonormal set. Indeed, if X is the matrix whose columns are eigenvectors, then the Matlab command $[Q, R] = qr(X)$ will yield an orthogonal matrix Q and an upper triangular matrix R such that $X = QR$. The columns of Q form a maximal set of orthonormal eigenvectors of A.

» v1 = [-1; 1; 0]; v2 = [-1; 0; 1]; v3 = [1; 1; 1];
» X = [v1 v2 v3]

```
X =
 -1   -1    1
  1    0    1
  0    1    1
```

» [Q, R] = qr(X)

```
Q =
 -0.7071   -0.4082    0.5774
  0.7071   -0.4082    0.5774
       0    0.8165    0.5774
```

```
R =
 1.4142    0.7071    0.0000
      0    1.2247    0.0000
      0         0    1.7321
```

Next we obtain an orthonormal set of eigenvectors of B.

Step 1 Enter the matrix B and find its eigenvalues.

» B = [2 0 0 0; 0 1 -2 2; 0 -2 1 2; 0 2 2 1]

```
B =
 2    0    0    0
 0    1   -2    2
 0   -2    1    2
 0    2    2    1
```
» eig(B)

```
ans =
 -3.0000
  2.0000
  3.0000
  3.0000
```

Step 2 Find the eigenvector(s) for eigenvalue -3. That is, solve the system $(B + 3I)v = 0$ for v.

» rref(B + 3*eye(4))

```
ans =
 1    0    0    0
 0    1    0    1
 0    0    1    1
 0    0    0    0
```

This gives $x = 0, y + w = 0, z + w = 0$. Choosing $w = 1$, we get $y = -1, z = -1$. Thus $(0, -1, -1, 1)$ is an eigenvector corresponding to the eigenvalue -3.

Step 3 Find the eigenvector(s) for eigenvalue 2. That is, solve the system $(B - 2I)v = 0$ for v.

» rref(B - 2*eye(4))

```
ans =
   0   1   0   0
   0   0   1   0
   0   0   0   1
   0   0   0   0
```

This gives $y ==, z = 0, w = 0$. Choosing $x = 1$ we get the eigenvector $(1, 0, 0, 0)$ corresponding to the eigenvalue 2.

Step 4 Find the eigenvector(s) for eigenvalue 3. That is, solve the system $(B - 3I)v = 0$ for v.

» **rref(B - 3*eye(4))**

```
ans =
   1   0   0   0
   0   1   1  -1
   0   0   0   0
   0   0   0   0
```

This gives $x = 0, y+z-w = 0$. Thus $y = -z+w$. Choosing $z = 1, w = 0$ and $z = 0, w = 1$, successively we obtain two eigenvectors $(0, -1, 1, 0)$ and $(0, 1, 0, 1)$ corresponding to the eigenvalue 3.

Step 5 To obtain the orthonormal set of eigenvectors, we use the matlab command $[Q, R] = qr(X)$, where X is the matrix whose columns are eigenvectors. The columns of Q form a maximal set of orthonormal eigenvectors of A.

» **v1 = [0; -1; -1; 1]; v2 = [1; 0; 0; 0]; v3 = [0; -1; 1; 0]; v4 = [0; 1; 0; 1];**

» **X = [v1 v2 v3 v4]**

```
X =
   0   1   0   0
  -1   0  -1   1
  -1   0   1   0
   1   0   0   1
```

» **[Q, R] = qr(X)**

```
Q =
        0     -1.0000      0.0000      0.0000
   0.5774      0.0000      0.7071      0.4082
   0.5774      0.0000     -0.7071      0.4082
  -0.5774     -0.0000     -0.0000      0.8165
R =
  -1.7321           0           0           0
        0     -1.0000     -0.0000      0.0000
        0           0     -1.4142      0.7071
        0           0           0      1.2247
```

Solution to Drill 8.6

Step 1 Enter the matrix A and find its eigenvalues.

» **A=[0 2 2; 2 0 -2; 2 -2 0]**

```
A =
   0    2    2
   2    0   -2
   2   -2    0
```

```
» eig(A)
```

ans =
   2
   2
   -4

Step 2 Find the eigenvector(s) for eigenvalue 2. That is, solve the system $(A - 2I)v = 0$ for v.

```
» B=A-2*eye(3);
» rref(B)
```

ans =
   1   -1   -1
   0   0   0
   0   0   0

Thus $x = y + z$.

By substituting $\{y = 0, z = 1\}$ and $\{y = 1, z = 0\}$ we obtain two independent eigenvectors for the eigenvalue 2, $\{(1, 0, 1), (1, 1, 0)\}$.

Step 3 Similarly, find the eigenvector(s) for the eigenvalue -4. That is, solve the system $(A + 4I)v = 0$ for v.

```
» B=A+4*eye(3);
» rref(B)
```

ans =
   1   0   1
   0   1   -1
   0   0   0

Thus $x + z = 0$ and $y - z = 0$.

By substituting an arbitrary value for $z = 1$, we obtain an eigenvector $(-1, 1, 1)$ for the eigenvalue -4.

Step 4 Let X be the matrix whose columns are eigenvectors of A. Invoke the Matlab command $[P, R] = qr(X)$. Then the columns of P form an orthonormal set of eigenvectors of A. Note that the right-hand side function is always qr.

```
» x = [1 1 -1; 0 1 1; 1 0 1];
» [P, R] = qr(x)
```

P =
   -0.7071   -0.4082   -0.5774
         0   -0.8165   -0.5774
   -0.7071   -0.4082   -0.5774
R =
   -1.4142   -0.7071   -0.0000
         0   -1.2247   -0.0000
         0   0   1.7321

Step 5 We show that $P^T A P$ is a diagonal matrix whose diagonal entries are the eigenvalues of A.

```
» P'*A*P
```

```
ans =
   2     *     0
   *     2     *
   0     *    -4
```

Step 6 To find A^{10} using $P^T A P = D$, we note that $P^T A P = D$ implies $A = P D P^T$ (since P is orthogonal, $P P^T = I$). Therefore $A^{10} = P D^{10} P^T$.

» **P*D^10*P'**

```
ans =
    350208      -349184      -349184
   -349184       350208       349184
   -349184       349184       350208
```

We can verify the result by directly computing A^{10} using Matlab.

» **A^10**

```
ans =
    350208      -349184      -349184
   -349184       350208       349184
   -349184       349184       350208
```

Solution to Drill 8.7

Step 1 Let the vector X and the matrix A be as follows.

» **X=sym('[x; y]')**

```
X =
  [ x]
  [ y]
```

» **A=[1 2; 2 3]**

```
A =
   1     2
   2     3
```

Step 2 Show that left-hand side of the quadratic form is equal to $X^T A X$.

» **transpose(X)*A*X**

```
ans =
  (x+2*y)*x+(2*x+3*y)*y
```

This is indeed $x_1^2 + 4x_1 x_2 + 3x_2^2$ (Note: $x_1 = x$, $x_2 = y$)

Step 3 Find the eigenvalues of A (say λ_1 and λ_2) and so by Theorem 8.3.1, $x_1^2 + 4x_1 x_2 + 3x_2^2$ can be transformed into $\lambda_1 x'^2 + \lambda_2 y'^2 = 4$.

» **format rat**
» **eig(A)**

```
ans =
  -305/1292
   1292/305
```

Since $(-305/1292)x'^2 + (1292/305)y'^2 = 4$ represents a hyperbola, the graph of the original equation also represents a hyperbola.

Solution to Drill 8.8

Step 1 Let the vector X and the matrix A be as follows

» **X=sym('[x; y; z; t]')**

X =

[x]

[y]

[z]

[t]

» **A=[2 0 2 -1; 0 2 0 -1; 2 0 3 0; -1 -1 0 2]**

A =

2	0	2	-1
0	2	0	-1
2	0	3	0
-1	-1	0	2

Step 2 Show that the left-hand side of the quadratic form is equal to $X^T A X$.

» **transpose(X)*A*X**

ans =

(2*x+2*z-t)*x+ (2*y-t)*y+ (2*x+3*z)*z+ (-x-y+2*t)*t

This is indeed $2x^2 + 2y^2 + 3z^2 + 2t^2 + 4xz - 2xt - 2yt = Q(x, y, z, t)$.

Step 3 The eigenvalues of A are

» **eig(A)**

ans =

1.2679

0.0000

3.0000

4.7321

Step 4 The transformation $X_1 = PX$, where P is an orthogonal matrix whose columns are the eigenvectors of A and $X_1 = [x_1, y_1, z_1, t_1]$, reduces the given expression to $(1.2679)x_1^2 + (0.0000)y_1^2 + (3.0000)z_1^2 + (4.7321)t_1^2$. This expression is always nonnegative.

Solution to Drill 8.9

Step 1 First find a basis for the column space of A. Clearly, the two columns of A are linearly independent (not multiples of each other). Therefore, we will construct an orthonormal basis using the Gram-Schmidt method. (Note: we write column vectors as row vectors for saving space.)

» **format rat**

» **v1=[1 0 1];**   % first column of the coefficient matrix

» **v2=[1 2 1];**   % second column of the coefficient matrix

» **u1=v1/norm(v1)**   % normalize the first column

u1 =

985/1393   0   985/1393

```
» w2=v2-(v2*u1')*u1;        % compute the second orthogonal vector
» u2=w2/norm(w2)            % normalize it

u2 =
    *    1    *
```

Step 2 Compute $b_1 = (u_1 b^T)u_1 + (u_2 b^T)u_2$

```
» b=[2 1 0];                      % find b1 such that Ax = b1 is consistent
» b1=(u1*b')*u1+(u2*b')*u2
b1 =
    1    1    1
```

Step 3 Solve the system $Ax = b_1$ for the least squares solution x. First find the reduced row echelon form of the augmented matrix $[A \ b_1]$ and solve.

```
» rref([v₁ v₂ b1'])
ans =
    1    0    1/2
    0    1    1/2
    0    0    0
```

Step 4 The reduced system is equivalent to $x = \frac{1}{2}$ and $y = \frac{1}{2}$, the least square solution to the system.

Solution to Drill 8.10

Step 1 First find a basis for the column space of A. We will write column vectors as row vectors simply for space saving reasons. We now check if three columns of A are linearly independent. Then $a(1, 3, -1, 3) + b(2, 5, -1, 5) + c(-1, 5, -2, 0) = (0, 0, 0, 0)$ implies

```
» eq1='a+2*b-c=0';
» eq2='3*a+5*b+5*c=0';
» eq3='-a-b-2*c=0';
» eq4='3*a+5*b=0';
» [a,b,c]=solve(eq1,eq2,eq3,eq4)
```

Warning: 4 equations in 3 variables.
```
» In C:\MATLAB5\toolbox\symbolic\solve.m at line 110
a =
  0
b =
  0
c =
  0
```

Step 2 Therefore, the three columns of A are linearly independent. We use the Gram-Schmidt method to construct an orthonormal basis.

```
» v1=[1 3 -1 3];
» v2=[2 5 -1 5];
» v3=[-1 5 -2 0];
```

```
» u1=v1/norm(v1)                    % first orthonormal vector
u1 =
   646/2889     646/963     -646/2889     646/963
» w2=v2-(v2*u1')*u1;                % second orthogonal vector
» u2=w2/norm(w2)                    % normalize it
u2 =
   1068/2263     179/2655     1426/1627     179/2655
» w3=v3-(v3*u1')*u1-(v3*u2')*u2;    % third orthogonal vector
» u3=w3/norm(w3)                    % normalize it
u3 =
   -129/524     387/524     129/1048     -653/1061
```

Step 3 Compute $b_1 = (u_1 b^T)u_1 + (u_2 b^T)u_2 + (u_3 b^T)u_3$

```
» b=[1 1 1 0]
b =
   1     1     1     0
» b1=(u1*b')*u1+(u2*b')*u2+(u3*b')*u3   % find b1 such that Ax = b1
                                        % is consistent

b1 =
   2/3     1     7/6     1/6
```

Step 4 Solve the system $Ax = b_1'$ for the least squares solution.

```
» A=[v1;v2;v3]'
A =
    1     2     -1
    3     5      5
   -1    -1     -2
    3     5      0
» rref([A b1'])
ans =
   1     0     0     -23/6
   0     1     0      7/3
   0     0     1      1/6
   0     0     0       0
```

Therefore, the least squares solution to the system $Ax = b$ is $(-23/6, 7/3, 1/6)$.

CHAPTER 9

Solution to Drill 9.1

Step 1 Write $(-1, 1, 0)$ as a linear combination of the vectors $(1, 2, 1)$, $(2, -3, 0)$ and $(0, 0, 1)$, say $a(1, 2, 1) + b(2, -3, 0) + c(0, 0, 1) = (-1, 1, 0)$, and solve the system for a, b, and c.

```
» format rat
» eq1='a+2*b=-1';
» eq2='2*a-3*b=1';
» eq3='a+c=0';
```

```
» [a,b,c]=solve(eq1,eq2,eq3)
a =
  -1/7
b =
  -3/7
c =
  1/7
```

Step 2 Therefore, we write $(-1, 1, 0) = (-1/7)(1, 2, 1) + (-3/7)(2, -3, 0) + (1/7)(0, 0, 1)$ Since f is a linear transformation,

$$f(-1, 1, 0) = (-1/7)f(1, 2, 1) + (-3/7)f(2, -3, 0)$$
$$+(1/7)f(0, 0, 1).$$

If the function values at $(1, 2, 1)$, $(2, -3, 0)$ and $(0, 0, 1)$ are f_1, f_2, and f_3 respectively, we get $f(-1, 1, 0)$ as follows.

```
» f1=[1 -1 0]; f2=[4 1 0]; f3=[0 0 1];
» value =(-1/7)*f1 + (-3/7)*f2 + (1/7)*f3
value =
  -13/7    -2/7    1/7
```

To compute $f(2, 5, 1)$ we proceed as above.

Step 1 Solve the linear system obtained by equating the corresponding entries of

$$(2, 5, 1) = a(1, 2, 1) + b(2, -3, 0) + c(0, 0, 1)$$

by first reducing the augmented matrix into reduced row echelon form.

```
» rref([1 2 0 2; 2 -3 0 5; 1 0 1 1])
ans =
  1    0    0    16/7
  0    1    0    -1/7
  0    0    1    -9/7
```

Step 2 Compute $value = f(2, 5, 1) = af_1 + bf_2 + cf_3$.

```
» a = 16/7; b = -1/7; c = -9/7;
» value = a*f1 + b*f2 + c*f3
value =
  12/7    -17/7    -9/7
```

It can be similarly shown that $f(3, 0, 1) = (33/7, -3/7, -2/7)$.

Solution to Drill 9.2

Step 1 Write $(2\ 3\ 4)$ as a linear combination of the vectors $(1, 0, 0)$, $(1, 1, 1)$, and $(1, 0, 1)$, say $a(1, 0, 0) + b(1, 1, 1) + c(1, 0, 1) = (2, 3, 4)$, and solve the system.

```
» eq1='a + b + c = 2';
» eq2='b = 3';
» eq3='b+ c = 4';
```

» [a,b,c]=solve(eq1,eq2,eq3)

a =
   -2

b =
   3

c =
   1

Step 2 We may write $(2, 3, 4) = (-2)(1, 0, 0) + 3(1, 1, 1) + 1(1, 0, 1)$. If the function values at $(1, 0, 0)$, $(1, 1, 1)$, and $(1, 0, 1)$ are denoted by $f1$, $f2$, and $f3$ respectively, we obtain $f(2, 3, 4)$ as follows:

» f1 = [1, 1, 1]; f2 = [0, 1, 1]; f3 = [0, 0, 0];
» value = (-2)*f1 + 3*f2 + 1*f3

value =
   -2   1   1

Solution to Drill 9.3

a. Step 1 First find the kernel of the linear transformation T by solving the homogeneous system $Ax = 0$.

» A=[1 2 3 -1; 3 5 8 -2; 1 1 3 0];
» rref(A)

ans =
   1   0   0   1
   0   1   0   -1
   0   0   1   0

Step 2 The reduced matrix is equivalent to the system, $x + t = 0$, $y - t = 0$, $z = 0$. This implies $x = -t$, $y = t$, $t = t$, $z = 0$. Therefore the kernel of T is given by the set $\{(-t, t, 0, t) \mid t \text{ is arbitrary}\}$ and a basis for the kernel is given by $\{(-1, 1, 0, 1)\}$.

b. We reduce A^T to reduced row echelon form and find a basis for the row space of A^T. This gives a basis for the column space of A and therefore a basis for the range of A.

» A=[1 2 3 -1; 3 5 8 -2; 1 1 3 0]
» rref(A')

ans =
   1   0   0
   0   1   0
   0   0   1
   0   0   0

This shows that there are three linearly independent columns, and a basis of the range is $\{(1, 0, 0)^T, (0, 1, 0)^T, (0, 0, 1)^T\}$.

c. The dimension of the range of T is 3 and the dimension of the kernel of T is 1. Therefore,

$$\dim \text{ kernel } T + \dim \text{ range } T = 4 = \dim V$$

This verifies the rank-nullity theorem.

Solution to Drill 9.4

Step 1 Find the matrix representation A of the linear transformation T as follows:

» b1=[1 0]; b2=[0 1];   % **choose two standard basis vectors**
» x1=b1(1); x2=b1(2);
» T1=[x1 x1+x2 x1+x2]   % **find the image vector of b1 under T**

T1 =
  1   1   1

» x1=b2(1); x2=b2(2);
» T2 =[x1 x1+x2 x1+x2]   % **find the image vector of b2 under T**

T2 =
  0   1   1

Step 2 Form a matrix whose columns are the vectors T_1 and T_2 and reduce the matrix to its row echelon form.

» A=[T1;T2]'
A =
  1   0
  1   1
  1   1

» rref(A)
ans =
  1   0
  0   1
  0   0

Step 3 By solving the homogeneous system $Ax = 0$, we get kernel of $T = \{(0, 0)\}$. Since $T : R^2 \rightarrow R^3$, by the rank-nullity theorem, dimension of range of $T =$ dimension of $R^2 -$ dimension of kernel $T = 2 - 0 = 2$.

Solution to Drill 9.5

a. Step 1 The standard basis for R^3 is given by $e_1 = (1, 0, 0)$, $e_2 = (0, 1, 0)$ and $e_3 = (0, 0, 1)$. First we find $T(e_1)$, $T(e_2)$, and $T(e_3)$. Enter the transformation T and find its value at standard basis vectors.

» T=sym('[a+b, b+c, c+a]')

T =
  [a+b, b+c, c+a]

» **T1=subs(T,{'a','b','c'},{1,0,0})**

T1 =

   1   0   1

» **T2=subs(T,{'a','b','c'},{0,1,0})**

T2 =

   1   1   0

» **T3=subs(T,{'a','b','c'},{0,0,1})**

T3 =

   0   1   1

Step 2 T_1, T_2 and T_3 can be expressed in terms of the standard basis as follows.

$$T_1 = (1, 0, 1) = 1(1, 0, 0) + 0(0, 1, 0) + 1(0, 0, 1),$$
$$T_2 = (1, 1, 0) = 1(1, 0, 0) + 1(0, 1, 0) + 0(0, 0, 1),$$
$$T_3 = (0, 1, 1) = 0(1, 0, 0) + 1(0, 1, 0) + 1(0, 0, 1).$$

Therefore, the matrix representation of T with respect to the standard basis is the matrix whose column 1 consists of the coefficients of T_1 in the above expression of T_1, column 2 consists of coefficients of T_2 in the above expression of T_2, and column 3 consists of coefficients of T_3 in the above expression of T_3.

» **A=[1 1 0; 0 1 1; 1 0 1]**

A =

   1   1   0

   0   1   1

   1   0   1

Step 3 The matrix A behaves like T, as shown below.

» **v=sym('[a; b; c]')**

v =

  [a]

  [b]

  [c]

» **A*v**

ans =

  [a+b]

  [b+c]

  [c+a]

That is, $Tv = Av$ for any vector v.

b. Step 1 First find the images of the basis vectors under T.

» **T = sym ('[a + b, b + c, c + a]')**

T =

  [a+b, b+c, c+a]

» **T1=subs(T,{'a','b','c'},{1,1,0})**

```
T1 =
   2    1    1
» T2=subs(T,{'a','b','c'},{1,0,1})
T2 =
   1    1    2
» T3=subs(T,{'a','b','c'},{0,1,1})
T3 =
   1    2    1
```

Step 2 Express the image vectors in terms of the given basis.
$(2, 1, 1) = a(1, 1, 0) + b(1, 0, 1) + c(0, 1, 1)$.

```
» eq1='a+b=2';
» eq2='a+c=1';
» eq3='b+c=1';
» [a,b,c]=solve(eq1,eq2,eq3)

a =
   1
b =
   1
c =
   0
```

Therefore, $(2, 1, 1) = 1(1, 1, 0) + 1(1, 0, 1) + 0(0, 1, 1)$. Write $(1, 1, 2)$ as a linear combination of the basis vectors: $(1, 1, 2) = a(1, 1, 0) + b(1, 0, 1) + c(0, 1, 1)$.

```
» eq1='a+b=1';
» eq2='a+c=1';
» eq3='b+c=2';
» [a,b,c]=solve(eq1,eq2,eq3)

a =
   0
b =
   1
c =
   1
```

Therefore, $(1, 1, 2) = 0(1, 1, 0) + 1(1, 0, 1) + 1(0, 1, 1)$. Write $(1, 2, 1)$ as a linear combination of the basis vectors: $(1, 2, 1) = a(1, 1, 0) + b(1, 0, 1) + c(0, 1, 1)$.

```
» eq1='a+b=1';
» eq2='a+c=2';
» eq3='b+c=1';
» [a,b,c]=solve(eq1,eq2,eq3)

a =
   1
b =
   0
```

c =
1

Therefore, $(1, 2, 1) = 1(1, 1, 0) + 0(1, 0, 1) + 1(0, 1, 1)$. Construct the matrix as in part (a), step 2.

» B=[1 1 0; 0 1 1; 1 0 1]'
B =
    1   0   1
    1   1   0
    0   1   1

c. Step 1 We express the new basis (nonstandard) in terms of the standard basis.

$$(1, 1, 0) = e_1 + e_2 + 0.e_3,$$
$$(1, 0, 1) = e_1 + 0.e_2 + e_3,$$
$$(0, 1, 1) = 0.e_1 + e_2 + e_3.$$

The matrix X can be constructed with its first column as the coefficients in the expression for $(1, 1, 0)$, the second column, as the coefficients in the expression for $(1, 0, 1)$, and its third column as the expression for $(0, 1, 1)$.

Solution to Drill 9.6

b. Step 1 Let A, B, and C be the rotation matrices corresponding to 60 degrees clockwise, 15 degrees counterclockwise, and 40 degrees clockwise.

» x=(-60)*pi/180;
» A=[cos(x) -sin(x);sin(x) cos(x)]   % clockwise 60 degrees
                                        % note that in Matlab cosx must be written
                                        % as cos(x)

A =
   0.5000   0.8660
  -0.8660   0.5000

» x=15*pi/180;
» B=[cos(x) -sin(x);sin(x) cos(x)]   % counterclockwise 15 degrees

B =
   0.9659   -0.2588
   0.2588   0.9659

» x=(-40)*pi/180;
» C=[cos(x) -sin(x);sin(x) cos(x)]   % clockwise 40 degrees

C =
   0.7660   0.6428
  -0.6428   0.7660

Note that negative angles are used for clockwise rotations.

Step 2 Enter the original point P and apply matrices A, B, and C respectively.

» P=[1;1]; A*P

ans =
    1.3660
    -0.3660

» B*ans
ans =
    1.4142
    0.0000

» C*ans
ans =
    1.0834
    -0.9090

Solution to Drill 9.7

a. **Step 1** Enter the coordinate matrix.

» A=[-0.4 0 0.4 -0.4; -1 1 -1 -1]
A =
    -0.4000   0   0.4000   -0.4000
    -1.0000   1.0000   -1.0000   -1.0000

Step 2 Plot the graph of the triangle.

» x=A(1,:); y=A(2,:);
» plot(x,y)

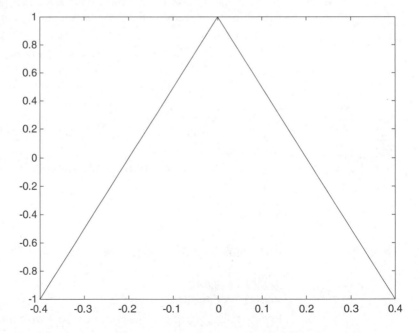

b. **Step 1** Enter the values of a and b and the translation matrix T.

» a=-0.2; b=0.3;

» **T=[1 0 a; 0 1 b; 0 0 1]**

T =
1.0000	0	-0.2000
0	1.0000	0.3000
0	0	1.0000

Step 2 To view translation operation as a matrix multiplication we need to work with homogeneous coordinates $(x, y, 1)$ corresponding to coordinates (x, y). The last row of A is used to store the homogeneous coordinates 1.

» **A=[-0.4 0 0.4 -0.4; -1 1 -1 -1; 1 1 1 1]**

A =
-0.4000	0	0.4000	-0.4000
-1.0000	1.0000	-1.0000	-1.0000
1.0000	1.0000	1.0000	1.0000

Step 3 Apply the translation matrix T to coordinate matrix A.

» **B=T*A**

B =
-0.6000	-0.2000	0.2000	-0.6000
-0.7000	1.3000	-0.7000	-0.7000
1.0000	1.0000	1.0000	1.0000

The rows of B (except the last row) contain the coordinates of the translated points. Let x_1, y_1 be the new coordinate vectors. Plot the graph of the translated points.

» **x1=B(1,:); y1=B(2,:);**

» **hold on**

» **plot(x1,y1)**

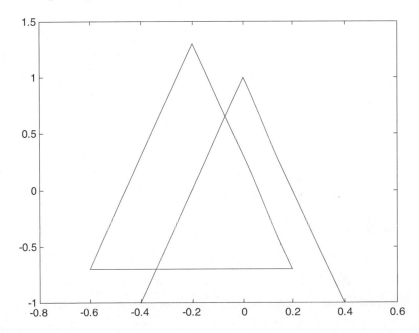

Some Basic Matlab Operations

General Notes

Throughout this section we shall use » to indicate a Matlab command statement. The command prompt » may vary from system to system. Command prompt » (or an equivalent prompt) is given by the system and you only need to enter the Matlab command. For example you type **eq1 = [1 − 1 2]**, not » **eq1 = [1 −1 2]**. Furthermore, a semicolon at the end of a Matlab command does not display the output.

How to Begin and Save the Work

- Begin your work in Matlab as in any computer algebra system (CAS).
- Save your work, say, on 'A' drive by giving the Matlab command

    save a: filename

  For example if you want to save work in file 'MyLinearAlgebraProject' on drive A, type

    save a: MyLinearAlgebraProject

  Note: This command saves each variable used in the work and its most recent value. This command does not save the entire workspace.

How to Use the Saved Work

- Open the file by going to the file menu as in any other computer algebra system. Some older versions of Matlab may require the command

    load a: filename

  Neither of these procedures will show any contents on the screen. However all the variables of the saved work will be loaded in the memory and can be used.
- To know the value of any variable in the saved work, type name of the variable.
- In order to know the list of all variables type '*who*'. This will list all the variables.

How to Save the Entire Work or Part of the Work with the Entire Command Sequence

- To save the entire workspace or part of the work with the entire command sequence, say, on 'A' drive, type

    diary a: filename

This command will start a diary of all the commands. For example if you want to save the command sequence in the file 'MyLinearAlgebraProjectCommands' on drive A, type

diary a: MyLinearAlgebraProject

At any point during the work, typing the command

diary off

will terminate the recording process. To resume recording, type the command

diary on

■ Exit Matlab session when finished.

Note: The work saved using the diary command during one Matlab session cannot be continued at another Matlab session. If the work is incomplete before exiting Matlab and intended to be used later, then it must be saved using the save command as explained above.

Note: Save command saves the work as *.mat* file. It is not necessary to give the extension *.mat* while entering the name of the file. In order to avoid confusion, it is recommended that the filename in the save command be different from the one in the diary command.

How to Give Comments in Your Work

It is possible to include comments in Matlab workspace. Typing % before a statement indicates a comment statement. Comment statements are not executable. Matlab ignores all such statements. For example, if you want to type 'This is my Project 1' for future reference, type

» % This is my Project 1

How to Get Help on a Topic

■ To get help on a topic, say rank, enter

» help rank

Chapter 1: Linear Systems and Matrices

(I) Equations

(A) Entering Equations

■ **List Form.** To enter an equation in the list form, enter all the coefficients separating them by a blank space and enclose the coefficients in square brackets. For example, to enter the equation $x - y = 2$, type $eq1 = [1\ -1\ 2]$ at the system prompt, that is,

» $eq1 = [1 - 1\ 2]$

Note: If a variable is not present in the equation, its coefficient must be entered zero. For example, to enter the equations $2x - 3y + z = 1$ and $x + z = 2$, type

» $eq1 = [2\ -3\ 1\ 1]$
» $eq2 = [1\ 0\ 1\ 2]$

- **Regular Form.** To enter an equation in the regular form, type the equation within single quotes. For example, to enter the equation $x - 3y = 1$, type $eq1 = $ '$x - 3*y = 2$' at the system prompt, that is,

» $eq1 = $ '$x - 3*y = 2$'

(B) Elementary Operations on Equations

- **Multiplying an equation, *eq*, by a constant, *k***

  » $eq = k*eq$

- **Multiplying an equation, *eq 1*, by a constant, *k*, and adding to another equation, *eq 2*.**

  » $eq2 = k*eq1 + eq2$

- **Interchanging two equations, *eq 1* and *eq 2***

  » $temp = eq\ 1; eq\ 1 = eq\ 2; eq\ 2 = temp;$

Note: In order to apply elementary operations, the equations must be entered in the list form.

(C) Substituting Values of Unknowns

- To substitute the value k of an unknown x in an equation eq, enter

  » $eq = subs(eq, 'x', k)$ or   » $x = k; eq = subs(eq, 'x', x)$

The later command assigns the value k to the unknown x and then substitutes that value in the equation.

(D) Solving Equations

- To solve an equation eq in the unknown x, enter

  » $solve(eq, 'x')$ or   » $[x] = solve(eq)$

Note: The former command will only display the value whereas the later command will give the variable name and its value.

- To solve more than one equation, write all the variables within square brackets on the left hand side and list all the equations on the right hand side. For example, in order to solve *eq1* and *eq2* in the unknowns *x* and *y*, enter

    » *[x,y] = solve(eq 1, eq2)*

and in order to solve equations *eq1*, *eq2*, and *eq3* in the unknowns *x*, *y*, and *m*, enter

    » *[m, x, y] = solve(eq 1, eq2, eq3)*

Note: In order to solve equations, the equations must be entered in the regular form.

Note: The unknowns *x*, *y*, and *m* on left-hand side must be written in alphabetical-numerical order.

Note: If the number of equations is fewer than the number of unknowns then *virtual* equations, such as *eq = '0 = 0'*, must be introduced in order to make the number of equations at least equal to the number of unknowns.

(II) MATRICES

(A) Entering Matrices

- **Entering a matrix *A* with numerical entries.** To enter a matrix, list all the entries of the matrix starting with the first row, separating the entries by blank spaces, separating two rows by a semicolon, and enclosing the list in square brackets. In other words, type the command

    » *A = [First row entries; Second row entries; and so on]*

For example, to enter the 3×4 matrix *A* with 1, 2, 3, 4 in the first row, 3, 2, 1, 4 in the second row, and 4, 1, 2, 3 in the third row, type

    » *A = [1 2 3 4; 3 2 1 4; 4 1 2 3]*

- **Entering a matrix *A* with symbolic entries (symbolic matrices).** Type

    » *A = sym ('[First row entries; Second row entries; and so on]')*

For example, in order to enter the symbolic matrix *A* with *a*, *b*, *c* in the first row and *d*, *e*, *f* in the second row, type

    » *A = sym ('[a, b, c; d, e, f]')*

Note: In the case of symbolic matrices, commas must separate entries of each row.

(B) Substituting Numerical Values

To substitute a value, say m, of a symbol a in a symbolic matrix A, type

> » $A = subs(A, \text{'}a\text{'}, m)$

For example, in order to substitute 2 for a, type

> » $A = subs(A, \text{'}a\text{'}, 2)$

In order to substitute for more than one symbol, replace *'a' with* the list of symbols, writing each symbol within quotes and separating the symbols by commas and replace m with the list of values, separating the values by commas. In other words, to substitute α, β, and γ for the symbols a, b, and c respectively, type

> » $A = subs(A, \{\text{'}a\text{'}, \text{'}b\text{'}, \text{'}c\text{'}\}, \{\alpha, \beta, \gamma\})$

For example, if A is a symbolic matrix that has the symbols a, b, and c then to replace a with 1, b with 0, and c with 2, type

> » $A = subs(A, \{\text{'}a\text{'}, \text{'}b\text{'}, \text{'}c\text{'}\}, \{1, 0, 2 \})$

(C) Elementary Operations on Matrices

- **Multiplying ith row (column) by a scalar k**
  Row:

  > » $A(i, :) = k * A(i, :)$

  Column:

  > » $A(:, i) = k * A(:, i)$

- **Multiplying jth row (column) by a scalar k and adding to ith row (column)**
  Row:

  > » $A(i, :) = k * A(j, :) + A(i, :)$

  Column:

  > » $A(:, i) = k * A(:, j) + A(i, :)$

- **Interchanging two rows**
  Rows:

  > » $A = A([\text{new order of rows separating the entries by commas}], :)$

For example, if the matrix A has three rows and we want to change rows *1* and *3*, type

> » $A = A([3, 2, 1], :)$

Note: The method can be used to change the order of any number of rows.

Columns:

> » $A = A(:, [new\ order\ of\ columns\ separating\ the\ entries\ by\ commas])$

For example, if the matrix A has four columns and we want to change columns *1* and *3*, type

> » $A = A(:, [3, 2, 1, 4])$

Note: The method can be used to change the order of any number of columns.

■ **Reducing a matrix A to reduced row echelon form**

> » $rref(A)$

(D) Selecting a Row, a Column, or Submatrix

■ **Selecting kth row of the matrix A**

> » $A(k, :)$

■ **Selecting kth column of the matrix A**

> » $A(:, k)$

■ **Changing (i, j)th entry of the matrix A to the value m**

> » $A(i, j) = m$

(E) Replacing Entries

■ **Replacing an entire row of a matrix A.** In order to replace the kth row of a matrix A set $A(k, :)$ *equal to* new entries of the row separated by a space and enclosed in square brackets, that is, type

> » $A(k, :) = [New\ entries\ of\ kth\ row]$

For example, to change the 2nd row of a 3×4 matrix A to 1, 1, 1, 1, type the command

> » $A(2, :) = [1\ 1\ 1\ 1]$

■ **Replacing an entire column of a matrix A.** In order to change kth column of the matrix A, set $A(:, k)$ equal to the new entries of the column in square brackets separated by semicolons, that is, type

> » $A(:, k) = [New\ entries\ of\ kth\ column]$

For example, to change the 2nd column of a 3×4 matrix A to $[1, 1, 1]'$, type the command

> » $A(:, 2) = [1; 1; 1]$

(F) Augmenting Matrices and Extracting Submatrix

- **Augmenting two matrices A and B having the same number of rows and saving the answer in the matrix C.**

  » $C = [A\ B]$

- **Extracting a submatrix**

  To extract a submatrix consisting of all columns of the given matrix A and rows $k_1, k_2, \ldots, k_r$, type

  » $A([k_1, k_2, \ldots, k_r], :)$

  For example, to extract the submatrix consisting of first two rows, type

  » $A([1,2], :)$

  To extract a submatrix consisting of all rows of the given matrix A and columns $k_1, k_2, \ldots, k_r$, type

  » $A(:,[k_1, k_2, \ldots, k_r])$

  For example, to extract the submatrix consisting of columns 1 and 3, type

  » $A(:, [1,3])$

  In order to extract a submatrix consisting of some rows and some columns, replace : with the list of rows (columns). For example, the command

  » $A([1], [1,2])$

  will extract the submatrix consisting of first row and first and second columns of the matrix A.

Chapter 2: Algebra of Matrices

(A) Additive and Multiplicative Operations

- **Multiplication of a matrix A by a scalar α and saving the answer in the matrix C.**

  » $C = \alpha * A$

- **Adding $m \times n$ matrices A and B and saving the answer in the matrix C.**

  » $C = A + B$

  For symbolic matrices A and B

  » $C = symadd\ (A,\ B),$ or   » $C = sym(A) + sym(B)$

- **Subtracting matrix B from matrix A and saving the answer in the matrix C.**

  » $C = A - B$

- **Multiplying an $m \times n$ matrix A with an $n \times p$ matrix B and saving the answer in the matrix C.**

  » $C = A*B$

  For symbolic matrices A and B

  » $C = symmul\ (A,\ B),$ or   » $C = sym(A) + sym(B)$

- **Powers of a square matrix A.**

  To find A^k and saving the answer in the matrix C, type

  » $C = A^\wedge k$

(B) Operations on a Matrix

- **Transpose of a matrix A and saving the answer in the matrix B.**

  » $B = A'$

  For symbolic matrix A

  » $B = transpose\ (A)$

- **Finding sum of the entries of a row, say kth, of a matrix A**

  » $sum(A(k,\ :))$

- **Finding sum of the entries of each row of a matrix A**

  » $sum(A')$

- **Finding sum of the entries of a column, say kth, of a matrix A**

  » $sum(A(:,\ k))$

- **Finding sum of the entries of each column of a matrix A**

  » $sum(A')$

- **Augmenting two matrices A and B having the same number of rows and saving the answer in the matrix C.**

  » $C = [A\ \ B]$

(C) Creating Special Matrices

- **Creating a zero matrix with *m* rows and *n* columns.**

    » $A = zeros(m, n)$

- **Creating an *n*×*n* zero matrix.**

    » $A = zeros(n)$

- **Creating an *n*×*n* identity matrix.**

    » $I = eye (n)$

- **Generating an *n*×*n* random matrix.**

    » $rand(n)$

- **Generating an *m*×*n* random matrix.**

    » $rand(m, n)$

- **Creating an *n*×*n* random matrix with nonnegative integral entries < *k*.**

    » $round(k * rand(n))$

- **Creating an *m*×*n* random matrix with nonnegative integral entries < *k*.**

    » $round(k * rand(m,n))$

- **Creating an upper triangular matrix from a given matrix *A*.**

    » $triu(A)$

- **Creating a lower triangular matrix from a given matrix *A*.**

    » $tril(A)$

- **Creating an upper triangular matrix from a given matrix *A* with zero diagonal.**

    » $triu(A, 1)$

- **Creating a lower triangular matrix from a given matrix *A* with zero diagonal.**

    » $tril(A, 1)$

- **Creating *n*×*n* Hilbert matrix.**

    » $H = hilb (n)$

- **Creating a Toeplitz matrix.**

    To create a Toeplitz matrix with a given column vector *C* as the first column and a given row vector *R* as first row

    » $T = toeplitz (C, R)$

Chapter 3: Subspaces

Graphing in Matlab

- **Two-dimensional**

  1. Divide the interval into subintervals of equal width. To do this, type

      » x = a: d : b;

  where a is the lower limit, d, the width of each subinterval, and b, the upper limit of the interval.

  2. Enter the expression for y in terms of x.

  3. Type

      » plot (x, y)

  For example, to graph $y = 2x + 1$ from $x = -2$ to $x = 2$ by dividing the interval into subintervals of width 0.1, type

      » x = -2 : 0.1 : 2;
      » y = 2*x + 1;
      » plot (x, y)

- **Three-dimensional**

  For functions of 2 variables, $Z = f(X,Y)$ and $3D$ plots, use the following procedure:

  1. Define scaling vectors for X. For example, to divide the interval $[-2, 2]$ for x into subintervals of width 0.1, enter

      » x = -2:0.1:2;

  2. Define scaling vectors for Y. In order to use the same scaling for y, enter

      » y = x;

  One may, however, use a different scaling for y.

  3. Create a meshgrid for x and y axis.

      » [X,Y] = meshgrid(x, y);

  4. Compute the function $Z = f(X,Y)$ at the points defined in Steps 1 and 2. For example, if $f(X,Y) = -2X + Y$, enter

      » Z = -2 * X + Y;

  5. To plot the graph of $Z = f(X,Y)$ in $3D$, type

      » mesh(X,Y,Z)

Chapter 4: The Rank

- **Rank of a matrix A.**

    » *rank(A)*

- **Reducing a matrix A to reduced row echelon form**

    » *rref(A)*

- **Basis of the null space of a matrix A**

    » *null(A)*

Chapter 5: Inverse, Rank Factorization, and LU-Decomposition

- **Inverse of a square matrix A and saving the answer in the matrix B.**

    » *B = inv(A)*

    For symbolic matrix A

    » *B = inverse(A)*

- **Augmenting matrices**
    See Chapter 1 Commands
- **LU-decomposition of a matrix A.**

    » *[L, U] = lu(A)*

    » *[L, U, P] = lu(A)*

    Note: The latter command gives *PLU*-decomposition of A.

Chapter 6: Determinants

- **Determinant of a matrix A.**

    » *det(A)*

    For symbolic matrix A

    » *B = determ(A)*

- **Elementary operations**
    See Chapter 1 Commands.
- **Extracting submatrices**
    See Chapter 1 Commands.

Chapter 7: Eigenvalue Problem

- **Characteristic polynomial of a matrix A.**

    » $poly(A)$

- **Eigenvalues of a matrix A.**

    » $eig(A)$

- **Entering a polynomial p.**
  If $p = 3x^2+5x\text{-}6$ is a polynomial, enter

    » $p = [3\ 5 - 6]$

- **Roots of a polynomial p.**

    » $r = roots(p)$

- **Eigenvalues and eigenvectors of a matrix A.**

    » $[U, D] = eig(A)$

Note: Here U is a matrix with columns as eigenvectors and D is a diagonal matrix with eigenvalues on the diagonal.

Chapter 8: Inner Product Space

- **QR-decomposition of a matrix A.**

    » $[Q, R] = qr(A)$

- **Singular value decomposition of a matrix A.**

    » $[U, D, V] = svd(A)$

Note: Here D is a diagonal matrix with singular values of A on the diagonal.

Miscellaneous Commands

- **Clearing a variable A.**

    » $clear('A');$

- **Clearing workspace**

    » $clear;$

Note: This command will remove all definitions from the memory. Be careful.

- **Size of a matrix A.**

    » *size(A)*

- **Square root of a matrix A.**

    To obtain a square matrix B such that $B^2 = A$, type

    » *B = sqrtm(A)*

    To obtain a matrix B with entries square root of entries of A, type

    » *B = sqrt(A)*

Programming in Matlab

- **Handling repetitions.**

    The repetitions can be handled in Matlab using a *for* loop or a *while* loop. The syntax is similar to the syntax of such loops in any programming language.

For loop

The syntax includes a counter variable, initial value of the counter, the final value of the counter, and the action to be performed written in the following format.

» *for counter_name=initial_value: final_value, Action; end*

For example, in order to create the 1×4 row vector v with entries according to formula $v[i] = i$, type

» *for i = 1 : 4, v(i) = i ; end*

Note: The comma after the final value of the counter is optional.

While loop

The syntax for a while loop is as follows.

» *while (condition) action, increment_action; end*

The loop is executed until the condition (the statement in the parenthesis) is evaluated as true.

Note: The counter variable must be initialized before using in the above command.

Note: The increment action gives the increment in the counter variable. For example, in order to display a matrix A four times, type

» *i = 1;*

» *while (i <= 4) disp(A), i = i + 1; end*

Nested for loops

In order to have a nest of for loops or while loops or for and while loops, each type of loop must have a separate counter. The syntax for two nested for loops is

> » *for counter 1 = initial_value 1 : final_value1,*
>   *for counter 2 = initial_value2 : final_value2, Action; end, end*

For example, to create a 5×4 matrix A given by the formula $A(i, j) = i + j$, type

> » *for i = 1: 5, for j = 1:4, A(i,j) = i+j ; end, end*

Decision Statements

Two types of decision statements are possible in Matlab—one-way decision statements and two-way decision statements. The syntax for the one-way decision statement is

> » *if (condition) , action, end*

and that for a two-way decision statement is

> » *if (condition), action, else action, end*

For example, if x and y are two numbers and you want to display the value of the bigger number, type

> » *if (x > y), disp(x), else disp(y), end*

Index